# Management of the Fuzzy Front End of Innovation

Oliver Gassmann • Fiona Schweitzer
Editors

# Management of the Fuzzy Front End of Innovation

*Editors*
Oliver Gassmann
Institute of Technology Management
University of St. Gallen
St. Gallen
Switzerland

Fiona Schweitzer
Innovation and Product Management
University of Applied Sciences Upper Austria
Wels
Austria

ISBN 978-3-319-34741-7     ISBN 978-3-319-01056-4 (eBook)
DOI 10.1007/978-3-319-01056-4
Springer Cham Heidelberg New York Dordrecht London

© Springer International Publishing Switzerland 2014
Softcover reprint of the hardcover 1st edition 2014
This work is subject to copyright. All rights are reserved by the Publisher, whether the whole or part of the material is concerned, specifically the rights of translation, reprinting, reuse of illustrations, recitation, broadcasting, reproduction on microfilms or in any other physical way, and transmission or information storage and retrieval, electronic adaptation, computer software, or by similar or dissimilar methodology now known or hereafter developed. Exempted from this legal reservation are brief excerpts in connection with reviews or scholarly analysis or material supplied specifically for the purpose of being entered and executed on a computer system, for exclusive use by the purchaser of the work. Duplication of this publication or parts thereof is permitted only under the provisions of the Copyright Law of the Publisher's location, in its current version, and permission for use must always be obtained from Springer. Permissions for use may be obtained through RightsLink at the Copyright Clearance Center. Violations are liable to prosecution under the respective Copyright Law.
The use of general descriptive names, registered names, trademarks, service marks, etc. in this publication does not imply, even in the absence of a specific statement, that such names are exempt from the relevant protective laws and regulations and therefore free for general use.
While the advice and information in this book are believed to be true and accurate at the date of publication, neither the authors nor the editors nor the publisher can accept any legal responsibility for any errors or omissions that may be made. The publisher makes no warranty, express or implied, with respect to the material contained herein.

Printed on acid-free paper

Springer is part of Springer Science+Business Media (www.springer.com)

# Preface

It is hard to escape the word "innovation" today. In most industries, the continuous creation of innovative new products and processes is considered to be of highest strategic relevance. Not only for companies, but also for whole nations, innovation is understood to be the key factor that enables competitive advantage, economic growth, and thus sustainable wealth. On the other hand, innovation draws scorching criticism. This year, the front page of *The Economist* bore the headline, "Will we ever invent anything (that) useful again?" To us as inveterate optimists, the answer to this question is clear: we will see a wealth of innovation, and the progress of science and innovation will continue to accelerate in the years to come. The question is *who* will win the global innovation race. For companies like GE, Siemens, IBM, Novartis, or Google, the technological world has become a global village. Innovation takes place not only within the classical triad, but also within the emerging markets, at an incredible speed. In the last 5 years, more than 80 % of all newly founded R&D labs have been founded in China and India and more than 90 % of all newly hired R&D staff has been hired also in these two countries. The world is flat, and the speed of the global innovation race has been increasing. New ideas which are bright, customer oriented, hard to imitate, and fast to implement will be vital for future competitiveness.

The key to this innovation ability clearly lies in the early innovation phase, the so-called fuzzy front end of innovation. Top managers tend to neglect this critical phase despite their knowledge of the leverage of the fuzzy front end of innovation for later product success. Much truth lies in the words an experienced project manager told us: "Tell me how the project starts, and I'll tell you how it will end." Empirical research has shown that the fuzzy front end of innovation is most critical for innovation success. Yet, most managers fail at designing an effective innovation process for the early phase. Real innovators, such as Larry Page, Steve Jobs, and Gottlieb Daimler, did not see this problem when they founded their companies. But innovation managers in established companies find it easier to focus on the late phases of innovation, where clear processes and rules can be defined and documented. This is a paradox: knowing that the factor decisive for project success lies in the start and early phases, managers focus their management attention on the late, highly structured phase.

Most top managers do not like "fuzziness"; they have been selected because they are good at planning and execution. Fuzziness is more difficult to address, and its outcome is hard to predict. But since the fuzzy front end of innovation is *the* critical factor for innovation and thus the competitive advantage of firms, we need to focus on it. Michael Dell's mantra on innovation – "fail earlier to succeed sooner" – contains much truth, according to our studies on the front end of innovation.

The cornerstone for innovation is laid in the very early stages of the innovation process – its fuzzy front end. The importance of this phase for overall innovation effectiveness has already been identified by many practitioners and researchers in the last few years. Yet, the front end is poorly understood, and managers experience a lack of knowledge on how to best organize the front end. Ideas about the intensity with which certain activities are to be carried out and the tools that assist their professional execution are vague, and the allocation of resources and top management attention are still ancillary in many companies as compared to other phases of the innovation process. These gaps all add up to an aura of fuzziness, which is often put on an equal level with unmanageability. This book addresses this sensitive phase by demystifying the front end and by providing practical information about activities and tools which boost front end performance.

The objective of this book is to show the way toward winning the innovation game at the front end of innovation by setting the course for successful innovation. To this end, the book is divided into two parts. In the first part, conceptual insights into key capabilities at the front end of innovation are provided. Each of the chapters offers a sound theoretical background and practical suggestions on how to develop these capabilities in order to maximize front end success. In the second part of the book, a multitude of case studies from different industry sectors illustrate how companies approach the fuzzy front end and manage to successfully navigate through it. Cases have been contributed by 3M, ABB, Autoneum, Bayer, BGW, BMW, Emporia, Evonik Industries, Google, Henkel, Hyve, IBM, Landis+Gyr, SAP, sprint> Radar, voestalpine Anarbeitung, and Volkswagen. These cases offer valuable insights into the challenges and opportunities of managing the fuzzy front end. Every case study concludes with clear lessons learned that help the reader to utilize the findings in his or her own business environment.

The book focuses on enabling professionals to broaden their understanding of *success factors* at the fuzzy front end of innovation and reflect on their innovation practices, and it aims to inspire the use of new tools and techniques to optimize the front end. At the same time, the readers of this book are not limited to innovation managers, as the book has been designed to allow master's students from product development, innovation management, and engineering management disciplines to gain insights into the front end processes and activities at leading companies. They will be enabled to critically reflect about the interplay of strategy, process, people, tools, and methods at the fuzzy front end of innovation.

We owe a debt to many for this book. Our appreciation goes to Julia Peherstorfer and Iris Gabriel for thoroughly formatting the book, to Herbert Gsottbauer for increasing the explanatory power of all the figures in the book, and to Elisabeth Hassek-Eder and Kurt Ubelhoer for their effective proofreading. This book was published within the Austrian project Front End (FFG No. 607404) funded by the Austrian Research Promotion Agency (FFG). Most of all, we would like to thank our authors, leading international academic thinkers and practitioners on innovation who have been willing to share their valuable time and insights through their contributions.

We wish the readers many new insights from reading this book and a successful design and implementation of their fuzzy front end of innovation.

St. Gallen, Switzerland　　　　　　　　　　　　　　　　　　　　Oliver Gassmann
Wels, Austria　　　　　　　　　　　　　　　　　　　　　　　　Fiona Schweitzer
April 2013

# Contents

**Part I  Conceptual Part**

**Managing the Unmanageable: The Fuzzy Front End of Innovation** ... 3
Oliver Gassmann and Fiona Schweitzer

**Structuring the Front End of Innovation** ... 15
Kurt Gaubinger and Michael Rabl

**Integrating Customers at the Front End of Innovation** ... 31
Fiona Schweitzer

**Out of Bounds: Cross-Industry Innovation Based on Analogies** ... 49
Marco Zeschky and Oliver Gassmann

**Trend Scanning, Scouting and Foresight Techniques** ... 59
René Rohrbeck

**Crowdsourcing as an Innovation Tool** ... 75
Oliver Gassmann, Sascha Friesike, and Michael Daiber

**Revolutionizing the Business Model** ... 89
Oliver Gassmann, Karolin Frankenberger, and Michaela Csik

**Managing the Intellectual Property Portfolio** ... 99
Martin A. Bader, Oliver Gassmann, and Nicole Ziegler

**Applying Cross-Industry Networks in the Early Innovation Phase** ... 109
Ellen Enkel and Sebastian Heil

**Accelerating Learning by Experimentation** ... 125
Stefan Thomke

**Dancing with Ambiguity: Causality Behavior, Design Thinking, and Triple-Loop-Learning** ... 141
Larry J. Leifer and Martin Steinert

**Leveraging Creativity** ... 159
Sascha Friesike and Oliver Gassmann

**A Design Perspective on Sustainable Innovation** .................. 179
Markus Kretschmer

**Part II  Practical Cases**

**3M: Beyond the 15 % Rule** ................................. 195
Stephan Rahn

**ABB: Integrating the Customer** ............................. 201
Patricia Sandmeier Kahmen and Petr Korba

**Bayer: Strategic Management of the Early Innovation Phase** ........ 207
Wolfgang Plischke, Jürgen Heubach, and Stephan Michael Maier

**BGW: Partnering the Outside-in Process – The Expert Innovation Journey** .................................................... 213
Christoph H. Wecht

**Emporia: The Merits of Online Idea Competitions** ................ 221
Fiona Schweitzer and Walter Buchinger

**Evonik Industries: Managing Open Innovation** ................... 227
Georg Oenbrink

**Case: Google Ventures** ..................................... 233
Sascha Friesike

**Idea Generation in the Consumer Business at Henkel** ............. 237
Thomas Müller-Kirschbaum and Juan Carlos Wuhrmann

**Crowdsourcing: How Social Media and the Wisdom of the Crowd Change Future Companies** .................................. 243
Johann Füller, Sandra Lemmer, and Katja Hutter

**Building a Bridge from Research to the Market: IBM's Industry Solutions Labs** ............................................ 251
Matthias Kaiserswerth

**The MINI Countryman: Successful Management of the Early Stage in a Cooperative Product Development Environment** ............. 257
Markus Seidel, Patrick Oberdellmann, and Antony Clayton

**Controlling the Early Innovation Phase at Autoneum** ............. 263
Javier Perez-Freije

**SAP: Bringing Economic Viability to the Front End of Innovation** .... 269
Uli Eisert

**Sprint Radar: Community-Based Trend Identification** .............. 275
Denis Eser, Kurt Gaubinger, and Michael Rabl

**Landis+Gyr: Designing and Analyzing Business Models in Value Networks** ............................................. 281
Amir Bonakdar, Branko Bjelajac, and Alexander Strunz

**Voestalpine Anarbeitung: Commercialization Framework for Technology Development Projects** ........................... 289
Kurt Gaubinger, Fiona Schweitzer, and Hans-Jörg Kirchweger

**Volkswagen: Open Foresight at the Front End of Research Innovation** ............................................. 295
Caroline V. Rudzinski and Gereon Uerz

**Fuzzy Front End of Innovation: Quo Vadis?** .................... 301
Oliver Gassmann and Fiona Schweitzer

**About the Authors** ............................................. 311

**About the Institutes** ............................................. 317

**References** ............................................. 321

**Index** ............................................. 337

# List of Figures

| | | |
|---|---|---|
| Fig. 1 | Activities at the fuzzy front end of product innovation | 7 |
| Fig. 2 | Drivers of front end success | 9 |
| Fig. 3 | Stage-gate process (second generation) | 18 |
| Fig. 4 | Three phase front end model | 19 |
| Fig. 5 | New concept development model | 20 |
| Fig. 6 | Holistic framework for the front end of innovation | 22 |
| Fig. 7 | Information need at the fuzzy front end of product innovation | 33 |
| Fig. 8 | Direct and indirect ways of customer integration | 34 |
| Fig. 9 | Aims of customer integration at the fuzzy front end of product innovation | 36 |
| Fig. 10 | Methods of customer integration at the fuzzy front end of product innovation | 38 |
| Fig. 11 | Customer types and their input at the fuzzy front end of product innovation | 43 |
| Fig. 12 | AlpineCo: from musical vibrations to ski vibration control | 52 |
| Fig. 13 | AluCo: developing robust crash boxes | 53 |
| Fig. 14 | Analogical thinking at TextileCo | 54 |
| Fig. 15 | Analogical thinking at PipesCo | 55 |
| Fig. 16 | Opening up the solution space through abstraction and analogies | 56 |
| Fig. 17 | $A^4$-cross-industry innovation process for new product innovation by analogical thinking | 57 |
| Fig. 18 | Errors in detecting signals of change | 61 |
| Fig. 19 | Expert search through pyramiding in the web 2.0 | 62 |
| Fig. 20 | Generic scouting network | 63 |
| Fig. 21 | Market foresight methods | 64 |
| Fig. 22 | Technology foresight methods | 65 |
| Fig. 23 | Gardner hype cycle 2012 | 66 |
| Fig. 24 | The two roles of corporate foresight in the front end of innovation | 70 |
| Fig. 25 | Key elements of the crowdsourcing process | 85 |
| Fig. 26 | Business model definition – the magic triangle | 91 |
| Fig. 27 | The business model innovation map: every node represents a revolution in an industry | 94 |

| | | |
|---|---|---|
| Fig. 28 | Pattern card set | 95 |
| Fig. 29 | The patent life cycle management model | 101 |
| Fig. 30 | Cross-industry network graph; *ICT* information and communication technology | 112 |
| Fig. 31 | Beiersdorf cross-industry network Pearlfinder | 113 |
| Fig. 32 | Implications of increased potential absorptive capacity on learning and cognitive distance | 115 |
| Fig. 33 | Network manager capabilities | 120 |
| Fig. 34 | Bayer MaterialScience cross-industry network future_bizz | 122 |
| Fig. 35 | Experimentation as four-step iterative cycles | 129 |
| Fig. 36 | A product development knowledge model based | 145 |
| Fig. 37 | Design process as iteration of divergence and convergence steps or prototype cycles | 150 |
| Fig. 38 | Details specifications and computer-modeled prototypes inhibit ideation and rough, low resolution, prototypes facilitate ideation | 151 |
| Fig. 39 | Tangible 3D prototypes, right, facilitate associative memory, analog thinking, and exploration better than 2D sketches | 151 |
| Fig. 40 | The role of generative design questions (GDQs) and deep reasoning questions (DRQs) in the design process | 155 |
| Fig. 41 | Analysis of designers' interaction for workspace process activities and purposes | 156 |
| Fig. 42 | *Top*: VCode interface showing a coded 16-s section. The squares in the *upper rows* indicate speaker turns. The *squares* in the lower part mark occurrences of negative and positive behaviors respectively. *Bottom*: Example point graphs of a regulated and a non-regulated programming pair. The graphs always show the emotion trajectories for each programmer separately. The *left graph* is drawn from a pair that scored amongst the lowest in the sample and the *right graph* is drawn from a pair that scored amongst the highest of the pairs studied | 158 |
| Fig. 43 | Procedure of the TILMAG method using the example of searching for new ideas for a portable, lightweight and attractive radio | 162 |
| Fig. 44 | Spider meeting | 163 |
| Fig. 45 | Morphological box | 170 |
| Fig. 46 | The different layers of strategic sustainable design in the fuzzy front end of sustainable innovations | 182 |
| Fig. 47 | The three decisive skills of designers | 188 |
| Fig. 48 | The strategic cycle at the Bayer Group and its interaction with Bayer's subgroups | 208 |
| Fig. 49 | Criteria during the strategic decision analysis and possible derived outcomes | 212 |
| Fig. 50 | Activities in the set-up and preparation phase | 215 |
| Fig. 51 | Selection parameters for internal and external participants | 216 |
| Fig. 52 | Overview expert innovation journey | 218 |

| | | |
|---|---|---|
| Fig. 53 | Established open innovation approaches in B2B companies | 228 |
| Fig. 54 | Example of an internet-based open idea competition | 230 |
| Fig. 55 | 4711 design contest flyer (*left*), Siemens sustainability contest platform (*right*) | 245 |
| Fig. 56 | The IBM Industry Solution Labs (ISL) as the starting point for innovative partnership projects | 254 |
| Fig. 57 | Simplified development process | 259 |
| Fig. 58 | Simplified cooperation model | 260 |
| Fig. 59 | Basic premises of Autoneum's control system | 265 |
| Fig. 60 | Stage-gate process at Autoneum | 267 |
| Fig. 61 | The approach which combines design thinking and business model innovation | 273 |
| Fig. 62 | Three-step procedure 'sprint > radar' | 276 |
| Fig. 63 | Campaign 'Trends in Mechatronics' | 278 |
| Fig. 64 | Eight step approach for the design and analysis of value network business models | 283 |
| Fig. 65 | Management Cockpit | 284 |
| Fig. 66 | Framework for TD commercialization (own illustration) | 291 |
| Fig. 67 | Overview of the different project stages | 296 |
| Fig. 68 | The basic information market mechanism | 297 |
| Fig. 69 | Overview of the balance between focusing and opening up in the process | 298 |
| Fig. 70 | Grasping opportunities at the front end of innovation | 302 |

# List of Tables

| | | |
|---|---|---|
| Table 1 | Pros and cons of front end process models | 21 |
| Table 2 | Value creation from corporate foresight in the front end of innovation | 71 |
| Table 3 | Factors that affect learning by experimentation | 132 |
| Table 4 | Possible outcomes from the use of incomplete models | 134 |
| Table 5 | Front-end process stages with customer and ABB involvement | 203 |
| Table 6 | Overview of the most common collaborative approaches in the pharmaceutical industry | 211 |
| Table 7 | Online idea competition versus focus group workshops | 223 |

# Part I

# Conceptual Part

# Managing the Unmanageable: The Fuzzy Front End of Innovation

Oliver Gassmann and Fiona Schweitzer

## 1 The Fuzziness of the Front End of Innovation

An increasing number of doomsayers, such as Peter Thiel, the founder of the internet payment company *PayPal*, criticize a lack of truly innovative ideas. They seem disillusioned because while individuals may have Facebook and smartphones, they still do not travel around in super-sonic flying cars, and because they still have to use refrigerators, microwaves, and ovens that might be more energy-efficient, but basically do not function really differently from what was around 40 years ago. While innovations at the turn of the last century, such as cars, planes, and phones, transformed the lives of billions of people, they critique that innovations today do not generate enough economic growth to improve income and welfare for a substantial number of individuals. Others hold that past and modern innovation is provoking many of the environmental and societal challenges that currently exist.

At the same time, managers consider innovation as vital. According to a recent CEO survey by *IBM* (2012) IBM global CEO study, innovation is one of the highest priorities of top management. But most managers work on the *late innovation process*, which is characterized by defined processes, clear procedures, and documented responsibilities and roles. Management prefers addressing the late phase of innovation despite knowing that the leverages are in the early phase. This is similar to the drunken man who lost his keys in the street and seeks them only under the streetlamp – because there it is bright. The real leverage in bringing

---

O. Gassmann (✉)
University of St. Gallen, Institute of Technology Management, Dufourstr. 40a, 9000 St. Gallen, Switzerland
e-mail: oliver.gassmann@unisg.ch

F. Schweitzer
University of Applied Sciences Upper Austria, Innovation and Product Management, Stelzhamerstr. 23, 4600 Wels, Austria
e-mail: fiona.schweitzer@fh-wels.at

up new ideas and improving the competitiveness of innovation lies in the front end of innovation, the so called 'fuzzy front end of innovation'.

The term *fuzzy front end* was first attributed to the early stages of an innovation project by Smith and Reinertsen (1991). They described it as the fuzzy zone between the time when the opportunity is known and the time when a serious effort is devoted to the development project. During the 1990s, when time to market became one of the greatest drivers in new product development, most companies started to speed up innovation by reducing the thinking time. As a result, many companies reached the milestone of product specifications faster – but this entailed failures, change requests and additional loops in the later phase. And this became very expensive. In the early 1990s, *Ford* used to have the same amounts of change requests as *Mazda*, but addressed the changes much later than Mazda. As a result, the impact of every change on costs, quality, performance and time was much higher at Ford than at Mazda.

Companies must energize the fuzzy front end of innovation in order to speed up the project as a whole. Moreover, the decision which opportunities and ideas the company selects and wants to pursue has to be taken fast. Instead of having long decision times before a project really starts and a short fuzzy front end, the project decision has to be made faster and the team has to be energized in the early phase. In turbulent environments – and most of our industries are much more turbulent today due to globalization and modern IT - speed is more important than the careful selection of the perfect opportunity. The opportunity costs of delay from starting a project late with subsequent deficiencies in time to market can outweigh opportunity errors in such markets. In some companies – often large companies operating under medium competitive pressure - it still takes months to years from the day when a compelling product idea lands on a manager's desk to the time when the first engineer starts developing technical solutions for the idea.

Today many companies seem to suffer from the same disease: Frontloading of specifications and requirement engineering as a discipline has to be re-learned. The great airplane A380 by *Airbus* is the most fascinating plane in the last decade, full of technology: Besides its huge size – a height of 24 m, a wing span of 80 m, and 560 t of weight with capacity for 520 passengers – it contains a vast amount of electronics: a sophisticated information system in the cockpit, two clicks with an over-dimensional mouse and every information is available to the pilot, and a 3-D-weather radar. But when the first customer Emirates changed its specifications in the infotainment area, the whole project nearly collapsed. Managers sometimes forget that this complex plane with 400 TV channels, 500 km of cables and 40,000 connectors cannot be changed in the last minute without consequences. The project was delayed and caused Airbus serious problems. Unplanned changes in the late project phases are dangerous. The motto is 'fail earlier in order to succeed sooner'.

*Boeing* was hit even worse with its Dreamliner project in 2012. For the first time in 34 years a plane was stopped from flying due to severe problems in many areas: oil leaks, broken batteries, and bugs in the brake system are only a few examples. Boeing did not integrate its suppliers in the early phase, instead Boeing only checked the specifications of its suppliers. This is not enough for a radically new system like the Dreamliner. The fuzzy front end of a project is also responsible for system integration functioning smoothly in the late phase of innovation.

It is said that in the early 1980s it would have taken *IBM* 3 years to ship an empty box if the correct procedures for the product development process had been followed. Developments in agile computing demonstrate that product development time can be reduced dramatically, and currently many industries are trying to use these principles to speed up their own 'shipment business' in order to prevent 'missing the boat' through delayed product development, a front end process which is too lengthy, and too many and too detailed activities of preparing, planning, analyzing and evaluating. Yet, 'sinking the boat' in the sense of developing inappropriate products through overly hasty actions is no minor challenge in product innovation. Setting the right course early on in the innovation process can save companies from expensive and time-consuming deviations in later stages of the development process. While it is undisputed that taking the right decisions as soon as possible is vital, many companies feel a lack of knowledge on methods and processes which improve front-end decision-making.

Decisions in the early phase are taken under *uncertainty* about technical feasibility can be reduced by virtual or real prototypes. The risk of market failure of the final product certainly remains throughout the innovation process and market acceptance can definitely be assessed only after the product has been launched. Nevertheless, there are methods that facilitate predicting likelihood of acceptance and that therefore ease decision-making regarding which ideas and concepts are worth pursuing. Risks always depend to a certain degree on factors that are beyond the company's control, e.g., competitors' actions or the economic climate. Yet, the right mix of methods and processes can help to identify drivers of risk, reduce uncertainties, and thus take some fuzziness out of the front end of innovation, while at the same time successful management of the fuzzy front end requires an entrepreneurial spirit that accepts risk and welcomes *risk-taking*. For these reasons, effectively managing the fuzzy front end of innovation is one of the most important, and simultaneously challenging, activities for innovation managers.

## 2  Time for Action at the Front End of Innovation

All actions that are taken between the first consideration of an opportunity and the decision whether to start product development make up the front end of innovation (Kim and Wilemon 2002). The *early front-end activities* include the identification of a problem or opportunity and the accompanying screening and evaluation processes. This phase can be described as strategic arena setting. *General Motors* also called this the 'bubble-up-process', where the strategic decisions for the new product development are made. Like many automotive companies, General Motors involves strategic procurement, advanced development and innovation marketing in this early phase. It is important that this phase is managed in a way which is as inter-functional and interdisciplinary as possible. The more perspectives are considered, the better are the strategic cornerstones for the project. *BMW* emphasizes the role of the customer and user in this phase: The task of the innovator in the fuzzy front end is to design a system which the customer desires. In the words of a BMW engineer: "Our task is to provide the customer with something that fills the customer with real excitement when he gets it, but that he never knew he was seeking in the first place." This is much more than a fancy marketing sentence. BMW does not ask the customer what he or she wants. In the early fuzzy front end, the usual way of questioning customers about their wants is hardly successful because the customer often only tells the engineers what they know already. BMW innovators have to know the customer better than he or she knows him or herself – this truly represents a very fine line between creating customer excitement and engineering happily in the wrong direction. Identifying latent customer needs is a key challenge in this phase, and approaches such as bodystorming, empathic design, netnography or observation can meet this target better than traditional surveying techniques. The fuzzy front end of innovation is the perfect phase for trying to anticipate customers' and users' future requirements and wishes. The adaptive headlights in the BMW X6, with sensors which constantly monitor the car's speed, yaw rate and steering angle, then calculate curve progressions accordingly to offer optimized illumination of the road ahead, were one of the results BMW developed from *customer insights*.

*Later front-end activities* comprise all the work that helps to specify the identified opportunities for innovating and to find possible solutions that seize these opportunities or meet specific problems. These activities include idea generation, idea evaluation, concept development, and concept evaluation for product innovations. The most difficult task is finding and specifying the right opportunity. Nobel prize winner Herbert Simon once said "Problem solving involves not only the search for alternatives, but the search for the problem itself." Engineers are very good at weighting and scoring alternatives for a given problem, they love cost-benefit-analysis. But it is much more difficult to find the right problem. The following small example will illustrate this issue.

Designers from the famous Stanford spin-off *IDEO*, meanwhile the most distinguished design company of the world with products like the iPod, the computer mouse and many more, always look for the *sweet spot* if they design a new product. The sweet spot is the point where the leverage between minimal efforts and maximum impact on user value is best. Once, the company worked for a train company which wanted to improve train riding. While typical engineers start collecting ideas on how to improve the train ride, these people started out in a more holistic way. They asked what makes the difference between taking the train and driving by car, the major competitive solution to going by train. And they found that for the customer the train ride is much more than just riding the train: It starts with planning and seeking information (which train leaves at what time), is followed by entering the train station, waiting for the train, boarding the train, the train ride itself, leaving the train at the station, and transferring to the final destination. The customer's perspective on train riding is more than just being on the train. Improving the waiting time for the train turned out to be the sweet spot; the team identified many ideas with a high potential of improving customer value at

# Managing the Unmanageable: The Fuzzy Front End of Innovation

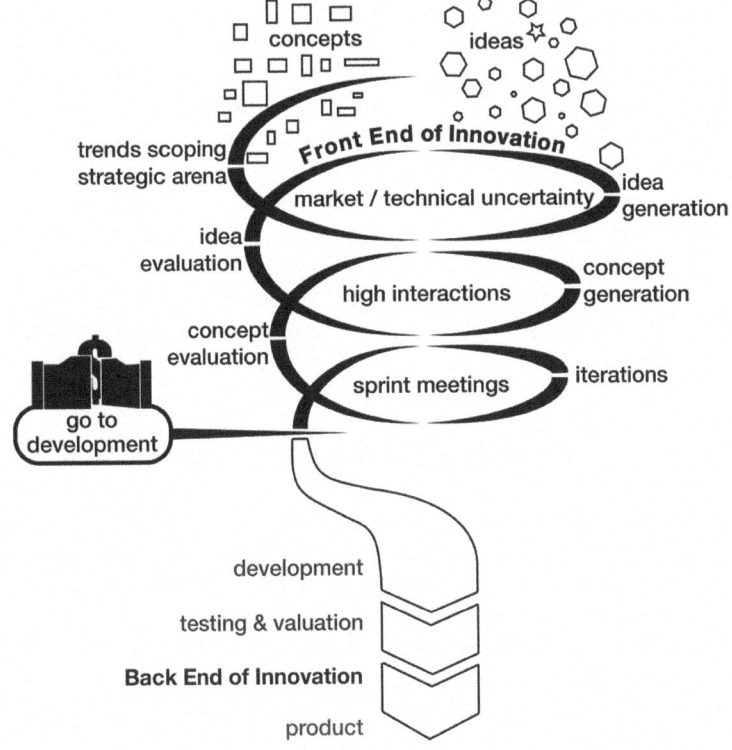

**Fig. 1** Activities at the fuzzy front end of product innovation

moderate cost. By directly jumping onto the train and discussing about its interior, the wrong problem would have been addressed.

Figure 1 illustrates the activities at the front end of innovation in a schematic form. The duration of these activities and the degree of detail with which each of the activities is executed for a certain new product idea may vary. Moreover, the number of iterations varies and concurrent fulfillment of activities can take place. Whether the activities can and should be followed in a sequential order will be discussed in later sections of the book. These activities are not limited to mere product innovation, but can be carried out in a more broadly-set technological pre-development phase or in a more product-focused early stage of the product innovation process as well as for process innovations and service innovations. The market and technical uncertainties, and with them the *scope of action*, are high at the very beginning of an innovation project and decrease throughout the innovation process through the diverse activities of strategic arena setting, idea generation/evaluation and concept generation/evaluation. While activities are taken to substantiate the innovation idea, time is consumed and costs are incurred. Therefore, efforts to optimize the whole innovation project are easier to effect at the front end. Front end decisions are considered to have high leverage for the whole innovation process, and wrong decisions at the front end may lead to costly and time-consuming deviations later on, as conceptual and empirical studies have shown (e.g. Kim and Wilemon 2002; Reid and De Brentani 2004; Verworn et al. 2008). Although in principle the ability to influence the direction of the innovation project is highest at the fuzzy front end, managers typically get more involved in a project after it has passed the so-called 'money gate', which is the time after the project has been transferred into the development phase and the time when innovation projects really start becoming expensive, as Harvard colleagues Wheelwright and Clark (1992) and their Finnish colleagues Poskela and Martinsuo (2009) have shown. However, as the seeds for these development projects are sowed at the fuzzy front end, it is of merit to take a closer look at the art of managing the front end.

## 3   The Art of Managing the Front End of Innovation

The painter Henri Matisse once said "The sign for which I forge an image has no value if it doesn't harmonize with other signs, which I must determine in the course of my invention and which are completely peculiar to it." Harmony refers to the ways in which elements are arranged. This balance need not necessarily be symmetrical in the sense of allowing each element an equal weight. Rather, the infinite variety of elements that make up the whole entity of an artwork must be composed in a way that allows no element to overpower the other, but to work together to produce fit. The art of painting involves balancing light and shades, vagueness and concreteness, sharp and soft lines, the positive and the negative. The art of cooking lies in mixing the right quantities and types of ingredients to make a dish tasty – not too sour or sweet, not too salty, but then again not stale.

Managing the fuzzy front end of innovation is a similar art: a balancing act between exploiting proven capabilities and dynamically exploring new ones, between stability and flexibility, between certainty and uncertainty, between formal and informal interaction, between market pull and technology push, between creativity and discipline, between free room and limitations. The art of managing the fuzzy front end is not the art of dictating what everyone has to do at what time. Nor is it the art of letting chaos reign. It is the art of identifying and understanding contradictory and complementary forces, supportive and counterproductive influences, and of providing the necessary framework, resources, and conditions to cope with these forces and influences. A late project phase in which hundreds of engineers in dozens of locations need to be coordinated, like the development of an airplane or a new car, often requires process experts and traditional managers. System integration and the convergence of schedules with dozens of cross-company teams necessitate clear gates and strict processes. Process management becomes key with all its instruments, milestones, stage-gate elements, and measures comparable to cockpit controlling.

In the fuzzy front end this is different: Process leadership is not unimportant, but the key capability is being good at managing people, i.e., finding the right people,

**Fig. 2** Drivers of front end success

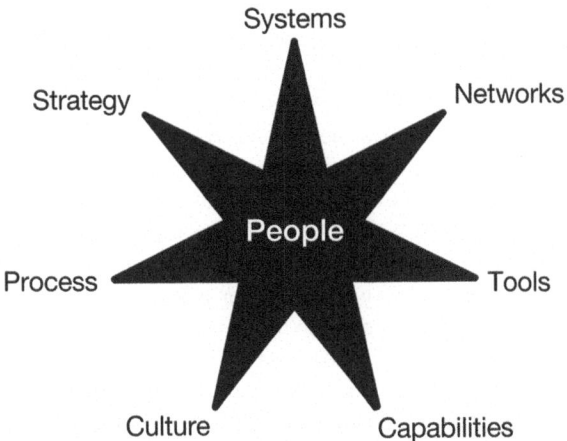

setting up a good network, coaching the teams, identifying the creative potential of the individuals and providing them with a strong vision and direction. Providing a meaning is extremely important in order to trigger the intrinsic motivation of researchers and innovators. Intrinsically motivated teams that work because of the interesting and challenging task are much more successful than teams that work for more status or more money. *Bombardier* has recently announced the maintenance free train as a vision in the company. This strong vision, its enforcement and genuine belief in it by the top management align all forces within a company and energize the fuzzy front end more than most of the project bonuses often used.

Factors that contribute to these managing capabilities are strategy, processes, methods and tools, interdisciplinary systems and networks, culture and people. At the end it is all about people, especially in the fuzzy front end of innovation. While the stage-gate system has been researched in all aspects and implemented in many companies in every detail and variant, the early innovation phase has been much less explored by academia and much less addressed by companies. One of the reasons is that it is much more difficult and complex. It is much harder to describe a good front end of innovation and consequently implement it compared to the later stage-gate process. Yet, the front end is where most companies have huge potential for improvement and the gap between best practice and average practice is enormous (Fig. 2).

## 4 The Course of the Book

Throughout this book we will take a closer look at the drivers of front end success. Take strategy: Textbook wisdom conveys that a strong strategy is vital for innovation success as it allows aligning all activities of all employees with the set target, and it is only in this way that the strategic goals can be reached (e.g. Cooper

2011). Without question, clear and focused strategies give overall direction and stimulate the target-oriented search for new innovation opportunities within certain search fields, but they intentionally hinder the emergence of initiatives that seem to be completely detached from strategies. Such ideas and activities are constantly filtered out, because they do not pass the value-laden filter of right and wrong as they do not conform to strategy. So strategy is a double-edged sword. Innovative opportunities can arise and a narrow strategy can prevent exploring them, as the Swiss company *Sulzer* showed. The former casting company had a clear strategy that called for exclusively pursuing its core business, which is casting and heavy metals. Once a surgery professor approached the head of prototyping and asked whether the latter could do him a favor and manufacture an implant for an artificial hip. Since the professor paid well and the engineer had become interested, he complied with the wish against all strategic directives. The business became bigger and the division developed into the largest division of the company – for a long time against Sulzer's explicit strategic directive. The user induced the strategic process bottom-up. Luckily the implementation of strategy had not been enforced rigidly at that time, so it was possible for the company to grow in the new area of business.

As the example shows, in dynamic environments, a company may have to change core competences to stay successful. Insisting on once successful competencies and strategies may lead companies into struggle. A capability that was once an asset can become a liability, if it is no longer appropriate and strategic pre-settings may encourage managers to run too fast in the intended direction without reflecting on whether the strategy is still adequate and ignoring ambiguities. In this sense, a rigid strategy can be compared with a creosote bush, whose roots gather every drop of water and do not allow any other plant in its surroundings to grow. The positive link between strategic focus and innovation performance recedes when the focus limits dynamic integration of emerging strategies and diversity (Burgelman 2002; Henderson and Cockburn 1996). The discussion in the next chapter, along with the presentation of *Strategic management of the fuzzy front end at Bayer* by Plischke, Heubach, and Meier, provides fruitful insights into the effective management of this balancing act.

In order to achieve the right strategic alignment, to justify and communicate decisions, and to control project progress, a structured process with clear decision points and control mechanisms for the fuzzy front end is worthwhile. Yet, such a process must not hinder flexibility and creativity. It has to permit iterations, concurrent engineering, improvisational approaches, experimentation slopes, and leaner process runs according to the necessities of the specific projects. These principles and their interplay are described in detail in the chapter "Structuring the Front End of Innovation" by Gaubinger and Rabl. The chapter on "Controlling the Early Innovation Phase at Autoneum" by Freije-Perez discusses how the facilitative role of controlling that focuses on knowledge creation, experimental non-linear operations, and a broad role definition allows to successfully steer through the front end of innovation in the automotive supply industry. The chapter "Voestalpine Anarbeitung: Commercialization Framework for Technology Development Projects " by Gaubinger, Schweitzer and Kirchweger shows an elaborate framework of a process for commercialization of a technology enriched with an overview of activities and tools that companies should focus on in each phase.

Given the quantity of different tools and methods available to identify trends, new product ideas and customer needs, to assess technologies and markets, and to assess technological feasibility, the appropriate selection and application of these tools becomes paramount. For this reason, in the chapter "Integrating Customers at the Front End of Innovation" by Schweitzer, we provide an overview and critical discussion of different tools to explore current or latent needs of customers and integrate them into activities of idea/concept generation and evaluation in the early stages. Sandmeier, Kahmen and Korba describe in detail how the customer was integrated into an innovative project at ABB for developing battery storage technologies in the energy sector. In "Building a Bridge from Research to the Market: IBM's Industry Solutions Labs", Kaiserswerth explains how these labs function as knowledge hubs, where trend information is exchanged and innovation issues are discussed between internal R&D experts and customers, and how these interactions lead to innovative joint projects.

Yet, not only customers can be integrated into the innovation process under the notion of open innovation, but web-enabled technologies allow the integration of a wide range of different pre-defined or even anonymous external actors. In the last few years, the use of such technologies for the purpose of seeking new product ideas or technical solutions has gained popularity under the term '*crowdsourcing*'. In the chapter "Crowdsourcing as an Innovation Tool", Gassmann, Friesike and Daiber systematize the possible types of crowdsourcing projects and explain step by step how a crowdsourcing project is carried out and which issues have to be addressed at each stage so that crowdsourcing is used in a successful way. An overview on different "Trend Scanning, Scouting and Foresight Techniques" is presented by Rohrbeck. The chapter provides clear assistance in choosing the right foresight method and highlights key issues that have to be borne in mind when integrating and using the results of foresight processes for decision-making at the front end of innovation.

In the chapter "Leveraging Creativity", Gassmann and Friesike stress that the ability to find creative solutions to challenges of new product development is not merely a personal gift only a few of us possess, but rather a capability that can be systematically built and used with the help of specific problem-solving techniques. A specific method of creative idea generation is cross-industry innovation, which is scrutinized in the chapter "Out of Bounds: Cross-Industry Innovation Based on Analogies" (Zeschky/Gassmann). In the chapter "Accelerating Learning by Experimentation", Thomke sets out in detail a four-step experimentation cycle that products evolve along, and he describes which chances and challenges teams usually encounter in technical experimentation and how factors such as the fidelity, cost, or iteration time of experiments affect front end success. The chapter includes several hints and tips that help to make the best use of experimentation at the front end of innovation.

A variety of tools for opening up the fuzzy front end of innovation to embrace input from customers and other external sources is discussed in the case studies: In

"BGW: Partnering the Outside-in Process", Wecht explains how an innovative tool called Expert Innovation Journeys works and how it can be used to generate ideas through active collaboration of corporate employees and external experts in a sophisticated workshop process. Another innovative open innovation tool that builds on the knowledge of external and internal experts is the "SPRINT Radar" which is presented by Eser, Gaubinger and Rabl. The case demonstrates how this multi-stage tool can be effectively used to explore trends and tendencies in the field of mechatronics. Further, Rudzinski and Uerz present an "Open Innovation approach to Strategic Foresight at Volkswagen" in their chapter, in which they highlight the importance of integrating different players from within the company to find broad acceptance in the company and to detect weaknesses in the current innovation strategy.

The chapter "Emporia: The Merits of Online Idea Competitions" compares focus groups to online idea competitions and provides insights into the strengths and weaknesses of the two approaches for the specific case of generating ideas for mobile phone solutions for senior citizens. Oenbrik provides an overview of the different open innovation approaches that Evonik Industries uses, from R&D co-operations, over technology scouts as agents for operative business units and for strategic topics to internet-based open idea competitions, and reflects on the strengths of these techniques for Evonik. Füller, Lemmer, and Hutter elaborate on such internet-based competitions in the chapter "Crowdsourcing: How Social Media and the Wisdom of the Crowd Change Future Companies" and present numerous examples of utilizing crowdsourcing for business innovation. They explain the principles of crowdsourcing which have to be considered in order to run successful crowdsourcing initiatives.

The increasing number of players who contribute to an innovation project, not only from inside the company, but also from the outside, turn *intellectual property* (IP) *management* into a core strategic issue. The quarrel about intellectual property infringement between *Apple* and *Samsung* that has lately received intense media coverage illustrates the difficulties that may arise for IP management when innovations are increasingly developed and produced in a collaborative way. While Samsung is a supplier of Apple and delivers for instance ASIC processors for iPhones, the two companies are fierce competitors, too. Sharing information with suppliers may contribute to innovation success, but if the supplier evolves into a competitor, this subject is sensitive. Crowdsourcing approaches may even intensify the challenges of properly managing intellectual property rights. For example, designers, technicians and inventors gathered on a virtual platform to spin ideas about cars. The ideas soon became so extensive and elaborated that *Local Motors*, a newly founded American open-source car developer, used them to develop the car model Rally Fighter. The US military has already ordered the development of a desert vehicle prototype by Local Motors and their co-creators. Further orders are in the pipeline. Ideas, designs and concepts were on the web for some time and everybody was allowed to contribute and to refine the ideas. The spirit behind this movement had been one of free access, sharing, and the openness of research and development. This idealistic demeanor is challenged when ideas actually materialize into products that make money. With thousands of co-developers that contributed to ideas, it is difficult to calculate fair shares in profit, and contributors might claim intellectual property rights, if not enough attention is paid to a good intellectual property management. In the chapter "Managing the Intellectual Property Portfolio", Bader and Gassmann address this intellectual property challenge and lay down a systematic model for managing patents along the lifecycle of technologies. This holistic life-cycle model allows companies to recognize the importance of aligned activities that are rooted in the corporate technology and product strategy, and it scrutinizes the decisions that have to be taken in each phase of a technology's life cycle.

As indicated in several of the above chapters, the tools and methods presented can only live up to companies' expectations when the right type of people are involved in their application. For example, Schweitzer highlights that different types of customers are important for the various tasks at the front end of innovation; while technically savvy users may prove vital for concept development, for early evaluation of the market potential of different concepts emergent users or early adopters may provide important information. As for trend forecasting, Rohrbeck explains the contribution that proponents and opponents of innovation projects make in the foresight process and suggests different methods of institutionalizing such roles at the front end of innovation. In the chapter "A Design Perspective on Sustainable Innovation" – *strategic design for sustainable innovation*, Markus Kretschmer focuses on the specific skills designers contribute at the front end of innovation with their particular problem-solving approach, their understanding of product culture, and their ability to think and communicate visually. In the same vein, the chapter "Dancing with Ambiguity: Causality Behavior, Design Thinking, and Triple-Loop-Learning" by Leifer and Steiner demonstrates how design thinking drives front end performance. In addition, Sandmeier and Korba describe the challenges of finding and motivating the right experts to contribute in the early phases of an innovation project. In "Google Ventures", Friesike presents an intriguing way in which Google looks for innovative ideas and the right people to realize these ideas. This is achieved by supporting start-ups, especially such start-ups whose aim it is to directly compete with Google's products. Google invests in these companies, provides them with marketing, managerial, and technical know-how, and ties them to the company so that finally these start-ups turn into co-operative partners rather than competitors.

In the chapter "Applying Cross-Industry Networks in the Early Innovation Phase", Enkel and Heil explain how innovation networks including companies and individuals from other industries fuel innovation and discuss the importance of selecting the right partners for such networks. If the cognitive distance between people from the company and outside sources is too extensive, the company may not be able to absorb external knowledge. Conversely, if the cognitive knowledge is too similar, the firm might not absorb vital new knowledge, because the knowledge the employees may obtain from the external sources equals their existing knowledge and thus is redundant. So again it is the right selection of people that determines the success of front end activities. Furthermore, the authors describe how the internal capacity of absorbing external knowledge can be enhanced and discuss the importance of cultural elements, such as trust and openness, shared norms, and common objectives, in creating multilateral innovation networks. The power of culture as an enabler of fuzzy front end performance is also demonstrated in the chapter "3M: Beyond the 15 % Rule", in which Rahn lays down the cultural cornerstones that make 3M such an innovation machine. Besides awareness of cultural aspects, the management of front end networks demands good coordination capabilities. If project participants and their roles do not remain the same throughout the project, as in the case of the *ABB* project, such coordination can become a critical challenge. For example, in chapter "The MINI Countryman: Successful Management of the Early Stage in a Cooperative Product Development Environment", Seidel, Oberdellmann and Clayton describe the difficulties arising from the shift in responsibilities from *BMW* in the concept phase to *Magna Steyr Fahrzeugtechnik* after target agreement.

While so far innovation research has mainly focused on product innovation, currently service innovation and system innovations are gaining increasing attention. Ingrained in this enhanced understanding of innovation is the concept of business model innovation. In "Revolutionizing the Business Model", Gassmann, Frankenberger, and Csik are set out to explain how the value proposition towards the customer, the value chain and the revenue model are created to profit from innovative business models, and they present the BMI Navigator, which enables the creation of business models in a three-step process, from initiation to ideation and integration. In the "Landis+Gyr Case Study", Bonakdar, Bjelajac and Strunz demonstrate how an analytical process can be followed in practice so that business models can be designed and analyzed systematically and how the main results of this process can be integrated into a management cockpit to allow easy monitoring of the basic indicators and easy comparison of key data between different business models.

The first part of this book is a profound source of conceptual contributions to the five key dimensions of managing the front end of innovation: strategy, processes, methods and tools, interdisciplinary systems and networks, culture and people. The second part of the book offers a rich selection of successful practice cases that demonstrate how these key success factors have been prosperously applied in different industries and organizational structures.

While reading through the chapters, the words of Paul Cezanne shall serve as a guiding rule: "To paint is not to copy the object slavishly, it is to grasp a harmony among many relationships." In this sense, the chapters of this book aim to inspire the reader to reflect on the ingredients needed to efficiently and effectively navigate the fuzzy front end of innovation.

# Structuring the Front End of Innovation

Kurt Gaubinger and Michael Rabl

## 1 Introduction

Structuring and managing the innovation process represents one of many critical process-related factors traditionally associated with innovation success. One of the principal objectives of process models is to structure typical tasks in the corresponding field to ensure the targeted application of work techniques, methods and tools. A well-defined process is transparent for all departments and a common understanding can be developed, which facilitates communication within the company (Gaubinger 2009). While the benefits of structured stage gate processes are broadly accepted for later stages of the innovation process, at the front end a broad variety of concepts and process models for structuring and systematizing the innovation process currently can be found at the operational level and in literature (Barczak et al. 2009; Cooper 2001).

## 2 Process Models for Formalization the Front End of Innovation

Managing the Fuzzy Front End is a continuous conflict between creativity and systematization (Verworn and Herstatt 1999). The early stages imply high risk and uncertainty, ill-defined results and an unclear way of setting and achieving goals. Therefore, it is essential for organizing the front end of innovation (FEI) in order to find the right balance between flexibility and creativity (weak-defined processes and targets) on the one hand and structure and bureaucracy (well-defined processes and targets) on the other hand. Too much structure kills creativity, while too little

K. Gaubinger (✉) • M. Rabl
University of Applied Sciences Upper Austria, Innovation and Product Management,
Stelzhamerstr. 23, 4600 Wels, Austria
e-mail: kurt.gaubinger@fh-wels.at; michael.rabl@fh-wels.at

structure negatively affects FEI-performance (Gassmann et al. 2006). This relationship between degree of formality and performance shows an inverted u-shape curve implying that too much as well as too little formality is negative. One of the main causes can be found in the turbulent environments characterized by increasingly dynamic and complex markets, rapid technological progress and shortened product life cycles which calls for adaptable and flexible processes (Calantone et al. 2003). Herstatt and Verworn (2007) also stress the importance of a situation-appropriate balance between structured processes and sufficient room for creativity.

There are a vast number of innovation process models which divide the front end into phases, stages, steps or elements, varying with regard to priorities, number of phases, perspective, definition of starting point and the ending point of the process and degree of detail (Verworn and Herstatt 1999). Sequential models follow a linear course, conduct one task after another and thus allow for an easy access of recommended actions, facilitating transparency and predictability (Khurana and Rosenthal 1998). However, they also run the risk of not corresponding to reality and of not adequately considering creative exchange and feedback loops among employees. To speed up innovation pace scholars suggest parallelizing development activities. As Cooper (2011) states, more activities are undertaken in an elapsed time by multi-disciplinary teams with parallel processing (rather than sequential). Generally, parallelism and integration of external stakeholders seems to be a central success factor not only in the NPD execution but also in the FEI.

In sight of turbulent environments some researchers advocate flexible processes in innovation management additional to parallelism. Models with flexible and dynamic processes, feedback loops and parallel actions are mainly referred to iterative process models in the literature (Sandmeier and Jamali 2007). Koen et al. (2001) for example support a circular shape of the front-end elements, which means that ideas are expected to flow and iterate between the sub-phases, because these sub-phases of the fuzzy front end are unpredictable, chaotic, informal and poorly-structured by nature. According to Ayers et al. (1997), flexibility, ambiguity and keeping a broad set of possible options open are especially vital to innovation success. Cooper und Kleinschmidt (2007) also point out that top companies conduct their innovation processes in a flexible and scalable way.

Never the less the actual implementation of flexible models in a company turns out to be difficult due to the abstract nature of these models not lending itself easily to deriving concrete recommended actions for employees. However, the developments in information and communication technology have the potential to enable the implementation in an efficient manner. Furthermore the use of linked IT-systems, simulation and rapid *prototyping* technologies, and comprehensive information systems continues to reduce development time and development costs. 'Electronification' of innovation processes thereby constitutes a decisive feature of the latest generation of innovation management processes.

## 2.1 Variable Degree of Structuring the Front End of Innovation

The term 'fuzzy front end' incorrectly suggests that the early stages of the innovation process have to be unstructured, fuzzy and chaotic by nature and cannot be managed because of all its unknowable und uncontrollable factors (Koen et al. 2001). However, creative problem solving needs not necessarily occur chaotically, but may very well be subject to certain structures and regularities. This rather speaks for the position of Steiner (2003), which holds that a deterministic chaos, where creativity is guided through certain formal processes, is advantageous as it enables employees to fully unfold their creative potential in the various steps without distraction and with clear goals and time-frames. Quinn (1985) also perceives 'controlling the chaos' as a potential way out of this dilemma. This approach does not imply suppressing the chaos, but just controlling it. Similarly, Brown and Eisenhardt (1998) point out the importance of a 'dissipative equilibrium' between chaos and bureaucracy. For Van Aken and Weggeman (2000), an ideal management regime contains approaches that both operate formal (tightly managed) and free (undirected exploration). Cooper (1994), for example, provides with his stage-gate-process with flexible gates and fluid stages (third generation) an approach, which manages well the straddle between chaos and bureaucracy. Hence, due to the pros and cons of both sequential and iterative models, many researchers look for a combination of these two approaches in order to find a process structure (Sandmeier et al. 2004).

Effectively managing the fuzzy front end of innovation represents one of the most important and simultaneously challenging activities for innovation managers (Kim and Wilemon 2002). In the later phases of the NPD process a structured stage gate process is widely accepted in theory and practice, whereas difficulties arise from the fact that the early innovation phase is mainly considered as dynamic, fuzzy, unstructured and hardly formalized (Murphy and Kumar 1997). Process models have been developed to structure the front end of innovation to reduce its uncertainty (Holtorf 2011) and to visualize and manage the process in its entirety (Rothwell 1994). Consequently, the following chapter is concerned with the evolution of process models for the front end of innovation.

Selected from the multitude of available models, the ones presented in the following section are the models that are frequently quoted in literature.

## 2.2 Stage-Gate Process (Cooper)

The Stage-Gate process divides the innovation process into stages separated by gates where go/no-go decisions are made based on information generated during the activities in the previous stages (Fig. 3). The new ideas collected during the discovery phase (stage 0) through internal and external sources are evaluated and filtered during stage 1 according to criteria like strategic fit, market attractiveness and technical feasibility. During the scoping phase (stage 1), a first rough elaboration of market-related and technical advantages is carried out, to be followed by

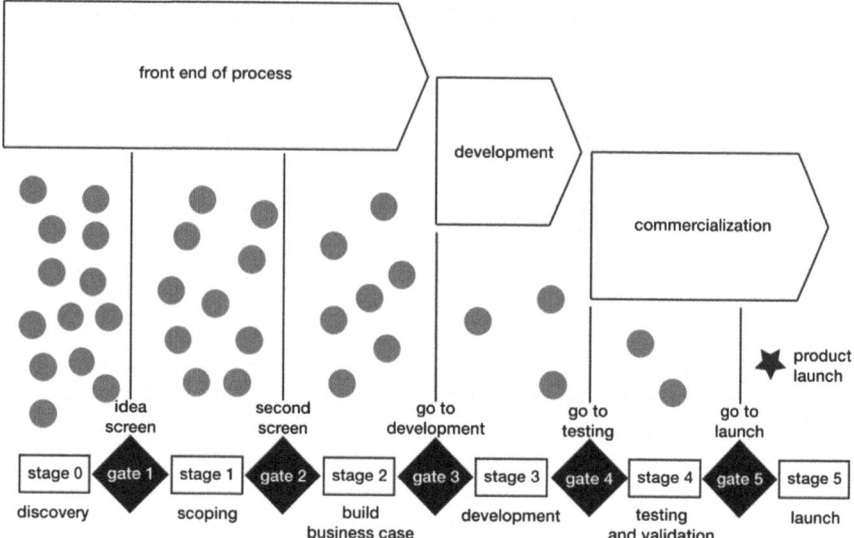

**Fig. 3** Stage-gate process (second generation)

further evaluation at gate 2. At the subsequent stage 2, detailed tests with regards to technology, market and competition are carried out, culminating in the draft of a business case depicting the route from ideas to product concept. At gate 3, which separates the front end of innovation from the development phase and is also referred to as 'money gate' since the firm has to decide if it is willing to allocate resources; an even more detailed assessment forms the basis for making a decision on the launch of a development project (Globocnik 2011).

Cooper's model has evolved over time through several generations and is one of the most frequently cited models. He expanded the above described so-called second generation model into the stage-gate-model of the third generation, characterized by 'four fundamental Fs' (Cooper 1994). Transitions between stages are fluid and activities can be conducted increasingly in parallel fashion (*fluidity*). Within the scope of a gate-decision, a project can be continued to some degree even if not all criteria for the respective stage have been met. Cooper talks about 'fuzzy gates' here. Likewise, tasks of a subsequent stage can also be carried out prior to a gate-decision. An optimal allocation of resources between different innovation projects is an increasingly important factor in determining gate decisions (*focused*). Also, in third generation stage gate models, projects only have to pass through certain process stages, depending on the respective project's degree of risk (*flexibility*). Processes are perceived as being scalable, hence those with a lower degree of risk can be processed in a 'leaner' way, i.e. in fewer process segments and gates. One of the drawbacks of third generation process models, though, is that flexibility is often achieved at the expense of robustness, with projects that are continued on condition often not being aborted on time. The last 'evolutionary stage' of innovation process management systems is subsumed by Cooper's terms *NexGen*

*Systems* (Cooper 2008a). In addition to an increased degree of scalability and flexibility, the most characteristic feature of the latter model is its openness in the sense of the open innovation approach.

## 2.3 Three Phase Front End Model (Khurana and Rosenthal)

Khurana and Rosenthal separate the front end of innovation in their sequential process model into the three sections pre-phase zero, phase zero and phase one. In addition to project-specific elements (such as project definition, respectively - planning, as well as the product concept), which continuously support the project, project-independent activities, so-called *foundation elements*, also influence pre-phase zero. The foundation elements can be considered as important push factors during the early phase and as influencing the quality of implementation as well as the efficiency of individual phases. They primarily comprise a clearly defined product-and portfolio strategy as well as clearly defined roles, norms and structures for the organization of product development. Over the course of pre-phase zero, innovation opportunities are being searched for, ideas are being generated via market and technology analysis and the new innovation project is launched, with an elaboration of the concept to follow in the ensuing phase zero. In phase zero, not only customer needs, but also market segments, competitive situations and business prospects are identified. Finally, in phase one, the technological and economic feasibility of the product concept is assessed and the product development concept is planned. The early phase of an innovation project eventually ends with a decision on the continuation or conclusion of the presented business case, presented as go/no-go decision (Khurana and Rosenthal 1997, 1998) (Fig. 4).

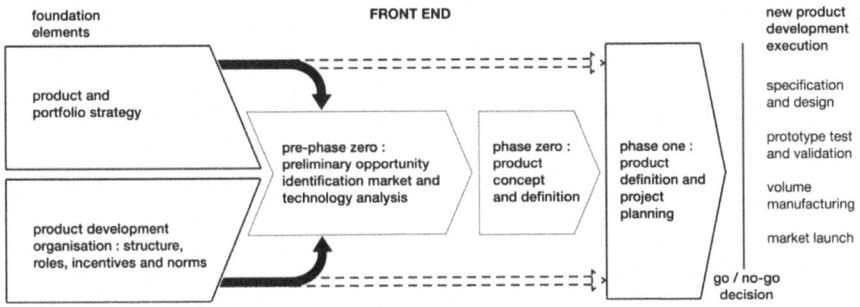

**Fig. 4** Three phase front end model

## 2.4 New Concept Development Model (Koen et al.)

The New Concept Development (NCD) Model from Koen et al. is intended to help people to better manage the early stages of the innovation process and to provide a common language on the front-end activities. It consists of the three parts engine,

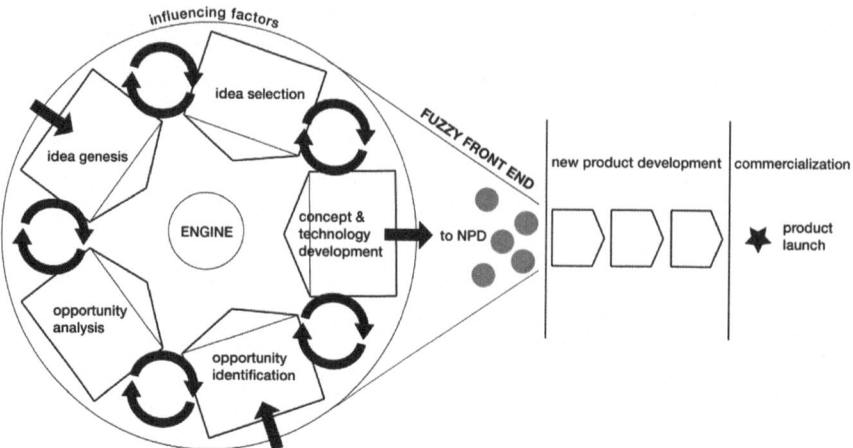

**Fig. 5** New concept development model

front-end elements and influencing factors. A characteristic feature of the NCD Model is the circular, iterative arrangement of the five front-end elements. They are not subjected to any particular order, but can be carried out at random, as often as desired, in parallel fashion or consecutively. In the course of opportunity identification, taking into account the goals of the company and resorting to tools and techniques (e.g. brainstorming) as well as problem solving techniques (e.g. causal analysis), potential chances respectively possibilities are being found, until finally at the stage of opportunity analysis, technological and market-related criteria are used to assess the question of whether the pursuit of an opportunity makes sense. In the phase of idea genesis, detailed ideas are developed in an evolutionary, iterative process. The most promising ideas are selected in the following process of idea selection. The engine of the front end elements comprises all factors that can be controlled and steered by the company (e.g. leadership, culture, business strategy) and create an environment for successful innovation. In addition, internal (organizational skills, technologies, strategy) as well as external strategies beyond the company's control (channels of distribution, customers, competitors) also influence the front end of innovation (Fig. 5).

In their NCD model Koen et al. put the focus on the product development aspect and integrate the technology process development only partially – if at all. The larger the investment into a technology development process is, the more resources are needed, the more structured is the way the decisions are made and the less likely is the integration of technology development into the framework of the NCD process (Koen et al. 2001, 2002).

Table 1 summarizes the advantages and disadvantages of the three described models.

**Table 1** Pros and cons of front end process models

| Model | Pros | Cons |
|---|---|---|
| Stage-gate process (Cooper) | Very famous and frequently cited model | Product concepts can be stopped to early |
| | Flexible to both radical and incremental innovations | Gatekeepers low level of knowledge can lead to wrong decisions |
| | Integrates both the market and technological perspective | Lack of flexibility due to sequential approach, except third generation model |
| | Activities are performed in parallel fashion | |
| Three phase front end model (Khurana and Rosenthal) | Additional consideration of elements of the organizational environment (foundation elements) | No feedback loops |
| | Useful tool to visualize and structure front-end activities, reduce the fuzziness and ease communication | No description of the preliminary opportunity identification and idea generation in detail |
| | | Tool lacks flexibility |
| | | Decision making could be enhanced by a more structured process (especially in the pre-phase zero and phase one phases) |
| New concept development model (Koen et al.) | Includes all company related factors | Abstract model that is hardly transferable to a business situation |
| | Stimulates innovation due to its non-sequential order of phases | Practitioners criticize the lack of application of these methodologies |
| | Flexible with regards to both radical and incremental innovations | Model mainly focuses on product development |
| | | Influencing factors are not controllable |

## 3 Conceptual Design of a Process-Oriented Framework for Structuring the Front End of Innovation

To balance the aforementioned conflict between structure and creativity respectively flexibility, a new scalable process-oriented framework for structuring the front end of innovation was developed. Figure 6 shows the fundamental structure of this framework and its four modules. The module *innovation strategy* encompasses three stages and is dedicated to strategic oriented opportunity identification. The integration of *technology development* (TD) as a main module of this framework is due to the fact that although TD projects represent a small proportion of a typical company's development activities, they are often vital to the company's growth and survival. Therefore TD projects have to be selected and managed in a systematic and focused manner throughout a well-defined process model (Cooper 2006). Because TD projects are quite different in terms of risk, uncertainty, scope and cost of typical *new product development* (NPD) projects, these processes have to be different from traditional NPD processes. Nevertheless this module is only relevant

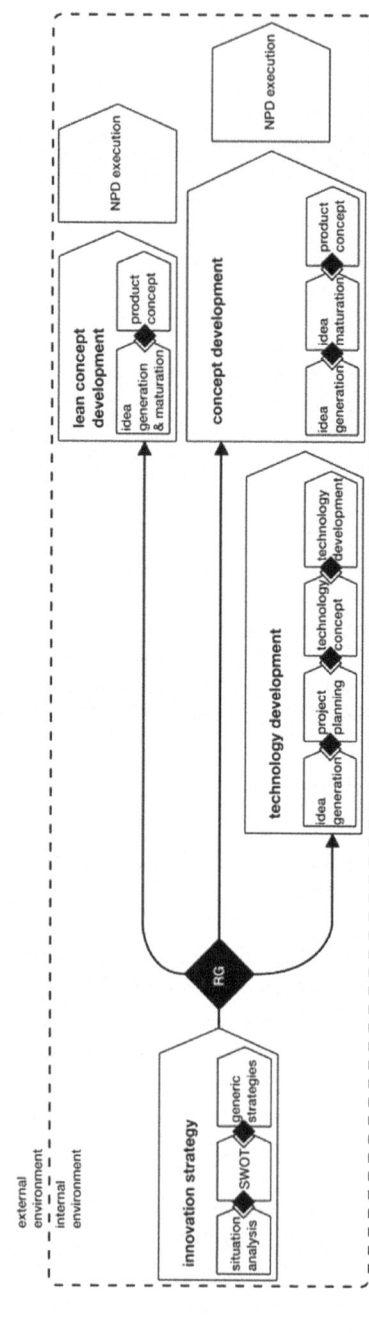

**Fig. 6** Holistic framework for the front end of innovation

when the output of the strategy phase aims to develop new technologies and is therefore an optional part of the framework.

Because theory and practice show that one innovation process model does not fit all projects, the developed framework is a scalable model and includes two different front end processes for concept development. In accordance with Cooper (2008b), major new product or platform developments have to go through a finely structured multi-stage front end process whereas moderate-risk development projects such as modifications, re-launches and extensions follow a leaner process with fewer stages and gates. Hence it is essential that the routing decision at *routing gate* (RG) for the type of process depends on the novelty degree and on the risk level of the potential project respectively. *Pöttinger*, a leading Austrian manufacturer of farm machinery, for example has implemented two different process models for predevelopment activities and series development depending on the novelty of the idea.

In detail, at the aforementioned gate RG a 'strategic courses of action related to innovation' which is the outcome of the strategic phase has to be assessed in a two-step procedure. First, it has to be evaluated if the potential project is targeting either towards new correlation effects between the natural sciences and technical advancement or towards the enhancement of an existing product or the development of a new product. In the first case the process technology development must be chosen. If the identified strategic courses of action is related to the improvement of an existing product or the development of an new product the second evaluation stage has to be executed. Ideally, thereby an utility analysis should be applied, which assesses every project regarding estimated development costs, potential payoffs, and the novelty degree regarding technological, environmental, organizational and market aspects.

Generally all activities encompassed by the model should preferably be carried out in a parallel manner at all stages, both within disciplines (e.g. technology, marketing, design) as well as across disciplines. Here parallel means that inside the stages iterating and overlapping activities of the multidisciplinary team members are typical. Moreover, two characteristic features of this model are its openness in the sense of the open innovation approach and its fuzzy gates, where projects can be continued to some degree even if not all criteria for the respective stage have been met.

## 3.1　Innovation Strategy

Without a clear innovation strategy, decisions at the front end of innovation become ineffective. An innovation strategy expresses a company's long-term innovation goals and primarily comprises all strategic statements on development and marketing of new products, technologies and procedures as well as on the opening of new markets. Innovation strategy is always part of a set of strategies. Its objectives are derived from the overall corporate strategy and it is linked particularly to marketing strategy (Trott 2012). A clearly defined innovation strategy determines where a company wants to focus its R&D efforts and therefore where it wants to search for ideas (Cooper 2011).

The process of formulating an innovation strategy can be set off by various impulses. Relevant literature distinguishes the two prototypes of innovations initiated by newly developed technologies (technology push) and innovations triggered by customer's needs (market pull) (Deppe and Kohn 2002). In the first scenario, innovation activity is mainly competence-driven, while innovations in the second scenario are being developed with the goal of satisfying specific or sometimes latent customer needs. In general, the two prototypes do not occur in pure form. Rather, we can observe multiple factors triggering innovation in parallel fashion. Regardless of the engine of an innovation, the decisive factor for its success consists in acknowledging customer needs and problems by matching offers, thereby generating value for the customer and also, in the final analysis, for the company.

Several steps have to be taken when implementing an innovation strategy. The starting point consists in comprehensive analyses of the company-internal and external situation. These analyses generate information on existing product offers and possible innovation potential, providing a basis for defining and implementing concrete strategies. In order to ensure long-term market success, innovation management has to carry out a situation analysis (Cooper 2011). This assessment focuses on the current and future economic and technological situation of the company and the relevant business environment (internal assessment).

Since development and changes of the business environment considerably impact the success of a company, it is essential to analyze current and future developments outside the company by means of a systematic *external assessment*. This procedure primarily serves the purpose of identifying new *innovation opportunities*. At the same time, it is meant to identify developments that constitute *threats* to the company's success. In order to identify opportunities and risks, one needs to select from all possible variables of the company's micro and macro environment those that are especially relevant to the company's specific decision-making situation.

When defining innovation and new product strategies, it is advisable to conduct a PESTEL analysis focusing also on technological developments, since a technology-oriented early warning system gains in significance as product life cycles become increasingly shorter (Gaubinger 2006). As a matter of fact, it should be noted here that the early identification of chances and risks is of central importance, since the resulting head start constitutes an essential competitive advantage in an increasingly dynamic and discontinuous business environment. However, when aiming for innovation opportunities within the framework of strategic planning, simply focusing on the business environment is not enough. The company also needs to have the necessary competence and potential in order to actually use those chances.

An *internal analysis* serves the purpose of identifying the strengths and weaknesses of a company in relation to its strongest competitors. In many cases this task is organized by evaluating each of the activities of the value chain. In a second step, the company's strongest competitors are assessed in terms of their potential. In a final step, a figure depicts the company's strengths and weaknesses in relation to its competitors in a polarity profile. The profile yields important information with regards to the relative potential for developing product innovations. In the course of the SWOT analysis, identified strengths and weaknesses as well as opportunities and risks are related to each other. In this context, the aim of the SWOT analysis consists in deriving concrete strategic courses of action related to innovation. Once the analysis of the strategic starting position has been completed, the results of the analysis can then constitute the basis for making long-term decisions with regards to innovations.

Thereby the following two generic archetypes of strategies can be used to define the fundamental orientation of the innovation activities:

**Product-Market-Focused Strategies.** Based on Porter (1980) a company has the option to pursue differentiation strategy, which focus on the development of distinctive offerings delivering superiors value to the target segment. Another option is to focus on cost leadership where a company competes by being able to lower its costs relative to the competitors. This can be achieved through downsizing, experience effects in volume production or organization improvement by leaner processes. The third approach targets a specific need of a market niche where the competition is less intense.

**Time-Based Strategies.** This set of strategies refers to the aspect, whether a company intends to be a first or early entrant into the market or a late entrant. In following a pioneering strategy, a company pursues the goal of occupying the position of innovation leader in the market, e.g. of being the first one to place new developments on the market. On the negative side of the balance sheet is the fact that it usually falls on the pioneer to build up the market for an innovation and to be the sole carrier of costs for communicating the perks of the innovation (Walker and Mullins 2011).

A second option with regards to market entry consists in pursuing an imitation strategy, i.e. observing the innovation activities of the competitors and imitating promising innovations. Pursuing this strategy, the company can either take on the role of the *fast follower* or that of the *late follower*. The fast follower enters the market shortly after the pioneer with a comparable product, while the late follower postpones market entry until the point when market developments and demands have stabilized. The fundamental chances following from the pursuit of an imitation strategy consist in minimizing risks and costs of market entry in relation to the pioneer, and in utilizing the pioneer's experience for the company's own product optimization. However, the strategy's drawback is the fact that it does not yield pioneering profits, only the status of a productivity or efficiency leader. If customer needs are already being met in a satisfactory way by the pioneer, this market scenario provides significant obstacles for successfully competing against the pioneer (Ahmed and Shepherd 2010).

## 3.2 Technology Development

Following Cooper (2006), the technology development process 'feeds the NPD process' and consists of the following sub phases: idea generation, project planning, technology concept and technology development. Especially in the field of technology development, idea generation is often done by members of the R&D department, but it should be also the result of other activities. As mentioned above, ideas should be proactively generated within the strategic areas, which are defined in the strategic planning phase. Furthermore, the results of the external assessment (technology forecasting, scenario planning, customer analysis) should focus the idea generating activities. Within these search fields a lot of technology ideas should be generated, derived from different internal and external sources of information. A structured suggestion scheme and the integration of lead experts and lead users ensure the appropriate direction of the idea-finding process. Finally, alternative ideas for new technologies are evaluated in interdisciplinary teams, which consider a certain market orientation also within the context of technology development projects.

In the phase of project planning it is necessary, that the whole development team creates a general state of knowledge (Slama et al. 2006). Essential activities in this phase are technical literature search, patent and IP search and a preliminary technical assessment (Cooper 2006). Based on these activities, an initial project plan is prepared. Since TD projects are usually based on a set of vaguely defined market information at project start, project planning must be specified with increasing levels of information in the ongoing phases.

In a next step the technology concept must be defined. Based on a detailed conceptual technological analysis the application potential of the new technology concept has to be evaluated. Since possible areas of applications of technologically induced innovation ideas are often unknown (Herstatt and Lettl 2000), promising areas of application and target segments for the new technology have to be identified (Bower and Christensen 1995). Start of this activity is the determination of the strengths and weaknesses of the new technology (Schwery and Raurich 2004) and the subsequent translation of these features into utility functions. Based on these results a list of potential industries can be narrowed down by means of a stepwise assessment procedure. Thereby industries with potential application fields have to be evaluated concerning their strategic fit and their attractiveness with checklists. With even more detailed analysis relevant target industries and target market segments can be determined. These activities are the foundation for the identification of a pilot customer, who ensures the application-oriented development of the new technology happening in the next phase.

In the technology development phase, the full experimental plan has to be implemented and the technological feasibility must be proved. Effectiveness of development activities can be improved if a potential user of the technology or a potential customer can be already integrated in this phase to evaluate and determine specific technological requirements and basic conditions. The inter-organizational planning and the execution of the project have to be carried out using sound project management tools.

## 3.3 Concept Development

Following the strategic phase or building on existing technologies, respectively, successful products have to be developed for the search fields identified this way. In the first step, concrete ideas have to be seized upon. According to the solutions' degree of novelty, the process of *ideation* can be divided into *idea gathering* and *idea generating*. The process of gathering ideas resorts to existing ideas from various sources, which can be internal or external. In contrast, the process of generating ideas uses an array of methods, e.g. creativity techniques, for generating new solutions. During the process of compiling and storing ideas, problem-solving strategies are systematically categorized and stored in a standardized form, to be evaluated in the subsequent phase of *idea screening*. In the sub-phase *idea maturation* a quick scoping and a further specification of the project is done. Based on these steps technical feasibility, prospective market success and the contribution for reaching the goal have to be evaluated. The ideas selected in this process constitute the basis for a detailed product conceptualization. Product conceptualization takes its starting point from the results of idea assessment, consisting in relatively abstract problem-solving strategies. Through multiple phases of filtering, ideas are being selected, starting with a *rough selection* characterized by a very low degree of specification and using tools such as oral assessment, checklists, utility analysis etc. This process should use criteria for evaluation that can be informative in a multidimensional way in terms of marketability and technical feasibility.

In an additional step, the product has to be further specified within the framework of *conceptualization*. In this step, all of the gathered information on target markets, target groups, competitive situations and potential for differentiation is condensed and recorded in the *product brief* (Werani and Prem 2009). This document essentially describes the product requirements in detail and usually contains the following items:
- Definition and description of the target market,
- Demands of the customers,
- Essential performance data of the product,
- Relevant external and internal restrictions,
- Estimated production costs and product costs and
- Deadlines and project milestones.

Once the product concept has been finalized, a business plan is drawn up, leading on to the clearance for product development. Thus the starting signal for product realization in the iterative cycles of design, prototype construction and testing has been given (NPD execution).

## 3.4 Lean Concept Development

For projects of moderate risk, such as modifications and improvements, the lean concept development process is appropriate. In this process, *idea generation* and *idea maturation* are performed on a reduced level without a formalized gate between these two phases. In contrast to the full concept development process shown above, the gatekeepers to the next phase *product concept* are typically not the senior management team, but mid-management (Cooper 2011). As in the previous phase, the activity list in the concept phase of low-risk development projects is reduced and also the specification sheet is more compact.

## 4 Implementation of Process Models

Due to the cross-functional relevance of the process model described above as well as the strategic importance of the activities encompassed in it, its implementation is a challenging task. Therefore, following the discussion in Cooper (2002), process implantation shall proceed in four steps with regards to the introduction of the stage-gate process.

### 4.1 Teambuilding

Since the introduction of a comprehensive process model requires the involvement of a large number of people, the designated project coordinator, in cooperation with management, has the responsibility of building a *project team*. This team should be representative of functions and product areas centrally affected by the introduction of the process. Thus, it is crucial for success that the team includes leading members of research & development, construction, design, distribution and marketing, in addition to product management. Before the project team takes on its actual tasks, a company-internal *workshop* can create awareness for the urgency for improvements and give interested company members the chance to participate.

### 4.2 Analysis

An essential task of the project team consists in *analyzing* current practices in the areas of product strategy development, product development, product program policy and product maintenance. This can be done by an internal study or an *analytical workshop*. The analytical workshop provides transparency and shared understanding for strengths and weak points of the ongoing formal and informal processes. However, there is the risk of participants influencing each other and of the status quo being only assessed vaguely due to time constraints. *Internal studies*, the second option of assessment, especially suit the purpose of soliciting input from different levels of hierarchy and experts with diverging viewpoints. At the operational level, a combined use of both methods is in order. For instance, weaknesses and potential for optimization can be ascertained by means of an internal study, to be verified and refined later by all respondents within a workshop. All *employees* affected by product-related planning and implementation should be included in the analysis. So a broad spectrum of perspectives on the processes and the challenges can be capture and also increases the assessment's degree of detail. All processes and activities assessed should be depicted graphically in process diagrams and extensively described verbally at the end of the analysis.

## 4.3 Process Specification

Building on the analysis of the company's status quo, the next step is the development of the company-specific process. An *abstract model* of the new FEI process has to be developed, using the model presented earlier as a blueprint. For each stage, its *purpose* and the *main activities* to meet this purpose have to be defined. Further precision and *adaptation* of the draft definitely requires the participation of affected company members and management to make sure that feedback from other company members involved in the process is included. Depending on the size of the company and the degree of employee participation, further rounds of feedback gathering can be planned. The process concept generated in this way has to be approved by management prior to the project team's further specifications (Cooper 2001). Once *approval* has been granted, it is the project team's responsibility to further *specify* the stages of the company-specific process model in terms of the organization and the instruments to be used.

Since most of the tasks in innovation and product management are interdisciplinary, there are many cross-sections between departments throughout the process phases. These cross sections have to be clearly defined and subsequently be complied with. Defining the process organization is closely connected to identifying an adequate embedding in the *organizational structure* of the company.

## 4.4 Process Implementation

Prior to the actual introduction of the company-specific process in integrated innovation and product management, all employees affected by the process, including those who have not been involved in its conceptualization up to this point, should be informed about its advantages. By means of *internal marketing strategies* (e.g. information sessions, intranet, brochures etc.), employees are to be informed on the impact of the new process in a comprehensive way. Once employees have been sensitized to the importance of the new process, training should convey the required *technical knowledge* as well as *personal knowledge*. Once a process has been introduced, it requires constant optimization and *adaptation* to change. Reasons often include the company's use of new technologies (e.g. the introduction of an engineering data management system) or the growth and development of the company itself (Andreasen 2005).

## 5 Checklist

To sum up, the following checklist may serve as a guideline for successfully implementing a holistic framework for the front end of innovation:
- Guarantee support and commitment of top management for implementing the new processes
- Comprehensively assess the micro and macro environment of the company
- Define a clear and transparent innovation strategy
- Communicate the utility of the developed framework for the front end of innovation within the company
- Ensure the use of synergies in respect to market and technological aspects
- Focus on comprehensive market orientation throughout the entire process
- Develop a commercialization concept for all innovation projects early on
- Install a performance measurement system and continuously control cost, quality and time of the projects. This ensures a goal-oriented budget planning.
- Form interdisciplinary teams at all stages and gates of the process
- Involve external stakeholders continuously during the entire process.

# Integrating Customers at the Front End of Innovation

Fiona Schweitzer

## 1 Introduction

Companies are increasingly opening up their innovation processes to absorb knowledge and gain insights from people and institutions outside the firm. Some years ago, *Procter & Gamble* introduced its Connect & Develop strategy with the aim of developing half of its innovations with external partners. Today this goal has been reached, with an average net present value of such innovations exceeding internal projects by 70 %. The new target is to triple the company's annual revenue to three billion dollars with the help of this strategy. In 2007, *Cisco* initiated an online idea competition in order to gain ideas for innovative solutions and future business development. Two thousand five hundred ideas from 104 different countries were blogged within 5 weeks. Based on the winning idea, a new business unit has been established with an investment of ten million dollars. This trend towards *open innovation* and *crowdsourcing* inspires new ways of thinking about customers' involvement in the innovation process. Instead of surveying customers to find out about their needs, companies enable customers to co-create products via toolkits and co-development workshops.

Furthermore, the fast development in the fields of information and communication technologies triggers the development of new and better methods of interactive customer integration. The UK-based company *Realeyes* combines eye-spying webcams and intelligent image processing algorithms to analyze people's facial expressions for negative and positive feelings when they watch virtual product concepts or web ads. Leading car manufacturers already use similar systems to gauge emotional impressions new car models leave on consumers.

---

F. Schweitzer (✉)
University of Applied Sciences Upper Austria, Innovation and Product Management, Stelzhamerstr. 23, 4600 Wels, Austria
e-mail: fiona.schweitzer@fh-wels.at

Although such new techniques of customer integration sound attractive, it remains challenging for an individual company to decide which method of customer integration to use and how to appraise its benefits. When Second Life was a popular virtual world, the light bulb manufacturer *Osram* encouraged Second Life residents to participate in an idea competition in virtual cooperation rooms, and *Toyota* launched a virtual car model and invited avatars to modify and customize it. Both projects failed through lack of customer response (Kohler et al. 2011).

In order to provide valuable customer insights, to explore latent consumer needs, or to find ideas for radically new products, the possible contributions of customers at the fuzzy front end of innovation have to be understood, the targets of customer integration projects have to be set, and the appropriate participants have to be encouraged to get involved. In the following, insights into these issues will be provided.

## 2  Customers as Providers of Needs Information and Solution Information at the Fuzzy Front End

At the front end of the innovation process, which comprises the time between the first mention of the idea within the company and the decision on its approval or rejection for the regular process of product development, uncertainty and equivocality are usually high. Therefore, this phase is often referred to as 'fuzzy front end' (Reid and De Brentani 2004). In order to drive the product successfully from the front end to the next innovation stage, managers need information that helps them reduce these uncertainties (Cagan and Vogel 2002) (Fig. 7).

Such information can be either *solution information* or *needs information* or. Solution information helps to reduce technical uncertainties by providing answers to questions such as: Is the product feasible? How can the product best be realized? Which technologies and materials should be used for the product? Needs information decreases uncertainties about market-related issues such as: Is there a market for the product? Which features are valued by which potential customers? Which needs do we satisfy with the product? How important are these needs for the customer currently, and how relevant will they be in the near future?

Traditionally, consumers were understood as providers of information for reducing market uncertainty. For many years, market researchers have used consumer insights into their needs and consumer feedback on products and product concepts so that manufacturers could optimize their products and adapt them to better meet the consumers' needs.

The challenge of gathering information on the needs and wants of consumers, users and potential users is that such information is sticky. High stickiness means that it is difficult for the manufacturer to access the information, because it constitutes implicit knowledge or is highly specific or coded information (von Hippel 2005). Lack of appropriate instruments and lack of qualification and prior knowledge may make it difficult for companies to absorb such sticky information or may make the absorption task a time-consuming and costly activity (Cohen and

# Integrating Customers at the Front End of Innovation

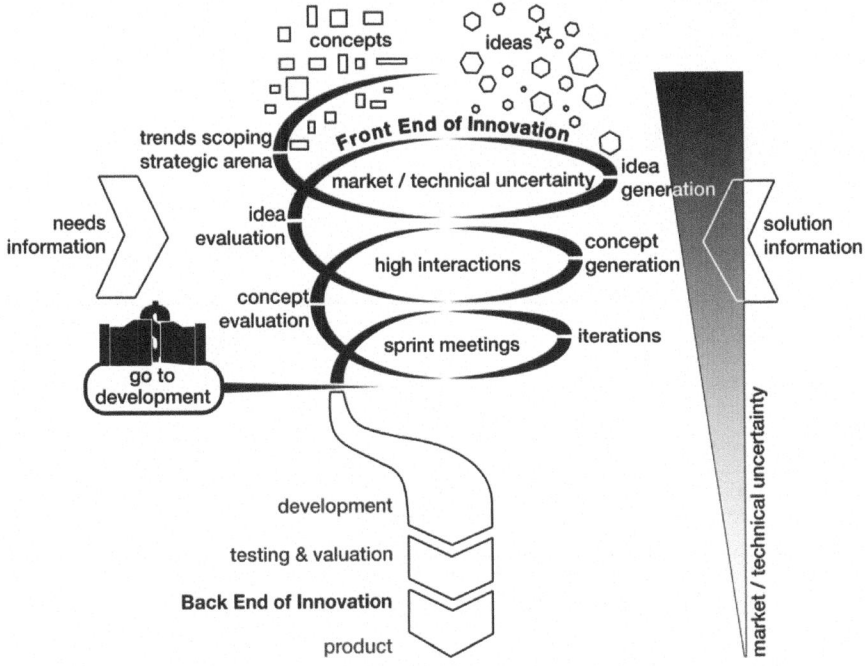

**Fig. 7** Information need at the fuzzy front end of product innovation

Levinthal 1990). A means of improving this resource-intensive *knowledge transfer* is to retrieve not only needs-based information, but also solution-based information from a customer, so that the needs of the customer and his or her knowledge of the product are already incorporated by the customer into an artifact, idea or product concept. Understanding (potential) users as *prosumers*, who are producers and consumers of ideas, supposedly leads to less misunderstanding in the translation of a need into a product requirement and consequently into a product feature. Several new *market research tools* focus on retrieving this sticky information, the most commonly used techniques are the lead user method, user toolkits, and netnographic approaches. The US company *Threadless* is an often cited example of retrieving sticky information through *co-creation*. The company has its customers create their own T-shirts. Instead of analyzing the next season's trends through surveying their needs and wants, Threadless enables consumers to design and upload their self-created drawings that are then produced individually (e.g., Ogawa and Piller 2006).

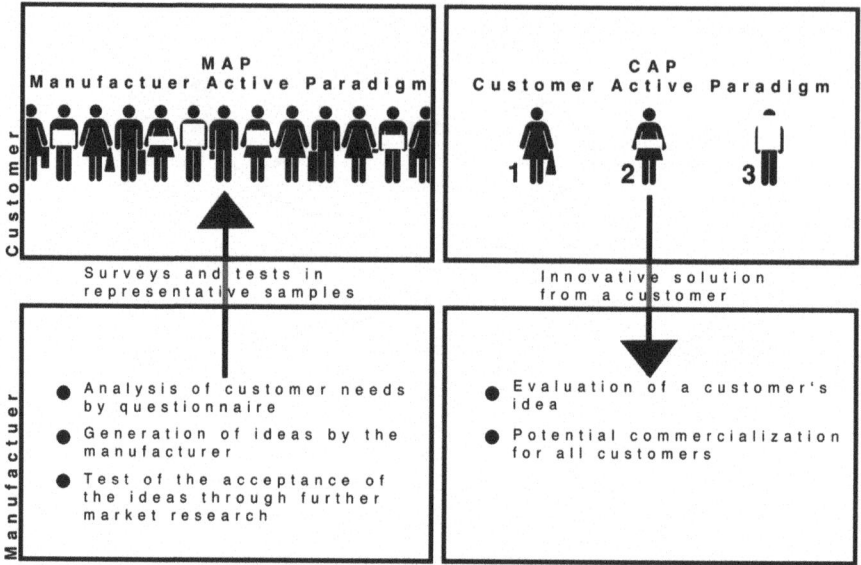

**Fig. 8** Direct and indirect ways of customer integration

## 3 Direct and Indirect Customer Integration

The traditional way of integrating the mind of the customer into the innovation process is characterized by customer orientation in the *manufacturer active paradigm* (MAP). Under this approach, the company indirectly integrates the customer by first retrieving information on needs from attractive customer segments, then developing a product that supposedly meets their needs, and testing acceptance of the product throughout the innovation process via concept tests, prototype tests, product tests, and market tests. In the course of the innovation process, the product is adapted step-by-step to finally meet the requirements of the targeted customers in the best way possible. The task of the manufacturer is to collect information on the customer, analyze it, and use it for developing new products. The task of the customer is to serve as a respondent who provides information when asked to do so.

Under the *customer active paradigm* (CAP), the customer has a more active and direct role in the innovation process. He or she is supposed to develop new product ideas. The idea of this approach is that the *creative potential* and the tacit knowledge of the user are directly used as innovation input in order to reduce the risk and cost associated with innovation. The task of the manufacturer is to discover user innovations, to select the most promising ones, to prepare them for production, and to produce and commercialize them. An extreme example of direct customer integration is the internet start-up *Quirky* that started with six million dollars of venture capital in 2009. Users first have to pay a registration fee and can then put their product ideas on the internet platform. All registered users may then refine the idea and are invited to filter and select the ideas that will be produced. The company owner,

Mr. Kaufmann, facilitates the commercial realization of the product in cooperation with factories in China, Taiwan and the USA within 80 days. The products are sold via the Quirky webpage, and the network of dealers and e-tailers is constantly improved. Thirty per cent of profits are distributed to the community. Besides the idea initiator, people who provide the idea for a name, a slogan, or the final design all participate in turnover and are all mentioned on the package of the product.

In Fig. 8, the difference between the manufacturer active paradigm and the customer active paradigm is illustrated. While under the first paradigm a company uses needs information from the customer through indirect customer integration into the innovation process, it directly uses solution information from the customer for new product development under the latter paradigm.

## 4 Aims of Customer Integration

Customer integration into the innovation process includes all activities designed to directly or indirectly use customer knowledge for process or product innovations. A myriad of different methods and techniques of integrating customers exists, and which one to choose depends on the actual knowledge that the company wants to retrieve (Fig. 9).

Customer integration can aim to gather knowledge on dissatisfaction with current product offerings and customer wishes for improvements of current value propositions (current needs information). To this end, customers are usually confronted with a current product, a product concept, or a product idea and are asked to evaluate its advantageousness compared to rival products and to assess their likelihood to purchase it. In addition, they are usually asked to provide information on points and elements that should be improved in order to make the proposed offering more interesting to them – the customer's task is therefore narrowed down to merely offering input for incremental optimizations and refinements of a company's product. A classic example of a method suitable for assessing this information need is prototype testing. A company carrying out prototype testing cannot expect the user to come up with creative or forward-thinking new ideas on future products, as this is not within the scope of this technique. Yet the method can be applied successfully when looking for advantages and disadvantages of certain features of the product in different usage situations, when aiming to eliminate teething troubles, and in estimating market potential.

Another approach is searching for latent and unarticulated customer needs and sensing developments that can turn into important future market trends (future needs information). In order to identify such trends, customers can be used as sensors. For example, users can be observed in different usage situations, and needs that these customers are unaware of or that are difficult to articulate can indirectly be derived from these observations. For obtaining early information on possible future needs, future workshops can be carried out. In such workshops, qualified customers, for example business-to-business customers who are market leaders in a certain industry, may provide information about new developments on

**Fig. 9** Aims of customer integration at the fuzzy front end of product innovation

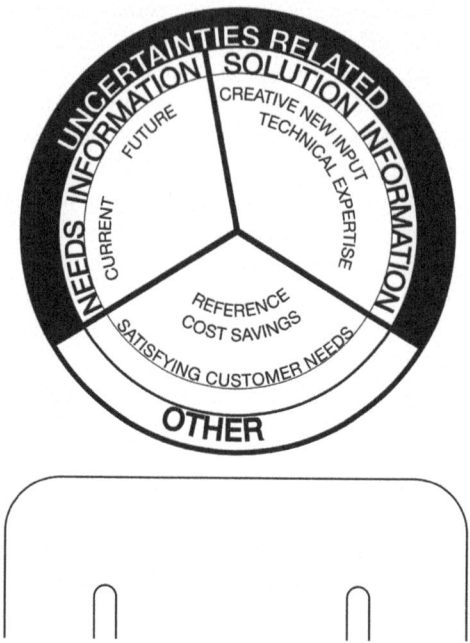

their markets and develop and discuss possible opportunities, trends and future scenarios together with corporate representatives. The main aim of such workshops is to identify weak trends, latent needs and new strategic arenas.

All these insights into customer needs help companies improve their understanding of their customers and of how these customers evaluate products, product concepts, product ideas or specific product elements. Therefore, this information assists them in optimizing the product, choosing the right product features, determining market potential, planning marketing strategies, and reducing market failure.

When solution information is sought, the company looks for information that helps meet actual or latent customer needs through product offerings. Solution information includes ideas on possible new product offerings and on ways in which they can be realized technically. To provide such information, customers have to be qualified in the sense that they need to have a certain level of creativity, technical understanding, and product expertise. Idea competitions, lead user workshops, and co-development are typical integration methods for this end.

Apart from the above-mentioned aims, which focus on reducing market and technical uncertainties, respectively, companies may have more sales-oriented reasons for integrating current or potential customers. They might integrate a certain customer in order to build and maintain trust and a good relationship so that this customer feels tied to the manufacturer, is more involved with the product, and consequently is more likely to buy from the said manufacturer. A further reason for early customer integration might be to gain a reference that facilitates selling to other customers. Moreover, customers who do not participate in the process themselves, but are aware of the participatory approach of other customers tend to be more favorably inclined towards this company, which leads to positive image effects for the company (Fuchs and Schreier 2011). These examples demonstrate that customer integration can be used as a pre-announcement and communication strategy for a new product. Some companies may also wish to involve customers for cost-saving reasons as customers who assume certain tasks also bear the associated risk and cost. As this chapter focuses on market research techniques at the front end of innovation, it primarily explores ways to reduce technical and market uncertainties relating to a planned product offering, while uncertainty-related targets and techniques suitable for reaching these targets through customer integration are not discussed further.

## 5   The Different Types of Customer Integration

Corresponding to the two main types of information companies look for in order to reduce technical and market uncertainties in the early stages of an innovation project, customer integration techniques can be divided into needs-focused methods and solution-focused methods. In Fig. 10, typical techniques representative of these two types of customer integration are illustrated.

**Needs-Focused Customer Integration Methods** include all traditional market research tools, such as in-depth individual explorative interviews, focus-group sessions, standardized surveys (face-to-face, telephone, written, online), or concept testing. Which method to apply depends on the insights that are to be gained. If exploratory tasks are to be performed – such as finding possible explanations for liking or disliking certain product concepts, spontaneous response to, and associations with, preliminary product concepts, or understanding motivations and needs – qualitative research tools (e.g., focus group sessions, in-depth interviews) are most suitable. In this context, psychological methods such as the means-end chain procedure can be deployed to discover fundamental beliefs and attitudes of consumers that drive their purchasing behavior. When relevant facts are to be described and recorded as accurately as possible, such as the relevance of a certain product feature to a target market or the percentage of potential customers who prefer one product version over another, conjoint measurement techniques or other more traditional quantitative methods are to be selected (Mohr et al. 2010).

Although these methods are well proven and widely used in practice, they have some deficiencies. One is the difficulty associated with engaging respondents. Refusals and discontinuation of interviews along with survey fatigue are typical phenomena. New data collection methods such as Securities Trading of Concepts (STOC) address this chink. In STOC, new product concepts are traded as financial securities or virtual shares. The participants can express their desires, needs, preferences and expectations about new product concepts and future events indirectly through the purchase and sale of virtual assets. Material rewards and cash prizes in these stock market games help to increase the participants' motivation to

**Fig. 10** Methods of customer integration at the fuzzy front end of product innovation

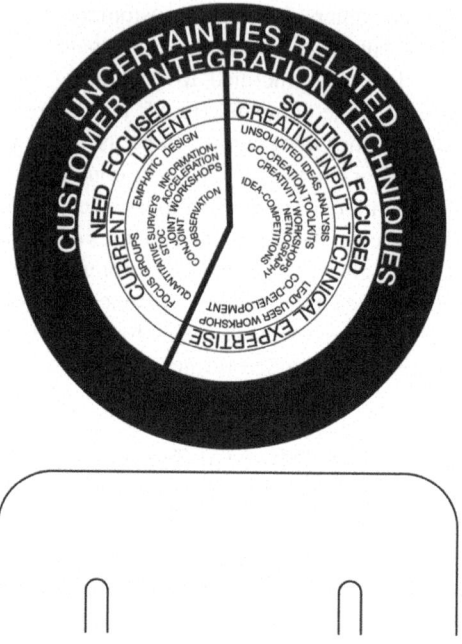

become involved and to reveal their true beliefs (Dahan et al. 2011). By the same token, digital animations and multi-media experiences have been included in focus-group sessions, or virtual communities have been activated with animated avatars to 'gamify' the process of collecting information and make it more stimulating, entertaining and interesting for participants (Kohler et al. 2011).

Furthermore, traditional market research is often criticized for malfunctioning when radical innovations are to be evaluated. Because of their alleged inability to conceptualize and evaluate radically new ideas that go beyond their immediate usage experience, average customers are said to have a tendency to turn down such ideas (Christensen 2000). To improve customers' evaluative capabilities in radical settings, information acceleration was developed. Information acceleration is a concept-testing method which works with interactive multimedia stimuli and uses virtual buying environments (Hauser 1996). Consumers are first immersed in future product environments through methods of a sensory nature, for instance through 3D modeling, and are then asked to provide feedback on virtual prototypes. As a result, many variations of a product can be tested at low cost at an early stage of the development process. This leads to accelerated product development learning and positively impacts on time to market. Data gathered on purchase intentions in this virtual environment allow simulation of consumers' future response to the new product and prediction of sales potential (Paustian 2001).

Many traditional research tools are also criticized for their limited suitability for detecting latent needs. In defense of these tools, it has to be mentioned that in-depth interviews can be a powerful method for this end when properly applied by experienced researchers with qualified knowledge in psychology, sociology and market research. Yet, they are often misunderstood as a fast and cheap tool for assessing customer orientation, and such cursory and half-hearted interviews will consequently detect only obvious trends and needs that the company already knows about. Another effective way of exploring needs which customers are unaware of or which are difficult to articulate are observational techniques, such as empathic design or ethnographic product studies. In both approaches, people are observed while using a certain product. Empathic design consists in short-term observation of product use in the natural environment of the user and focuses on observing several different use situations of many different users. *Viking*, an Austrian manufacturer of high-quality robotic lawn mowers, lawn tractors, garden shredders, scarifiers and tillers, used empathic design to analyze the core processes and activities of private gardening throughout the four seasons and to detect user problems with gardening products in order to develop new products and product features, respectively to improve existing products and features. Empathic design enabled the company to find several ways of improving the ergonomic design of its products, which are now being discussed internally.

Longitudinal ethnographic product use studies usually include not only observational, but also interrogative techniques and aim at immersing the researcher in the life of a certain user or a small user group for a longer period. Researcher and researched subject usually live closely together, which enables the researcher to obtain a profound understanding of contextual and habitual use patterns along with lifestyle and values of users (Leonard and Rayport 1997). When *Volkswagen* encountered difficulties in meeting their sales targets in the US, the group carried out the Moonraker project, in which an interdisciplinary team of 23 employees in turns lived in California for 13 months to increase their understanding of the U.S. market and the American consumers, especially their distinctive tastes and needs. In search of marketing insights, and in order to determine and experience the vehicle needs of VW in the world's biggest auto market directly, the Moonraker team drove across the desert, struggled through snowstorms, hung out with surfers – team members accompanied their subjects at every turn in their daily routine. Finally, the process of translating the findings into new products for the U.S. market took several months to years and led to successful product and service innovations (Kurylko 2005).

Another method that was developed especially for tracing latent needs is *outcome driven innovation* (ODI). The outcome driven innovation method aims at revealing, assessing and analyzing customer needs systematically. Using ODI, customers are not asked how a product could be improved: in qualitative interviews, they are required to indicate which functions and benefits they want from a product. For example, customers do not want a cell phone, but they want to communicate over a distance. The qualitative interviews hence focus on gathering customer requirements. In a second step, a quantitative survey is carried out, in which respondents state the levels of satisfaction and importance which they attribute to the elicited customer requirements, which leads to a clear ranking of the most important, but hitherto unmet, needs of customers (Ulwick 2005). Yet another approach consists in not asking for needs, but for solutions.

**Solution-Focused Customer Integration Techniques** use special skills or know-how of customers to develop solutions that will satisfy needs or solve problems that have been discovered or specified through needs-focused customer integration techniques. Theoretically speaking, the special competence that is looked for can be varied, but usually it centers either on the customer's creativity or on his or her technical expertise.

In order to activate and use customers' *creative potential*, it is common to use idea competitions – notably online idea competitions. Another method is to hold creativity workshops with customers. In such workshops, a task, question or problem is presented, participants are introduced to one or more creativity tools, and these tools are applied step-by-step to develop creative ideas under the instruction of a skilled moderator. In the last few years, combined creativity workshops, in which selected customers and employees participate, have increasingly gained popularity. Through these workshops, the company can obtain access to its customers' knowledge on technologies, technological trends, current and future markets, and market trends (Jungk and Müllert 1987). At *voestalpine Stahl*, a leading Austrian manufacturer of high-quality steel, a future conference was used to engage with forward-looking customers in order to sense new opportunities. An interdisciplinary team of 60 people from the customer company and the steel company met for a 3-day joint conference to discuss relevant trends and future challenges for the year 2020 and beyond. Through this conference, five new search fields for innovation were identified, and a host of new ideas were developed that are now under internal evaluation. Additionally, the conference had a very positive effect on the relationship between the two companies (Peruzzi 2011).

Toolkits are increasingly common for consumer innovations, notably in the fashion industry. With toolkits, manufacturers no longer need to identify upcoming design trends and select the right designs for the next season going through costly and time-consuming iteration slopes, but it is left to the customers to configure their own products, and they are offered a new dimension of shopping experience. Customers can be creative in a controlled way by choosing from different motifs and colors and by up-loading their own designs. The manufacturer provides the mass-customized products and can decide whether or not to include the most popular and attractive designs in its product line. *Lego* – a provider of toys, experiences and teaching materials for children – uses a kind of co-development in its Lego Factory, where children are able to design new Lego models using a digital designer and can submit them to competitions. These ideas represent a primary source for new Lego products. Similarly, *Adidas*, *Nike*, or *Ralph Lauren* have provided toolkits for mass-customization on their web pages for several years and have been able to charge a premium of 30–50 % on comparable standard items. The trend toward user toolkits is also partially picked up by B2B companies. *International Flavors and Fragrances* (IFF) supplies flavors and fragrances to the food and beverage industry and offers customers the possibility to mix and alter flavors from a model kit, carry out iterations themselves, and eventually have IFF produce the final combination. In this way, consumers become co-producers within a pre-set solution space (von Hippel 2005).

While competitions and toolkits are manufacturer-driven initiatives, companies are also reporting that the number of unsolicited ideas supplied by customers is increasing. While this can provide a fruitful pool for inspiration, it poses a challenge at the same time, namely when a customer sends a potential new product idea to a company and the company is actually currently working on this idea: as soon as the product is launched, the customer who sent the idea might have the feeling that his/her idea has been stolen and might claim property rights. For this reason, several big companies have put a policy against unsolicited ideas in place (Shippey 2009).

If a company wishes to benefit from the *technical expertise* of customers, lead user workshops and co-development rank among the best suited techniques of customer integration. The *lead user method* was developed by von Hippel (1986) and is a method of involving users who are ahead of their time and hence feel certain needs earlier than other users. Central elements of the lead user process are the identification of relevant trends, the identification of appropriate lead users as regards these trends, and workshop sessions with lead users for the purpose of developing or refining ideas and product concepts. Although lead users were found to contribute to innovations in several industries (e.g., outdoor sports equipment, medical equipment, *open source* software), their integration may entail negative effects, such as developing innovations that serve only a niche market (Gassmann et al. 2010a).

Co-development, a method of joint product development, represents the highest level of customer integration into the innovation process. Selected customers with high expertise are integrated as cooperation partners, work together with the company's own researchers, and jointly develop innovations (mostly radical ones). Due to the intensive cooperation between the company's researchers and the customers, shared tacit and explicit knowledge can be established.

The diffusion of *web 2.0* technology and virtual communities is leading to the development of new methods of integrating customers. For example, *netnography* – composed of the terms 'internet' and 'ethnography' – is a new tool with broad areas of application. By using netnography, implicit and explicit needs, desires and attitudes of consumers with regard to particular products and brands can be identified (needs-focused application). Unveiled information can even include product prototypes that users have designed and published on the web (solution-focused application). Netnography is regarded as an exploratory method of subtle and unobtrusive observation of the (online) communication flow and the social interaction of a virtual group. The results of this analysis are condensed into so-called consumer insights, followed by a translation into customer-oriented products and services (Kozinets 1998). This technique makes use of the fact that some users who have needs that are not fulfilled in current product offers either criticize these products and exchange pros and cons of products with others via online communities, or become active themselves and develop prototypes or product supplements and share their inventions with other consumers through specialized virtual communities. *Hyve*, a company specializing in screening and analyzing these activities of virtual communities, has already helped companies in such diverse industries as sport shoes, water treatment, credit cards, and food and beverages to gain new product and service ideas and detect interesting user innovations (Füller 2010).

## 6 The Different Types of Customers and Their Role at the Front End of Innovation

In customer integration practice, current and potential customers can be integrated. Such customers include persons, companies, or other entities which bought or will buy goods and services from an organization. Customers are not limited to direct customers, but include all downstream market players, such as intermediaries and customers' customers. Not all of these customers are equally qualified for any customer integration method. Hence a factor critical for obtaining the required information is selecting the right respondent. To provide a guideline, different types of customers and their primary suitability for different types of customer integration are highlighted in Fig. 11.

In order to collect current needs information, the general user in a targeted market or market segment has to be reached. In qualitative research, typical sampling and theoretical sampling are viable sampling methods. In quantitative research, quota sampling or random sampling techniques are used. Convenient samples are to be avoided (Malhotra 2009).

Where more radical changes to products are sought, lead users have turned out to be of more use than the average user. Lead users have needs long before these become relevant to others, and they benefit significantly from having these needs satisfied. Therefore they have a strong self-interest in contributing to solutions that meet their needs. Pyramiding has proven an effective method to find such lead users (von Hippel et al. 2009), along with self-selection procedures such as crowdsourcing mechanisms (Piller and Walcher 2006). In practice, B2B companies often use important current customers as lead users and then wonder why these customers do not provide the information they have actually been looking for (Gassmann et al. 2010a).

Emergent users are another potential group of customers with whom innovative, new products can be co-developed. Their personal characteristics include optimism, intellective, reflective self-focus, openness to new experiences and ideas, and high levels of creativity. Furthermore, they have superior experiential and rational information-processing skills, both verbally and visually (Hoffman et al. 2010). These capabilities put emergent users in a position to envision how concepts have to evolve in order to be of interest to the mainstream marketplace.

For B2B companies, customers who are market leaders in an attractive industry can also function as channels through which they can absorb information on trends, future needs, and future opportunities in these markets. In addition, companies can also use qualitative research techniques (such as focus groups or in-depth

**Fig. 11** Customer types and their input at the fuzzy front end of product innovation

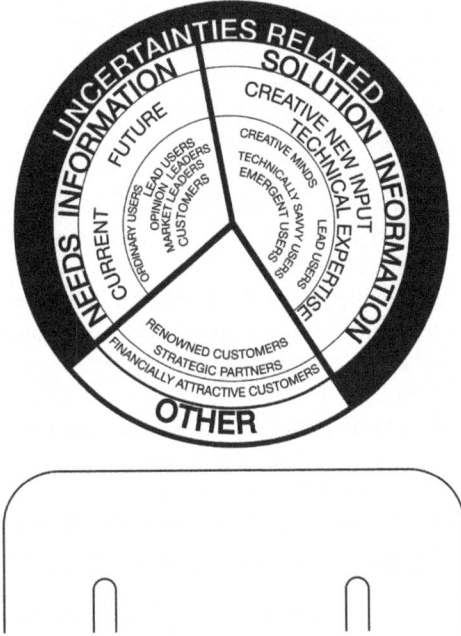

interviews) with customers' customers who have the ability to think ahead in order to retrieve information on future needs.

While the necessity to obtain needs information from customers is basically widely accepted in marketing research and practice, consumers' ability to provide useful solution information is the subject of intense discussion.

Opponents hold that articulated ideas are limited by experience and that the functional fixedness inspires only familiar product versions, rather than really radical and creative solutions (Christensen 2000). Moreover, the technical, procedural and intellectual knowledge and expertise of R&D personnel is said to put these specialists in a better position to solve consumer problems and provide solutions than customers (Ulrich and Eppinger 2008). Yet, empirical studies – for example in the global robotics industry – have found that a firm's professional use and re-use of existing knowledge leads to a certain point of exploitation beyond which additional internal knowledge results in reduced new product output (Katila and Ahuja 2002).

A recent experiment by Poetz and Schreier (2012) provides additional evidence that user ideas do not necessarily have to be less innovative. On the contrary, the researchers found that ideas from users outscore ideas from professionals in terms of novelty and customer benefits. Users created more original mobile phone services, which might be due to users harboring fewer concerns regarding technical feasibility. Professionals might be too focused on convergent thinking (i.e., on how an idea can be transferred into a service offer) to unleash divergent thinking mechanisms, which are necessary for developing completely new ideas. Yet it is likely that the ability of users to create valuable ideas differs on an industry basis and might be more pronounced with customers that have a certain amount of technical understanding in high-tech industries and other industries with complex technical products.

As far as solution information is concerned, it is hence likely that customers have to fulfill certain requirements to be attractive for customer integration projects. Basically, they should be creative and interested in tinkering and experimenting to deliver creative new input, and they should be technologically savvy to contribute to technical problem-solving tasks. Competence and experience of solution providers are deemed essential to deliver high-quality solutions. Lead users and emergent users reportedly have these strengths (Magee 2005).

When customers are integrated mainly for reasons other than reducing market and technology uncertainty, such customers may need to fulfill other requirements to qualify as partners. Financially attractive customers may be interesting when reducing the risk and cost of innovating is a central driver, a customer's reputation will be important if the customer is to be used as a reference (Gruner and Homburg 2000). For example, the Swiss lighting expert *Zumtobel* integrates renowned architects into the innovation process as co-developers to design high-quality lamps that are usually used in highly prestigious building projects and captivate through their innovative forms and functionalities. Some of these lamps enter the company's standardized line of products, most, however, are limited editions only. They are usually very challenging from an artistic point of view, but only incrementally new from a technical perspective, and not really attractive financially. Yet the company pursues such projects, as art and culture are important corner posts in the company's culture, and because they elicit positive publicity and image effects. Criteria for selecting architects as strategic partners include innovativeness, high standards of quality, and a track record of top-end building experience (Gassmann et al. 2010a).

## 7 Motivating Customers to Participate

Knowing about motivations for customers to participate and choosing the benefits offered for participation accordingly are central issues for successfully integrating customers into the fuzzy front end of product innovation. If a company looks for innovative, new product ideas, but is unable to attract qualified creative minds, the ideas collected might be of little value. If a company develops a sophisticated questionnaire, but does not receive representative customer response, results may be meaningless. Benefits that attract participants can vary from one participant to another according to their respective motivation. Basically, customer motivation can be classified into intrinsic, internalized extrinsic, and extrinsic motivation (Myers 2004).

Participants who are intrinsically motivated take part in a customer integration activity for the sake of the activity itself, for example, because they enjoy it or because they are eager to know what it is about. Although it is known from psychology that internal motivation is a central driver of performance, often the playfulness and enjoyability of the interaction experience itself are neglected when the interaction process is designed. New integration techniques hence focus on increasing fun and involvement.

Internalized extrinsic motivation to participate includes altruism, the wish to build and develop one's own skills and competences, acquisition of valuable information, and self-efficacy understood as a sense of accomplishment that results from one's contribution. Receiving credit for one's contribution, such as being officially named as a co-developer, is important in this respect as it helps participants to signal competence, to gain peer recognition, and to satisfy the longing for ego gratification (Lerner and Tirole 2005). Internalized extrinsic motives may not only play a role in solution-based integration activities, but also in needs-focused endeavors, for example, when participants in focus group sessions reveal opinions and ideas whilst looking for approval of their competences, positive feedback, and recognition from the moderator or other participants (Gassmann 2001).

In cases where contributions are made owing to intrinsic or internalized extrinsic motivation, consumers participate without expecting any tangibles in return. In contrast to such free-revealing behavior, externally motivated customers participate because they are interested in the outcomes of their efforts. They might either expect a certain advantage (e.g., a reward) or wish to avoid a certain disadvantage. Compensation, be it in-kind or monetary rewards, is often used to assure participation through external incentives. For example, customers can be compensated for filling out a questionnaire with free give-aways, small monetary refunds, or the possibility to win in a raffle, and the winners of online idea competitions can be rewarded financially (Kohler et al. 2011). Innovative approaches to making inventive customers profit from the outcome of their contributions are creative common licenses in software development (Achtenhagen et al. 2003). Other external motivators may be personal dissatisfaction with current products, the benefits of having a new product designed according to one's own needs, being among the first customers to benefit from the product, or having exclusive rights of use. For example, users' outright benefit from a programmed code has been identified as a key motivator for a user's willingness to invest time and effort in code development in the fields of open source software (Müller-Seitz and Reger 2009). In the same vein, fear of not being included in further innovation activities, seeing the company develop products that meet the interests of others (e.g., competitors) better than one's own interests, or losing a long-standing relationship with current suppliers can be important motivators, notably in business-to-business relationships.

According to these basic motivations to participate, customers can be divided into the following five key groups: win-oriented, intrinsically interested, curiosity-driven, need-driven, and refund-driven participants. The first two groups are particularly interested in solution-focused customer integration techniques. They are highly skilled, creative, and innovative persons who like to solve problems and to toy with possible solutions. While win-oriented customers are mainly interested in winning the monetary prizes offered for solution-based activities and in actually solving problems, intrinsically interested individuals are highly innovative novelty seekers who love experimenting and are driven by feedback and recognition. Curiosity-driven customers like to experience and test new products and want to get hedonic and inspiring experiences from customer participation activities (Füller 2010). Need-driven customers only participate, because they want to obtain a solution to their specific problems, and are usually more useful in needs-focused customer integration techniques than in solution-focused settings, as they are low in innovativeness and novelty-seeking behavior, an exception being need-driven lead users. Refund-driven participants take part in needs-focused customer integration activities such as surveys because they want the financial reimbursement, online panel members being a typical example of this group of informants.

## 8 Using Results of Customer Integration Projects

As discussed above, efficient and effective customer integration calls for selecting those customers that can really contribute, finding ways to motivate them to participate, and selecting the right procedure to integrate them. Companies should be careful when carrying out all these steps and should be aware that selecting and implementing the right procedure necessitates time and financial resources (Littler et al. 1995). Wrong partners, wrong tools, and improperly applied integration tools can be harmful. Frankly, the correct application of the method and analysis of the results is a challenge that should not be neglected. For this reason, a company that applies a specific customer integration method for the first time should consider obtaining professional assistance from a market researcher or consultant with the necessary methodological competence.

Moreover, companies have to handle IP issues concerning the ideas and concepts of users with care. As the customer is involved in the manipulation and creation of sensitive data, trustworthiness and commitment of the customer are of importance in any customer integration process, but are even more important in connection with solution-focused than with needs-based information. Companies fear opportunistic behavior of the customer, loss of proprietary information and know-how, and problems in assigning proprietorship. Non-disclosure agreements, building strong interpersonal relations, high embeddedness of the partners, selecting partners with past experience in co-development, and using long-standing contacts are only some of the most common methods to reduce such risks (Littler et al. 1995). The trend towards using well-known customers may, however, come with a cut in novelty of information (Rindfleisch and Moorman 2001).

As a prerequisite for successful customer integration, companies must furthermore ensure that the company's innovation culture and process are ready for customer integration. For example, the *not invented here (NIH) syndrome* – a negative attitude towards the acquisition of external knowledge – can lead to ideas being rejected and de-emphasized by an organizational unit if they stem from consumers. The NIH problem is often caused by fear of losing direct control of the development process. Hence, sufficient openness of the employees and a company culture that shows respect for adopting external ideas are critical factors for successful customer integration (Katz and Allen 1982).

For example, allowing customers to alter one's product and to add new functionalities as in the case of open source software requires high technical skills and specific knowledge of legal issues and business models: empirical data indicates that companies that embrace customers into their innovation process even have more highly skilled employees than more traditionally-minded competitors (Harison and Koski 2010). The fear harbored by employees that their knowledge will no longer be of use when customers are integrated into the innovation process therefore seems unjustified.

Last but not least, the expectations, requirements and needs of customers are often neither understood nor interpreted in the right way, in particular in the case of needs-focused integration techniques (von Hippel 2005). Among the main reasons for this phenomenon are failures and omissions in information gathering and analysis. The case of 'New Coke' illustrates this challenge. In the 1980s, *Coca Cola* lost US market share to its competitor *Pepsi*, allegedly because of *Pepsi*'s better taste. *Coca Cola* developed a new recipe. Blind tests revealed that its taste was preferred over both the traditional formula and Pepsi's formula by 60 % of respondents. When *Coca Cola* re-launched the product, it failed, and consumers protested that they wanted their original Coke back. *Coca Cola* had ignored the fact that American consumers saw *Coca Cola* as an American institution and symbol and wanted it the way it was. The new taste should rather have been introduced as a new version, such as cherry coke or coke zero (Kotler and Keller 2006). Another problem that frequently arises in needs-based integration is improper inter-functional knowledge transfer. The extracted needs have to be transformed and translated into clear parameters and specifications for product development. This task is often performed improperly due to friction caused by different ways of thinking and disparate terminologies used by marketers and technicians (De Luca and Atuahene-Gima 2007).

## 9 Checklist

To sum up, the following checklist may serve as a guideline for successfully integrating customers at the fuzzy front end of product innovation:
- Definition of goals: the objectives of integrating the customer have to be clarified. What result is expected? Should new insights into needs be retrieved, will product concepts be evaluated, or are new solutions required that address well-known needs?
- Selection of appropriate method: Corresponding to the integration goal, the most suitable method has to be selected. Attention has to be paid to internal requirements. Will traditional market research tools be applied? If so, are these tools of a qualitative or a quantitative nature? Are internet-based tools an option? Will sticky information be retrieved through toolkits or netnography? Can

observation be an interesting method of detecting latent needs? What experience with customer integration already exists in the company? Which employees have to be involved to carry out the project? Can the customer integration process be completed with the help of internal resources only or will external consultants be involved?

- Selecting the right customers: Customers who are willing and able to contribute have to be chosen. What are to be the selection criteria used for finding these customers? How can the customers be motivated to participate? What can the customer expect in return, e.g., reputation, financial reward, potential final outcome?
- Application, knowledge transfer and commercialization: The selected method of customer integration has to be implemented, and the results have to be analyzed and used. How is the information gathered best used for current and future innovation projects? What is the best way to share the gathered knowledge with different members of an innovation team? Who will interact with the customer, and who will perform the analysis? Which internal employees should receive the results of the customer integration project? How can the findings be implemented within the company?

While observing these guidelines will not necessarily lead to smooth customer integration in all cases, it will serve to address important issues to be considered in order to increase effectiveness of customer integration at the front end of innovation.

# Out of Bounds: Cross-Industry Innovation Based on Analogies

Marco Zeschky and Oliver Gassmann

## 1 Introduction

This chapter focuses on generating new product ideas in the early phase of the innovation process. This phase is characterized by individual and collective creativity as vital preconditions for new ideas, and therefore, for any physical product development activities. Creativity allows individuals or groups to link distinct pieces of knowledge together to form new combinations, which may take the form of a new solution to an existing problem or a new application to an existing technology. As such, innovation entails reassembling elements from existing knowledge bases in a novel fashion (Hampton 1998).

The *cross-industry innovation* process as described in this chapter is an effective approach to stimulating creativity and making systematic use of already existing solutions from other applications or industries in order to develop entirely new ideas (Gassmann et al. 2010b). Although creativity is primarily immanent in individuals, a large number of mechanisms exist which stimulate creativity and allow it to become effective in groups as well. For example, among the more popular creativity techniques are brainstorming, brain writing, the gallery method, or De Bono's six thinking hats, which make use of the individual's creativity and leverage it within a group to generate new ideas.

A more recent approach to fostering creativity and producing creative ideas is the abstraction of a specific problem and the subsequent search for analogies – a process which often occurs in cross-industry innovations (Herstatt and Kalogerakis

---

This is based on an earlier article by the authors in *Creativity & Innovation Management*

M. Zeschky (✉) • O. Gassmann
University of St. Gallen, Institute of Technology Management, Dufourstr. 40a, 9000 St. Gallen, Switzerland
e-mail: marco.zeschky@unisg.ch; oliver.gassmann@unisg.ch

2005). Cross-industry innovations are innovations where existing solutions are taken from one industry and are applied in a creative way to generate a novel solution in another industry. Analogies are a central mechanism in many creativity techniques (De Bono 1990a), and using analogical solutions, particularly across industry boundaries, can significantly contribute to the development of highly novel innovations while simultaneously limiting the risks of uncertainty.

For example, when the *BMW Group* introduced its ground-breaking man–machine interface iDrive in 2001, it took advantage of an analogous solution from a non-automotive domain and integrated it into a single controlling device. The iDrive is a device for controlling manifold functions in luxury cars, which until then had been manipulated by up to 200 different knobs and switches. The analogy was found in the joystick as an important device in the video game industry, and the respective knowledge was transferred and adapted to the specific requirements in the course of the development process.

Therefore, drawing analogies from an initial problem to distant but similar problem settings reduces uncertainty as potential solutions have already proved to function in a similar context. Another advantage of non-obvious analogies is that they often entail highly novel solutions, because the combination of more distant pieces of knowledge often results in a higher degree of novelty (Hargadon and Sutton 1997). In fact, divergence and lack of shared experiences are sometimes necessary preconditions for developing truly novel ideas at all. Thus, the use of analogies in product innovation entails many benefits; however, such analogies are not easily applied, rather uncovering them typically requires systematic effort. Because of that, the successful development of product innovation depends on the interplay of several factors on the firm, the business unit, and the individual level. Although individual creativity is a vital precondition for innovation, even the best ideas have little chance of survival when supportive structures and processes are absent. To increase these chances of survival, successful companies have implemented an innovation organization consisting of a clearly defined innovation strategy, an innovation process with stages and gates in which initial ideas are continuously redefined and improved, and organizational units which are responsible for pursuing either incremental or radical innovation. In short, innovation is no longer left to happenstance, but is subject to clear strategies and structures. The following sections will focus on the question how *analogical thinking* may contribute to more creativity in the early phase of the innovation process, and how analogical thinking may increase the chances of idea survival.

## 2 Analogical Thinking in Problem Solving

The role and importance of analogies for innovation has mostly been investigated in product design and psychology literature (Dahl and Moreau 2002). However, scholars have recently also started investigating the role of analogical thinking for strategy development in the firm (Gavetti and Rivkin 2005). Analogical thinking is a creative method applied to a problem that needs a solution and takes place if a familiar problem is used to solve a novel problem of the same type. For analogical thinking to occur, the problem has ideally been specified rather clearly. In general, analogies can be drawn in different settings and directions. In some cases, a solution is found in one industry and applied to solve a problem in another industry, as is the case with cross-industry innovations. In other instances, the analogy is drawn from a solution which is 'looking for a problem'. Again, in all cases, the search for a solution is stimulated by a rather specific problem. Within this 'problematic search', analogies to settings quite similar to the original problem can be drawn which potentially provide a solution.

Cognitive scientists agree that innovation entails reassembling elements from existing knowledge bases in a novel fashion (Gagne and Shoben 1997). Thus, analogical thinking is a mechanism underlying creative tasks in which people transfer information from a familiar setting and use it for the development of ideas in a new setting (Gentner and Rattermann 1993). Furthermore, *similarity* of concepts (such as problems or situations) at any level of abstraction facilitates analogical thinking (Ross 1989). Thus, the similarity of some basic elements between the origin of the problem (i.e., the problem source) and origin of the analogy (i.e., the solution source) is a vital precondition for the identification of analogies. Similarity has also been described in terms of a continuum ranging from 'near' or 'surface' analogies to 'far' or 'structural' analogies (Dahl and Moreau 2002). Near analogies are much easier to identify than far analogies, as near analogies often entail obvious surface similarities, such as similar design, while far analogies typically entail similarities in the structural relationships between source and target attributes. For instance, Dahl and Moreau (2002) illustrate the case of designing a new freeway system. A near analogy would imply looking at an already existing freeway system in another city, whereas a far analogy would entail arriving at a solution by considering the human circulatory system. The distinction is important, because near and far analogies require different types of information to be mapped and transferred. With near analogies, both surface-level attributes (e.g., roads) and relations between these attributes (e.g., the flow of cars through the freeway) are mapped and transferred, while the lack of surface-level attributes with far analogies leaves the mapping to occur between shared structural relations. The example intuitively shows that far analogies are more difficult to identify and require more cognitive effort. The identification of far analogies requires the identification of similarities in the relational (vs. surface) structure between the problem and the solution source, which is often difficult when surface similarities are completely absent. However, if successfully implemented, far or structural analogies serve as the basis for 'mental leaps' and can lead to radical innovation (Holyoak and Thagard 1995). On the other hand, if source and target share the same surface qualities, they often come from the same or from close conceptual domains (Ward 2004), which would result in rather incremental innovation. However, surface and structural similarities are two ends of a continuum, and a clear distinction between them is difficult to make. In this chapter, the term 'surface similarities' is used when there are similarities in features, such as product design and product features, and 'structural similarities' is applied when there are similarities in the principal technological function and architecture of the products.

**Fig. 12** AlpineCo: from musical vibrations to ski vibration control

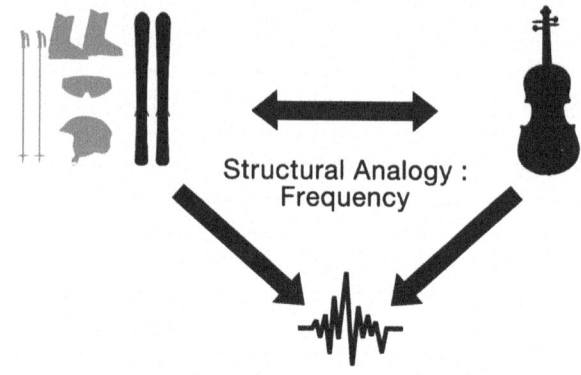

## 3 Some Real-Life Cases on Cross-Industry Innovation

**AlpineCo: Designing Skis by Looking at Musical Instruments.** AlpineCo had the problem that its downhill skis were difficult to control at certain speeds. Analyzing the cause, R&D found that the ski was coming into its resonance frequency at high speeds, which caused the ski to vibrate. During the phase of intense analysis of the problem, the head of R&D and three colleagues delved into the question of how the vibration could be damped or eliminated. From his background as a mechanical engineer, the head of R&D knew that vibrations were a recurring problem in settings such as machine or building construction. With the terms 'vibration', 'damping', and 'cushioning' unconsciously in mind, the team then decided to search for industries and applications where damping or elimination of vibrations were a problem: "we were actively looking for analogous solutions". However, initial search efforts were in vain because the scope of the search had been defined too broadly, as the R&D team was searching for anything that had to do with vibrations. The search was only successful when one team member proposed limiting the scope to include frequencies only above 1,800 Hz, as this was the range of frequency found in the vibrating ski. The team realized that this frequency is typically encountered in acoustics, and AlpineCo ultimately found a viable solution with an inventor who for years had conducted research on the elimination of undesirable frequencies in bowed instruments. The solution also proved to be easily transferable, as the material used to filter undesired frequencies in bowed instruments could easily be adapted to the skis. According to the head of R&D: "It's a simple idea and easily applicable, and did not require any additional investments." AlpineCo then applied the solution to its own demands by developing an extra layer in the ski, the structure and material of which were similar to the bowed instruments, and incorporated it into the ski. This technology is termed 'frequency tuning' and is found in virtually every ski today (Fig. 12).

**Fig. 13** AluCo: developing robust crash boxes

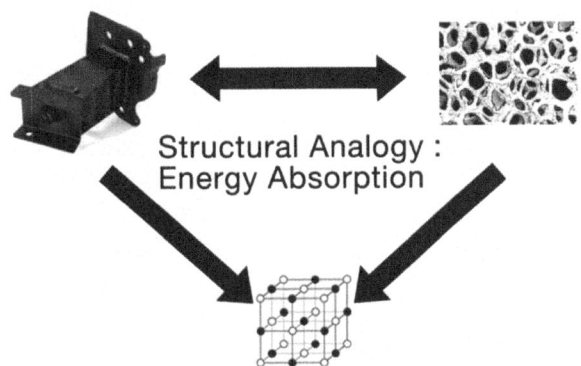

**AluCo: How to Transform Kinetic Energy in Crash Boxes?** For a long time, AluCo had been looking for alternative approaches to improving its crash management system (CMS), which consists of the front beam and two crash boxes that are mounted to the longitudinal chassis beams of a car as one module. Somewhat frustrated with the hitherto 'conventional' approach, AluCo's management realized that mere optimization of materials and tweaking geometric designs would not result in the major advancement it had hoped for. As the head of Future Technologies put it, "we have been doing this for decades now, and I believe our engineers have become too short-sighted to look beyond their own noses." Before 'prematurely jumping to solutions', to quote the head of Future Technologies once more, a team of four R&D employees engaged in an in-depth investigation of the current crash box. They particularly focused on gaining a detailed understanding of the product function, both from a technological and a customer utility point of view. In subsequent workshops, the team analyzed and described the technological function of the crash box first in terms such as 'protecting the car's longitudinal carrier from damage', and later in terms such as 'gliding grid structures in the material'. In the course of the analysis, AluCo developed key terms, such as 'energy absorption' and 'transformation of kinetic energy'. With these terms, AluCo built associations to different kinds of technologies, applications and industries where the absorption of energy was crucial. AluCo's R&D then started searching the internet with a focus on the key terms identified previously. In this way they identified several promising technologies new to their industry, which today are subjects for potential further development (Fig. 13).

**TextileCo: Using Computer Mouse Sensor Technology in Sewing Machines.**
TextileCo faced the problem that the speed of the material displacement was different from the speed of the sewing foot, which resulted in inhomogeneous stitch-lengths and spaces. Thus, first activities aimed at synchronizing the speed of the material displacement with the speed of the sewing foot. Analyzing how the displacement could be gauged under the given spatial constraints, TextileCo's R&D concluded that the displacement of the material had to be gauged with high precision because of the high speed of the sewing foot. As gauging was outside the

**Fig. 14** Analogical thinking at TextileCo

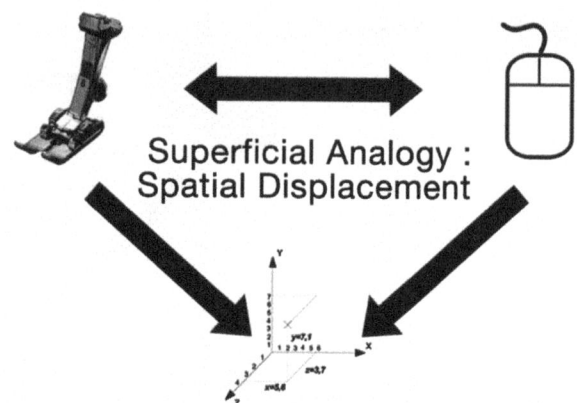

firm's competence, TextileCo decided to look for external solutions. A team of five R&D members started searching for solutions that were related to what TextileCo called 'real-time gauging'. In the course of their search, they approached an external technology service provider, who ultimately supplied TextileCo with a solution in the form of the optical sensor of a conventional computer mouse. The service provider had previously worked on another project where feedback loops played an important role and where a very similar sensor technology was applied. As the head of R&D said, "without the service provider we would never have come up with such a brilliant and simple solution, it took us only 18 months from problem formulation to market introduction, which is about half the time we usually need." TextileCo adapted the mouse sensor chip to its specific requirements and enhanced it so it would even recognize very smooth or dark fabrics. As a result, because of the automation accomplished, even beginners are now able to quilt genuine artwork of high quality. This had previously been the domain of only experienced quilters, and implementing the new technology allowed TextileCo to tap into a new and fast growing market (Fig. 14).

**PipesCo: How to Learn While Watering Your Flowers.** The piping division of PipesCo has profound know-how in production techniques, such as welding or gluing, in combination with material optimization for the joining of pipes. As the industry is characterized by long product life-cycles, the conventional strategy has consisted of constantly improving existing technologies and products. One day, an R&D employee was watering the flowers in his garden and realized that the hose and the sprinkler head were connected via a plug connection: "It was a lucky accident. The basic principle is the same; it's about a medium flowing through a pipe, only the way the pipes are connected is different." He introduced the idea in the company, and preliminary assessments convinced the CTO to pursue the idea, both because of the simplicity of the technology, which would tremendously facilitate the connection of large pipes in construction, and because of the enormous cost savings associated with the new technology. As the CTO put it, "it was a

**Fig. 15** Analogical thinking at PipesCo

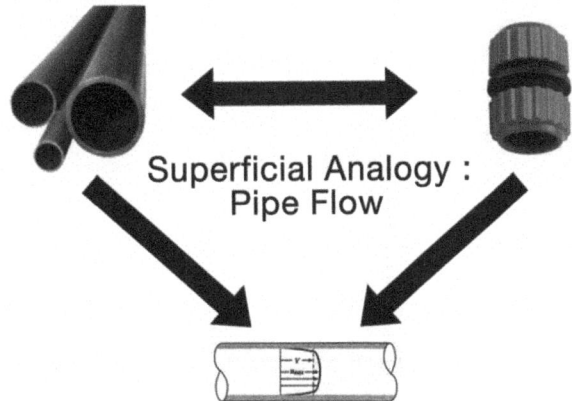

revolutionary development, but actually we simply incrementally advanced what had already been known in another industry. The biggest challenge was to adapt the solution to the existing requirements in terms of pressure, safety, and durability." Today, the plug connection technology has prevailed and led to significant competitive advantage for PipesCo (Fig. 15).

## 4  The $A^4$-Cross-Industry Innovation Process

The examples demonstrate the potential of analogies in the development of technological breakthroughs and radical innovations. Although similar in their highly innovative character, the cases reveal differences in how the analogies were identified and in the way analogical thinking is enabled ranging from pure cognitive abilities to very systematic efforts.

Furthermore, it is apparent that the mere identification of any similarity is not sufficient. Rather, particularly in the case of structural analogies, firms need strategic intent, i.e., the will to question their own technologies and the will to adapt new knowledge. Thus, beyond the mere identification of the analogous solution, firms need to transfer the relevant knowledge and adapt it to their own problem context – process steps which are vital for the 'idea' to become an innovation.

By abstracting from the original problem to its structural relationships, the space for potential solutions is opened up (Fig. 16), and the use of cognitive abilities is facilitated.

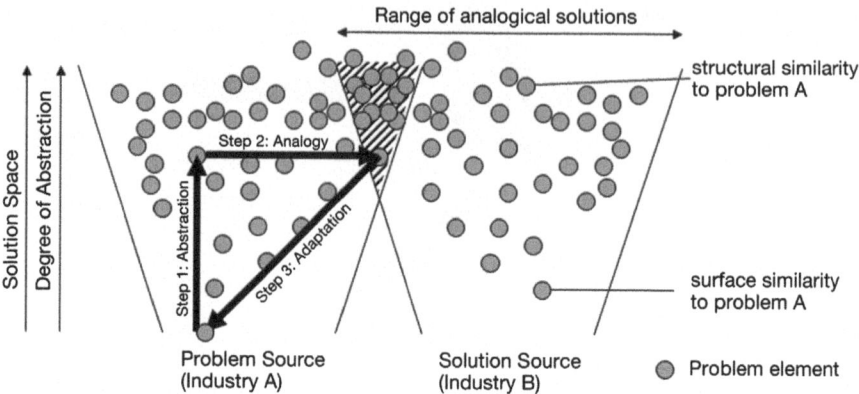

**Fig. 16** Opening up the solution space through abstraction and analogies

Problem abstraction as carried out by the firms described above may be an effective means of arriving at a proper problem formulation and has been found to be vital for successful product innovation. By using abstraction, problems can be redefined; the more abstract the terms in which the problem is redefined, the more familiar the problem seems.

The cases show that analogical thinking does not happen merely by accident, but is supported by a systematic approach. Based on the insights from the cases presented earlier, the $A^4$-*innovation process* for cross-industry innovations (Fig. 17) is proposed. Its purpose is to provide firms with a structured approach to facilitating analogical thinking and applying it to achieve *breakthrough cross-industry innovation*.

### Conclusion

The aim of this chapter was to show how firms facilitate and use analogical thinking for the development of cross-industry innovation. Thus, it
(a) Aims to extend literature on analogical thinking by providing empirical insights on how firms facilitate and use analogies, and it
(b) Aims at providing managers with a practicable process for achieving cross-industry innovation.

It has been found that firms must be open-minded to external solutions and willing to challenge their own technologies so that analogical thinking can be employed effectively. Therefore, top management must foster the search for external solutions and be willing to cannibalize established products and technologies. On this premise, analogical thinking can be a powerful approach to identifying new and non-obvious technological solutions with limited risk and cost. Apart from firms' strategic decision to be open for external innovations, the following aspects have been found to be particularly important:

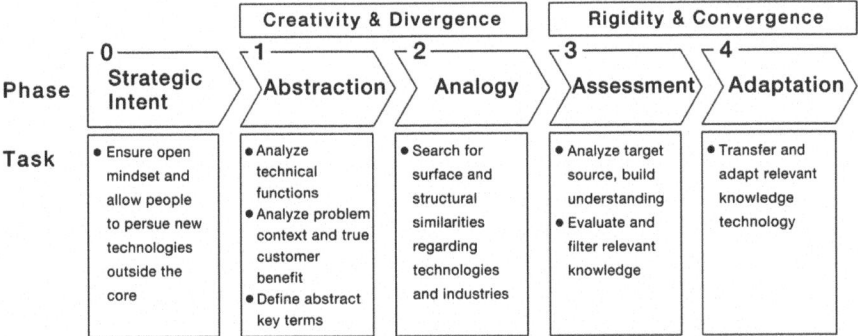

**Fig. 17** A⁴-cross-industry innovation process for new product innovation by analogical thinking

- Firms must establish a deep understanding of the problem and the context in which the problem is set. This requires an in-depth analysis of the problem, both from a technological and a contextual perspective. Such analysis leads to subsequent abstraction of the problem, which in turn allows for abstract search terms to be generated. These tasks might be difficult to fulfill for firms with long-established products, as existing technologies, competencies and conventional mindsets are not easily overcome.
- Since the identification of both surface and structural analogies between different settings is facilitated when there has been prior exposure to both settings, the firm must establish ways to explore domains which differ from its own application context. This is particularly true where R&D employees have not been previously exposed to different settings. In such cases, particular attention must be paid to establishing ways of exploring domains unrelated to a given industry.
- Firms must understand the context of the analogous solution in order to evaluate what knowledge is valuable and thus eligible for transfer. Failure to do so might lead to an analogy being prematurely designated as valuable, which ultimately results in the adaptation of useless knowledge.

To ensure successful cross-industry innovation, the company has to consider a few critical aspects:

**Strategy Level**
- Create an open-minded organization which allows external ideas to enter R&D.
- Seek to understand ideas and technologies which could cannibalize existing products/competencies.
- Demonstrate and live innovation culture on senior management level.
- Dedicate slack resources to pursuing ideas beyond current strategic alignment.
- Conduct regular creativity meetings to question existing products/solutions and foster an open-minded innovation culture.

**Individual Level**
- Critically question the true customer benefit of your product by abstracting from the original problem.
- Foster analogical thinking by looking at similar solutions from other applications or industries.
- Ensure that analogies are not superficial, but fundamentally connect to your problem. The $A^4$-cross-industry innovation process provides a structured approach for the identification of analogical solutions to develop breakthrough cross-industry innovation. The process targets the early and fuzzy front end of the innovation process and tries to support the search for highly novel solutions. The practical experience gained from applying this process in different industries is very positive and encouraging. In fact, many firms have found that by applying the approach they have arrived at better solutions earlier compared to their 'conventional' problem-solving approach. The particular strength of the $A^4$-cross-industry innovation process outlined above lies in the combination of existing knowledge in the problem source with experience with the solution source for creating new solutions in a given industry.

# Trend Scanning, Scouting and Foresight Techniques

René Rohrbeck

## 1 Introduction

The front end of innovation has earned the adjective 'fuzzy', particularly as it is considered unstructured, non-linear, and highly iterative (Khurana and Rosenthal 1998; Koen et al. 2001; Verworn et al. 2008). But this should not be misunderstood as a need to rely on hope or chance encounters to drive innovation.

Beating competition in the innovation game will require developing the abilities to innovate on the basis of early signals in trends, involve internal and external partners in discussing insights into the future, and to build an organization that is able to grasp opportunities in a timely manner.

This is by no means easy for any firm, and to make matters worse, building foresight capabilities involves working partly against organizational reflexes that are useful and critical. For example, the critical ability to focus on the current business can easily be damaged if the firm engages excessively in scanning its environment and entering new fields of business. Thus, building *corporate foresight* capabilities will always imply an important balancing act.

Corporate foresight comprises all activities that are aimed at identifying changes, creating a consolidated future outlook, and using these insights into the future in ways useful to the organization. These activities include developing a strategy, creating innovations, managing risk, and exploring new markets (Rohrbeck 2010a; Slaughter 1997).

If successful, a corporate foresight activity fulfills a dual role: it creates useful insights into the future and at the same time triggers and facilitates organizational response. The second role is crucial, as innovating in new business fields is

R. Rohrbeck (✉)
Aarhus University, School of Business and Social Sciences, Department of Business Administration, Bartholins Allé 10, 8000 Aarhus C, Denmark
e-mail: rrohr@asb.dk

particularly challenging for established firms (Chandy and Tellis 2000). New product categories often cannibalize existing ones, and while innovations in current products are blessed with a high level of planning certainty, innovating in new product categories requires organizations to deal with a high level of uncertainty (Nijssen et al. 2005). Using traditional techniques of evaluating innovations, such as discounted cash-flow analysis, to decide which new innovation project to fund would practically always lead to the ones in established product categories being chosen over the ones in new product categories.

To discuss the challenge and the solutions proffered by foresight at the front end of innovation, this chapter is structured along a generic process model of corporate foresight (Daft and Weick 1984):
- *Scanning:* How to detect signals of future change?
- *Interpretation:* How to facilitate the discussion about the impact of changes?
- *Action:* How to trigger new innovation initiatives?

## 2    Detecting Future Changes

Many firms still believe that relying on a list of the mega trends (i.e., trends with high impact and longtime horizons) makes them sufficiently oriented towards the future. They consider such mega trends sufficient for guiding innovation efforts towards promising future markets. The first bit of bad news for these firms is that the majority of their competitors will most likely already have done the same. The second bit of bad news is that by betting on mega trends, they can easily end up innovating in an area where uncertainty, and therefore the number and size of business opportunities are low. Thus firms willing to outcompete rivals on the basis of innovation face a first challenge:

**First challenge: How to detect signals of change that still yield a competitive advantage?** The first answer to that challenge lies in focusing on trends that are not as easy to perceive as mega trends. In the 1970s, Igor Ansoff proposed not to wait until everyone can spot changes, but to start by detecting 'weak signals' (Ansoff 1975). This is, however, challenging in itself as such a search is like searching for a needle in a haystack, without knowing what one is looking for. At the outset, every weak signal could – but in most cases will not – develop into a major change that can be the basis of a promising innovation. When scanning their environments for weak signals, firms should therefore only spend an appropriate amount of time and money to assess a potential trend. That appropriate amount is small at the beginning and increases only if the change proves relevant in the first phase of the analysis. Choosing the appropriate amount of effort is, however, very difficult as the potential relevance of a given signal is not known a priori.

Translated into practice, this means that, in the first step, a weak signal should only be identified and recorded. After an initial screening, only the weak signals that give signs of potential relevance should then be investigated further and/or be discussed with internal experts. This way the firm will only commit a significant amount of resources to a limited amount of candidates for relevant future changes.

**Fig. 18** Errors in detecting signals of change

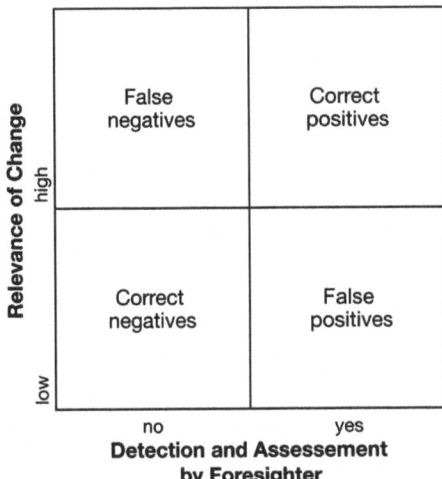

Assigning the scanning task to experienced employees can further enhance this process of detecting weak signals. Employees with a long track record in a given industry, a strong internal and external network, and a broad, rather than narrow, educational background are much more likely to reduce the number of errors in the detection process. In this context, two kinds or errors are relevant (see Fig. 18).

The false negatives result in the firm's inability to respond to a relevant change. This would open the door to competitors who may grasp the innovation opportunity. Or, in case of a change that carries risk, the firm will be taken by surprise and would suffer the consequences. But equally bad are the false positives errors. They carry the important negative effect that the firm wastes valuable resources on chasing after ghosts, i.e. opportunities that in the end do not materialize, or threats that never occur.

The second answer to the challenge of detecting change relevant to a firm's competitive position lies in using a broader scanning scope. As discussed above, most firms focus only on high-impact trends with a long timeframe. This might be justified for firms in industries with slow clock speed, but for firms in fast-changing markets, scanning the environment for trends with short time horizons is at least equally important.

The fashion business is an exemplary industry, where many firms have mastered the art of spotting changes in trend-defining subcultures and translating these into products in the market within 3–6 months. But also technological change can be swift, and, for example, many consumer electronics firms have built excellent sensors that generate insights into emerging technologies and new customer needs in the last decades.

Today, scanning for such change can often be partly automated through intelligent data mining (also called bibliometric) approaches. Online social networks provide an abundant amount of data on consumer behavior and needs and are easy to scan. In this context, both global and broad social networks as well as

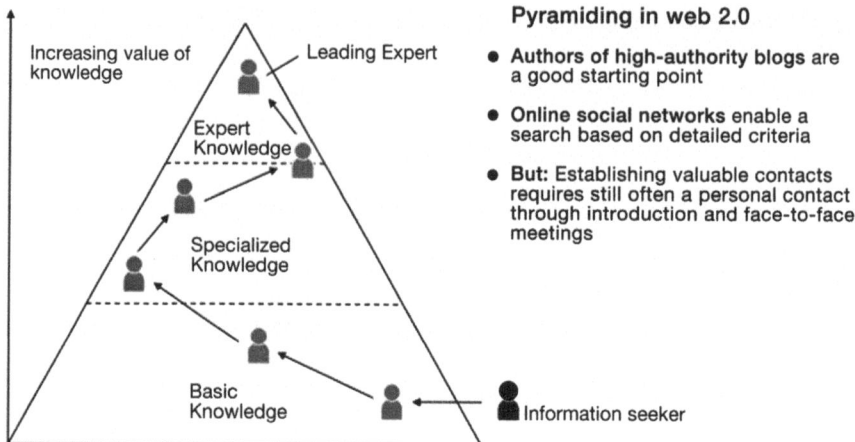

**Fig. 19** Expert search through pyramiding in the web 2.0

niche communities can be equally relevant, depending on the individual need of the firm. For instance, a company working in medical devices would be much more interested in scanning a community of medical doctors than a broad social network.

Data mining can be performed in different ways, ranging from a keyword-based search for monitoring a certain development, to exploring new changes through scanning for associated terms. The latter way of gathering data is particularly useful to identify innovation opportunities in converging industries.

For example, if 15 years ago, a firm dealing in cameras had searched for terms associated with digital photography, it might have been surprised to find a relation to the term 'mobile phone'. By identifying that link, the firm might have been in a position to proactively initiate a partnership with a manufacturer of smart phones and to profit from jointly innovating on image capture, storage and viewing.

These two ways of searching, i.e., monitoring (directed search) and scanning (broad search), can be used in the same way for technology change. The most relevant data sources in this field are still publication and patent databases. But more recently, social networks have also been used for identifying and contacting key experts in technological domains. Such techniques often build on the pyramiding principle, where the leading expert is identified through a series of contacts with experts in the field in that each expert is asked to name another expert whom they would regard as more knowledgeable than themselves (von Hippel et al. 2008).

Figure 19 shows how pyramiding can enable an information seeker to identify the leading experts in a field in a few steps. This technique allows the information seeker to efficiently work through basic and specialized knowledge to reach the true expert knowledge level. The new user generated content services on the internet further facilitate the search for experts, for example through enabling the identification of influential blog authors as starting points for the expert search. In addition, social networks can be accessed and searched by keywords to identify experts with the sought-after knowledge and skills.

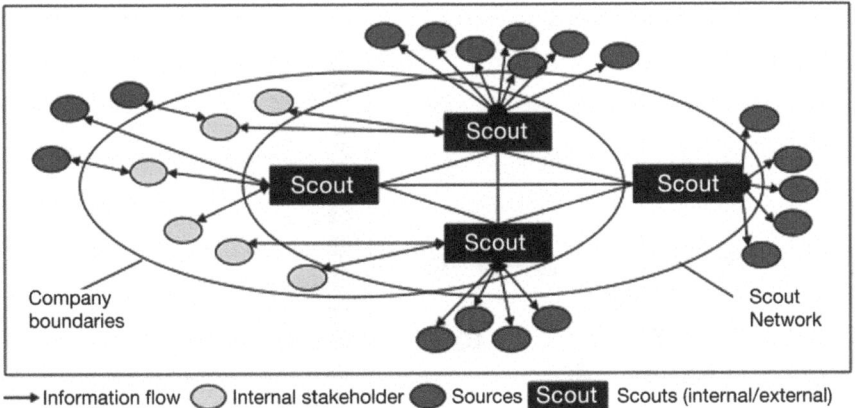

**Fig. 20** Generic scouting network (Rohrbeck 2010b)

Choosing an approach that puts people (scouts/foresighters) center stage, rather than databases, also helps to tackle another challenge:

**Second challenge: How to detect change when terminology is unclear and patchy?** Relying on experts rather than databases for identifying change makes dealing with immature terminology easier. Humans are able to update their terminology as it develops around them. To illustrate the challenge, let us assume that a firm has obtained information on a new phenomenon that involves user generated content and the phenomenon that people are increasingly willing to share their knowledge for free on the internet. At the outset, it would have been difficult to define good keywords to analyze this new phenomenon. For example, starting with 'crowdsourcing' and 'user generated content' would have provided good insight into part of the overall phenomenon at best.

However, through talking to experts, a scout might have been able to quickly link 'crowdsourcing' and 'user generate content' to other associated terms such as 'blogs', 'wikis', 'social networks', etc. This would have resulted in a much more comprehensive and rich understanding of the phenomenon, when compared with database search based on keywords alone.

This is also why in fast moving environments many firms are building their foresight activity on networks of scouts that gather information through direct communication with the persons that are leading the change. These can be researchers who develop certain technologies in leading companies, universities, and research institutes. But *scouting* works equally well with scouts that gather their information from leading thinkers that are behind socio-cultural changes.

Figure 20 shows a generic *scout network*, where scouts work like neuronal nodes in the brain, connecting external experts with internal stakeholders. However, equally important is the network between the scouts that provides a platform for early validation of signals and triangulation to ensure that a signal points at a relevant change.

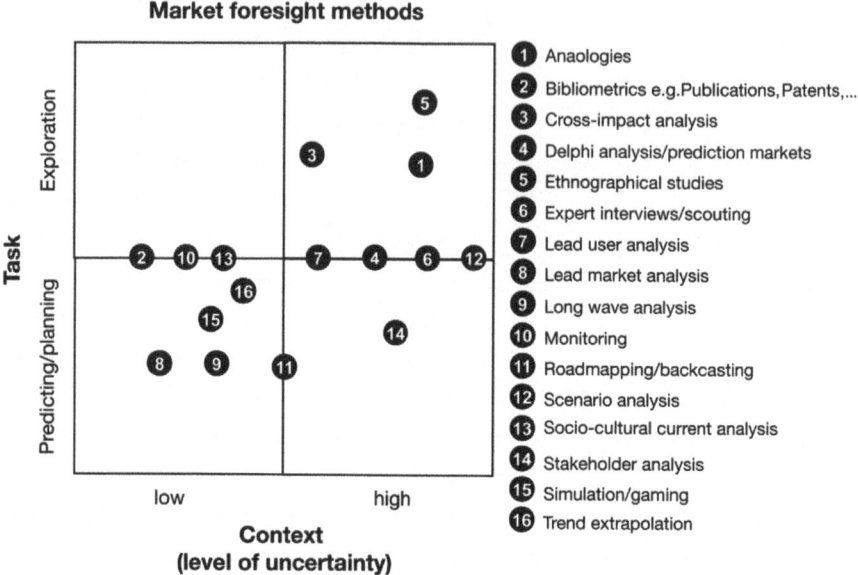

**Fig. 21** Market foresight methods

An additional advantage of scouting networks is that they facilitate the follow-up steps right up to developing the innovation. The personal contact they afford allows the gathering of additional information, if required to convince colleagues of the importance of the change. Also, experts' statements often carry more weight for decision makers than a good database analysis would. Going through multiple iterations with an expert and collecting valuable statements is further supported by the trusted relationships that the scouts normally maintain with their sources.

As scouting is only one possible method of establishing foresight, it points at another question that arises when designing foresight approaches:

**Third challenge: How to choose a foresight method that is appropriate for a given task and in a given context?** To guide this choice, two types of methods can be distinguished, i.e., methods that are more suitable for an exploration (finding options, new approaches, new customer needs, etc.) and methods that are more suited for a prediction/planning task (Porter et al. 2004; Reger 2001). Regarding the context, the dominant factor that guides choice is the level of uncertainty. Uncertainty includes complexity and volatility of the firm's environment.

A further distinction should be made between foresight methods suitable for exploring the future on the market side and methods more suitable for the technology aspect.

On the market side (see Fig. 21), at least 16 important methods can be identified that offer support in the front end of innovation. Methods such as ethnographical studies are particularly useful for exploring new customer needs even in product domains that are unknown to the firm and have a high level of uncertainty.

# Trend Scanning, Scouting and Foresight Techniques

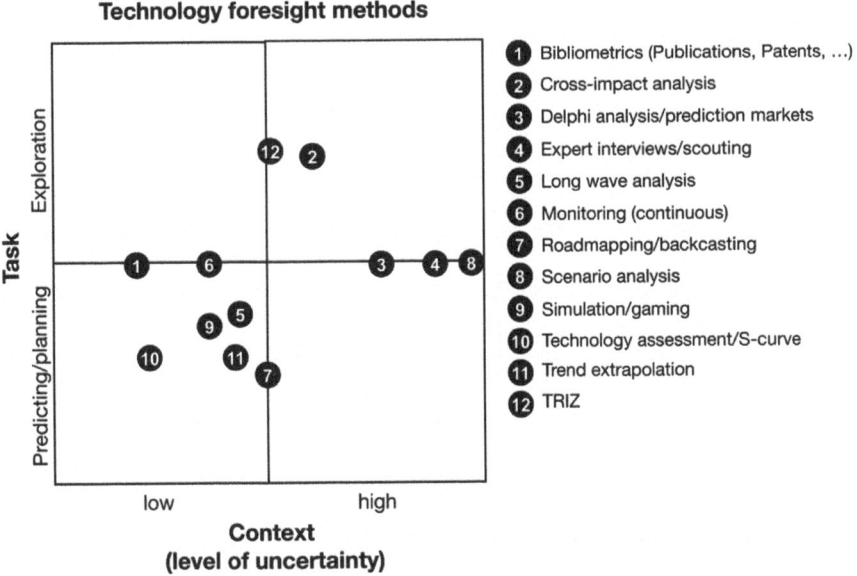

**Fig. 22** Technology foresight methods

On the other end of the spectrum, methods such as lead market analysis or simulations can be used, but only if the factors that influence the market are known and information about direction and rate of change is available. Thus, they can only be employed in low uncertainty environments. At the center of the portfolio are methods that can be employed for either prediction or exploration. These methods also include expert interviews or scouting networks. The outcome can be determined by the way in which the interviews are conducted. Roadmapping is also an example of a method that can be used for either exploration or planning. Some firms use it primarily as an internal planning tool to ensure collaboration between product and technology planning. But roadmapping can also be used as a central method in a workshop for identifying and discussing innovation opportunities.

On the technology side (see Fig. 22), classical methods such as S-curve analysis (where performance increase of a technology is plotted against time) are still being employed with the aim of predicting the right moment for switching from one technology to the next. An increasing usage of the internet for bibliometric analysis and targeted expert search can, however, also be observed in this context.

Other methods, such as TRIZ, are used only by few firms, even though they yield a high potential for exploring development trajectories or identifying the next technology generation for a given product. Traditional trend extrapolation is increasingly supplemented by other methods of assessing emerging concepts.

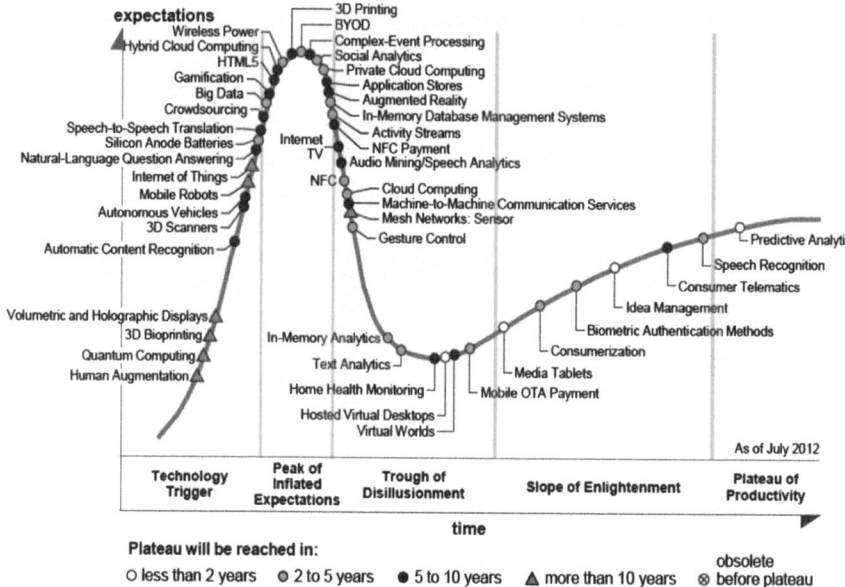

**Fig. 23** Gardner hype cycle 2012 (Source: Gardner Newsroom, 16 August 2012)

Such a method is the Gardner hype cycle, which is a model that proposes that the popularity of a given technology follows a certain typical development cycle (see Fig. 23). This cycle is composed of five phases. In the first phase, the technology gains popularity fast as the media quickly catches on to the promised performance gains as compared to those of current technologies. In the second and third phases, the technology fails to deliver on the performance its inventors have promised or on the performance the media has described and consequently loses popularity as fast as it gained it in the first phase.

If the technology survives this 'trough of disillusionment' and investments in technology development retain a sufficient level, it has the opportunity to reach the 'slope of enlightenment', where technology developers and potential users have aligned their expectations and the technology starts to deliver according to these expectations. In the final phase, the technology reaches the 'plateau of productivity', where it remains until it is substituted by another technology.

Some firms use the framework of the hype cycle, but do not rely on the assessment by Gardner and instead form their own judgment through internal expert panels. Such an assessment is typically made through workshops and thus provides the additional benefit of triggering an internal discussion about the potential of future technologies and innovations. Such a participative approach also provides strong support for later development of the innovation.

## 3   Interpreting the Impact of Future Changes

After a potential future change has been detected, the firm needs to establish whether it is relevant and how it might impact its business. This is referred to as organizational interpretation or translation into the organizational discussion. Much of the interpretation work can be facilitated by corporate foresight methods. For example, cross impact analysis applied to technologies can reveal which key enabling technology needs to be developed to allow the building of given product families. Other changes, for example those that have been identified through a bibliometric analysis, often need a dedicated translation effort, for example in the form of a workshop, where foresighters and innovation managers work together to identify innovation opportunities.

Before linking foresight to innovation is discussed, it should be clarified what is meant by change. For the interpretation phase, and also later in the response phase, it is important to know what kind of future change an organization is dealing with. In this context, the distinction between trends and uncertain dynamics applies.

Trends are characterized by a high level of certainty at least about the direction of change, and in many cases also good information on the likely rate of change. Most mega trends fall into this category. For example, that society is aging can be predicted rather confidently, as young high-school graduates entering the workforce take at least 17 years to 'develop'. This means that the current number of births predetermines the number of graduates that will enter into apprenticeships in 17 years. At the same time, the number of elderly persons is predetermined to an even stronger extent and can thus be predicted easily. In addition, history suggests that birth rates are difficult to alter, particularly in a stable external environment. These factors of predetermination make the aging society a trend which is particularly predictable.

By contrast: Who would be comfortable making predictions about the future of the European monetary policy? Questions, such as, "Will the EURO still be the primary currency in Europe?", or "What will be the role of the EURO in the international monetary system?", are questions to which the answers are notoriously difficult to predict. Therefore, it can be concluded that the uncertainty about direction (increase or decrease in the number of participating countries) and rate of change (speed of growth or decline of memberships) gives this change an uncertain dynamic.

Of these two categories, trends are much more popular in innovation planning. The simple reason is that most firms rely on management systems that work with plans, which in turn are built on assumptions about future states of influencing factors. Uncertain dynamics are highly inconvenient for such management systems, as they would require developing multiple plans for the different possible states of uncertain dynamics in the future. This leads to the following challenge:

**Fourth challenge: How to plan and innovate on the basis of uncertain dynamics, where direction and rate of change are unknown?** The first answer to this challenge lies in thinking in scenarios. While working with good, medium and worst case scenarios is standard practice in financial planning, this useful technique has not spread much in innovation management. The central proposition is that instead of trying to predict uncertain factors and making matters worse by combining these into predictive models, firms should rather acknowledge the uncertainty and identify a number of possible and consistent scenarios.

Managerial action on the basis of scenarios can than follow three routes (Gausemeier et al. 1998):
- *Planning-oriented:* Actions based on the premise that the most likely scenario will occur
- *Responsive:* Combine actions intended to move towards the scenario that yields the greatest opportunity and actions meant to avoid the scenario with the greatest risk
- *Trend-setting:* Actions designed to create a desired scenario, for example by influencing key actors in a given market to take joint actions.

All three routes can lead to success. Whether one route is more preferable depends on the strategy of the firm. For example, if the firm aims to achieve innovation leadership, the trend-setting route would be most appropriate. If a company strives to limit its exposure to risk, the responsive stance would be preferable. For large firms that need to ensure that all relevant internal units act in an orchestrated fashion, the plan-oriented route is often favored.

But even if a firm plans and acts on the basis of a most likely scenario (thus choosing to abandon the plurality of futures in its planning), it will have benefited from having engaged in scenario thinking. Identifying scenarios and discussing alternative routes of development trigger both explicit and implicit contingency planning and will make the firm or the individual innovation projects more responsive and robust.

## 4    Triggering New Innovation Initiatives

In this context, the broader question arises whether planning is the only way to trigger innovation. Research has shown that innovations often go through phases of stagnation until a new insight or individual creative idea unlocks the barriers that have previously prevented the further development (Backman et al. 2007; Gemünden 2001). Therefore, the assumption that all innovation can be planned through enough pre-development analysis should be abandoned. A better assumption would be that organizations will always lack some future insights (i.e., emerging promising market and technology opportunities) at a given moment and that they will be able to enhance their innovation in a later phase when more insights have been gathered. This is true throughout the whole innovation process, but particularly important in the front end of innovation. Therefore, an additional challenge should be taken into account when designing foresight approaches:

**Fifth challenge: How to use foresight methods in the highly iterative front end of innovation?** By studying serial innovators (individuals that have repeatedly and successfully driven innovations in large firms), Griffin et al. (2007) find that these innovators follow two cycles until a satisfactory outcome is achieved. Of these two cycles, only invention and validation are of relevance in the context of the front end of innovation. The second cycle, which deals with bringing the invention to market, is less relevant in the present discussion. The first cycle consists of three phases:

- Finding the right problem
- Understanding the problem
- Inventing and validating

This cycle can be triggered by any of the phases. For example, it might start with the discovery of an unfulfilled future need detected by using an ethnographical consumer research method, such as a 24 h stay in a user family. Such stays can uncover tasks that customers want to perform, but cannot with their current products. For example, one might have discovered that families would like to synchronize their weekly timetables. A large pin board mounted in the kitchen might currently be the preferred method.

But such a solution might be judged unsatisfactory, particularly if schedules are updated regularly and sometimes at short notice. Such an initial unfulfilled need could trigger the search for a better understanding of the 'problem', or a search for initial ideas for the invention. In this case, it could have been discovered that smart phone usage is spreading increasingly to the 'young consumer' segment, giving rise to the idea that a successful invention might involve using electronic devices and the internet to synchronize the calendars.

The first prototypes of such devices might then be validated through lead user workshops, leading to the discovery that features such as push messages to parents, when the children's appointments of the next day are being altered, would be extremely useful. That would lead to a better understanding of the 'problem' and trigger a new round in the three-step cycle, leading to an enhanced innovation concept.

This example illustrates how foresight methods can be selected and deployed throughout an iterative process of generating innovation. The ideal selection and deployment will depend on the firm and the context in which it operates.

The usefulness of foresight methods, however, goes beyond triggering innovation initiatives (Rohrbeck and Gemünden 2011). Successful innovations often emerge from the interplay of an *innovator* (working on and promoting his/her innovation) and an *opponent,* who through challenging the innovation ensures continuous enhancement and finally a superior product quality. Consequently, two roles of foresight can be distinguished in the front end of innovation; the role of the innovator and that of the opponent. Both roles are illustrated in Fig. 24. It also shows which elements of the two roles fuel the front end of innovation. For the illustration, the conceptualization of the front end as set out in chapter "Managing the Unmanageable: The Fuzzy Front End of Innovation" of this book (see Fig. 1) has been used.

To institutionalize an opponent role, a firm can create specific foresight units that engage in workshops with the project teams of ongoing innovation initiatives. This should be done regularly, for example every 3 months (Rohrbeck and Gemünden 2011). It is, however, important to ensure that such workshops have a constructive nature, for instance by emphasizing that the foresight team is not only expected to

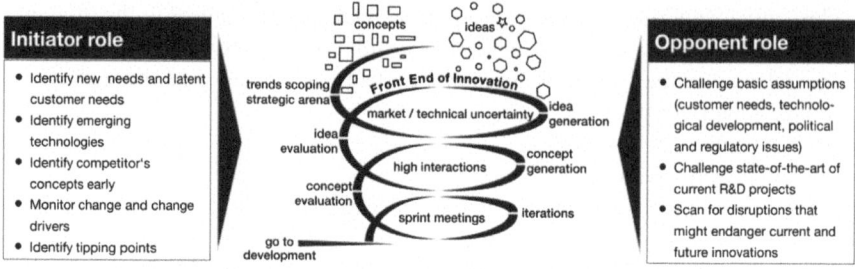

**Fig. 24** The two roles of corporate foresight in the front end of innovation

identify weaknesses of the current concept, but is also required to bring insights (e.g. on alternative technologies) to the table.

In addition, it is important to ensure that the workshops are taken seriously and that they can have real consequences. One way of doing this is to implement a process by which the innovation initiatives can be terminated if they fail to demonstrate that they will produce a state-of-the-art product.

A successful way of institutionalizing such a termination process is illustrated by the 3M so-called 'death parties'. They are real parties to which all R&D employees are invited to celebrate the termination of an R&D project. Organizing such events to create a positive and joyful atmosphere ensures that every employee understands that terminating a project is not a failure, but rather a liberation of a team on its way to developing an unsuccessful innovation. At the same time the death parties also mark a new beginning as the project staff and the budget can now be deployed to explore and develop a more promising innovation opportunity.

## 5 Value Creation of Corporate Foresight in the Front End

The value contributed by corporate foresight could be classified into 13 potential, positive impacts (Rohrbeck and Schwarz 2013). Of these, four relate to strategic management outcomes and will thus not be discussed here. The remaining impacts can be clustered into three categories: (1) those related to perception, (2) those related to interpretation and triggering actions, and (3) those that contribute to overall value. For this chapter, some value contributions have been rephrased to fit the perspective of the front end of innovation. The result can be seen in Table 2.

The perception category is the most obvious. By applying foresight methods, firms are able to channel more future insights into their front end of innovation and thus increase the likelihood of discovering interesting opportunities. In addition, a continuous scanning activity contributes to the discovery of potentially disruptive technologies that might endanger ongoing innovation initiatives (Reger 2001).

To maximize the value created in the perception phase, it should be continuously monitored and fine-tuned. This is best achieved by monitoring the innovation

**Table 2** Value creation from corporate foresight in the front end of innovation

| Group | Potential value contribution of strategic foresight |
|---|---|
| Perception | Gaining insights into changes in the environment |
| | Contributing to a reduction of uncertainty (e.g., through identification of disruptions) |
| Interpretation and triggering action | Identifying new key and disruptive technologies |
| | Enhancing the understanding of customer needs |
| | Identifying potential customers |
| | Enhancing the understanding of the market |
| | Identifying opportunities and threats regarding the product and technology portfolio |
| | Reducing the level of uncertainty in R&D projects |
| Overall | Facilitating organizational learning |
| | Shaping the future (e.g., by influencing other parties, such as politics and other companies) |

candidates. This facilitates assessing how many of the opportunities discovered fall into the four categories:
- Correct negatives
- Correct positives
- False negatives
- False positives

If, for example, the firm discovers that it misses important opportunities (high number of false negatives) it could encourage its foresighters to discard opportunities only after a more thorough search has been completed, or it could organize additional interdisciplinary review panels that screen the future insights before they are judged to be irrelevant.

In the second category, success could be monitored by simply counting the number of new key technologies or technologies that yield a disruptive potential for the firm. An analysis of this number over time shows whether the 'fitness level' of a scanning activity increases or decreases. The same is true for the market-related scouting successes, such as new customer needs, new customer (groups), or new insights into changes in the market or industry. In addition, the opportunities and threats identified that have an impact on the level of existing product and technology portfolios could be counted. These elements of value creation can also be classified as throughput measures that monitor how well the foresight system operates.

The second category also contains one output measure that is primarily related to the opponent role. This measure indicates if the foresight activities related to challenging ongoing R&D projects are successful in reducing the overall risk level of the R&D project portfolio.

The third category consists of two firm-level value drivers that have only an indirect link to the front end of innovation. They are, however, of high importance for the firm and can thus provide good arguments for the funding of the foresight activities as a whole.

The facilitation of organizational learning is important to mention, but unfortunately difficult to measure. It relates to the ability of a firm to break away from old mindsets and routines. This ability to embrace new business opportunities, create new organizational structures and processes is particularly important for responding to highly disruptive changes in the environment. Foresight methods can be expected to contribute to this ability in various ways, for example by opening the debate about multiple, possible futures, based on a scenario analysis.

A second value that could be created in this third category is an enhanced ability to influence others in order to shape the future. This is particularly important in cases when a firm aims to develop a systemic innovation, where multiple firms and sometimes also governmental and societal actors need to work together to create a new market.

An example of such an innovation is electric mobility. In this field, car companies are developing technologies jointly with their suppliers, while governmental bodies and interest groups develop the regulatory framework and a platform framework through which the battery charging infrastructure is built and operated. Corporate foresight can help in such systemic innovation developments with joint scenario development, joint roadmapping, and collaborative business modeling (Rohrbeck et al. 2013).

## 6   Future Outlook

Overall, the usage of foresight techniques in the context of business firms is rising. This can be ascribed to two main causes. First, many industries have recently experienced disruptive change, which has led to the bankruptcy of dominant firms that had previously been perceived as too big to fail. In the respective industries, this created a strong motivation to build foresight capabilities to avoid such a fate. Second, firms whose competitors have built foresight capabilities are interested in following suit as they struggle to respond to change as fast as their more capable peers.

These two effects have led to the increasing popularity of corporate foresight practices and the creation of practitioner cycles and will eventually result in corporate foresight being established as an academic field and a recognized practice domain. The coming years will show how many firms implement foresight techniques, build foresight units, and design tailored foresight processes.

Most likely corporate foresight will take a similar development route, as did innovation management 30 years ago. At the beginning, it was implemented as part of other firm functions, and later it grew independent into a function in its own right.

In addition to the overall growth of corporate foresight, the practice of corporate foresight is expected to develop in three directions in the future:
- **Corporate foresight and open innovation**
  As an increasing number of firms is exploring ways to innovate with external partners, interest in teaming up for corporate foresight is also growing. *Deutsche Telekom*, for example, is exchanging future-related information with both value

adding partners and competitors to enhance the company's future outlook (Rohrbeck et al. 2009). Others are using foresight techniques to support open innovation by identifying promising external technologies to be implemented in their own internal innovations (Veugelers et al. 2010).

- **Corporate foresight and Web 2.0/Web 3.0**
  The internet and particularly the emergence of the Social Web (2.0) have enabled instant expert identification and supported scouting approaches by providing powerful tools to discuss insights and jointly create knowledge (Gordon et al. 2008). It can be expected that the emergence of the Semantic Web (3.0) will yet again enhance foresight capabilities, for example by facilitating more intelligent patent analysis (Bergmann et al. 2008) or enhancing the ability to identify systemic patterns that open up innovation opportunities.

- **Corporate foresight and systematic exploration of new business fields**
  Large firms in particular are under increasing pressure to move to new business fields, as the time in which firms can enjoy innovation-leader premium profits is decreasing. The iPhone provides a clear example of where *Apple* is struggling to counteract the fast approaching threat of other vendors, such as *Samsung*, and the time in which Apple was able to enjoy high margins appears to be running out. Therefore, firms are increasingly looking towards corporate foresight to propose approaches that integrate multiple methods and allows the systematic exploration of new business fields (Heger and Rohrbeck 2012).

# 7 Checklist

To sum up, the following checklist may serve as a guideline for successfully integrating foresight into the fuzzy front end of product innovation:

- **Build additional sensors** to identify weak signals of change. Use a mixture of bibliometric and people-centric search mechanisms.
- **Experiment with methods and processes** until an approach has been found that works for the given task, the context and the company.
- **Provide foresight capabilities** that can be integrated **on demand** in the iterative process of the front end of innovation
- **Ensure a high level of interdisciplinarity** to ensure that the full extent of the impact of a change is perceived and that enough complementary perspectives can contribute to defining the innovation idea.
- Use methods that **enable a systemic observation of change** (such as scenario technique), rather than relying on methods that build on linear cause-effect relationships.

# Crowdsourcing as an Innovation Tool

Oliver Gassmann, Sascha Friesike, and Michael Daiber

## 1   The Idea Behind Crowdsourcing

Being more innovative than their competitors is a key challenge to many enterprises. Being the first to explore new paths and to realize new ideas is often a crucial success factor in a globalized economy. The networked world with omnipresent internet-connectivity and endless collaborative options has turned the previous war over talents into the present war over ideas. And in this context *crowdsourcing* – a term that only came to life in the last 5 years – plays an important role. The term is a neologism from the words 'crowd' and 'outsourcing'. While outsourcing assigns a given task to a predefined group of actors, crowdsourcing turns to a mostly anonymous crowd. The crowd then presents possible solutions, and the principal uses the one that fits its needs best (Dawson and Bynghall 2011). In 2011, to provide an example, the website humanrightslogo.net sought a logo for human rights. The project attracted an astonishing 15,300 submissions from 191 countries.

The use cases for crowdsourcing are diverse, and new platforms are constantly emerging (Howe 2008; Sloane 2012). The website crowdsourcing.org and the German site *crowdsourcingblog.de* provide extensive overviews on the current

O. Gassmann (✉)
University of St. Gallen, Institute of Technology Management, Dufourstr. 40a, 9000 St. Gallen, Switzerland
e-mail: oliver.gassmann@unisg.ch

S. Friesike
Alexander von Humboldt Institute for Internet and Society (HIIG), Bebelplatz 2, 10099 Berlin, Germany
e-mail: friesike@hiig.de

M. Daiber
ABB Turbo Systems Ltd, Bruggerstr. 71a, 5401 Baden, Switzerland
e-mail: michael.daiber@ch.abb.com

crowdsourcing landscape. In general, crowdsourcing projects can be divided into the following four groups:

## 1.1 Intermediate Platforms

With this format, a platform provider helps a principal to access a group of *solvers* to help with its challenge. Intermediate platforms generally maintain a group of solvers by constantly providing new challenges and thus keeping the crowd entertained. These platforms have different purposes and as such attract different followers. Mainly intermediate platforms fall into three categories: (1) R&D platforms, (2) ideation platforms, and (3) marketing and design platforms. *R&D platforms* help firms to answer research questions. These platforms attract researchers and companies usually pay fairly high prices for the solution to complex problems. Such platforms are for instance: *TekScout, InnoCentive, ideaconnection, allied mindstorms,* or *Innovationskraftwerk*. Some focus on already established R&D findings and work as matchmakers for intellectual property (e.g., *yet2.com, pharmalicensing.com*). A second group of crowdsourcing intermediate platforms are so called *ideation platforms*. Other than R&D platforms, the results of ideation platforms are more general and inspirational and less technically detailed. These platforms are especially interesting in the early phases of an innovation project as the outside view of the participants often helps to broaden the firm's solution space. Ideas provided range from only a few sentences to descriptions that fill several pages. Examples of such platforms are *atizo.com, jovoto.com, unseraller.de, crowdINNO.com, tricider.com, bonspin.de,* or *yutongo.com*. A third group of intermediate platforms fall into the category of *marketing and design*. These platforms attract design-savvy users and help firms to develop new logos, websites, brochures, or marketing campaigns. Examples include *99designs.com, crowdSPRING.com, quaxter.de, a-better-tomorrow.de, logoarena.de,* or *brandsupply.de*. They are mostly platforms for freelancers. These platforms offer environments where single tasks can be efficiently outsourced to a group of users. Such tasks could be finding dead links on a website, selling a product, performing local micro jobs, or translating or transcribing texts, to name but a few. Examples are *amazon mechanical turk, clickworker.com, textmaster.com,* or *AppJobber*.

## 1.2 Free Solutions

The second group of crowdsourcing projects can be described as 'free solutions'. Here, a group of people forms in order to jointly build a solution that can be used free of charge afterwards. In contrast to intermediary platforms, it is not a company that defines a problem; instead, users who share a common problem/need team up and find a solution. Whereas the first type of platforms corresponds to a top-down problem solving strategy, this second category represents a bottom-up approach. Essentially, free solutions appear either as websites or as software programs.

In software, such solutions are labeled 'open source' and represent tools such as the web browser *Firefox*, the operating system *Linux*, the web server *Apache*, or the video player *VLC*.

Most *open-source* software projects revolve around one single developer who occasionally receives help from others. Few projects attract the attention of a large number of programmers as in the examples mentioned. The second group of 'free solutions' is made up of websites. In most cases, a group of volunteers comes together to establish a free alternative to a proprietary solution. The best known example in this category is *Wikipedia*, but there are many similar projects of a lesser size. One such project is *Openstreetmap*, a free map that allows users to generate layers for special information, such as public transport, hiking, or wheelchair accessibility.

## 1.3 Corporate Platforms

Corporate platforms are used by firms if they feel the need to interact with a company-specific group of users. These platforms help companies to engage with customers, spot trends early on, and harvest feedback. However, they are labor-intensive, as users who provide feedback want companies to give something back to them. Constant interaction is necessary to keep the user group motivated and engaged. Two forms of corporate platforms are in use:

- **Product ideas and solutions:** In this group, companies look for new products that could be offered in the future. *Tchibo* asks designers and customers for ideas for their weekly specials, *Starbucks* asks what customers wish for, and *BMW* is looking for features of future cars in their *Virtual Innovation Agency*.
- **Branding and design:** These platforms tap into the design and the marketing understanding of their customers. At *burdastyle.com*, nearly 300,000 members share cutting patterns, at *Muji* customers can overhaul old products, and *OSRAM* is a community built around lights that are lit by LEDs.

Often corporate platforms are white-label (only the corporate logo is visible) versions of platforms used in other settings as well. In this field a small industry is developing that focuses on helping corporations establish their own crowdsourcing presence. The following two examples are based on such white label solutions:
- The leading Italian bank *Intesa Sanpaolo* launched the platform *europeanictchallenge.com* in 2011. Thirty ICT start-ups were selected from 140 applicants and invited to a multi-day start-up boot camp.
- *Skipso* developed *grameenfoundation.applab.org* for the *Grameen Foundation*. *Viathe* platform, mobile apps and services are designed for and tailored to developing countries. Expectant mothers receive information about a healthy diet, farmers can make better decisions based on predictions of weather and market prices or learn something new about caring for sick animals.

## 1.4 Marketplaces

The fourth group is labeled 'marketplaces'; such platforms offer creatives or inventors an outlet to reach an audience. These marketplaces either sell already established or designed goods, such as *dreamheels.com*, where anyone can become a pumps designer, or *threadless.com*, where the same applies to t-shirts, or they are marketplaces for ideas that still need funding. These funding brokers are called *crowdfunding* platforms and have received considerable amounts of media coverage in the last 2 years. On websites such as *kickstarter.com* or *sellaband.com*, movie makers, musicians, writers, or entrepreneurs ask for funding for their upcoming projects. If the project is interesting enough to create as much funding as has been asked for, the entire sum is given to the project. Most platforms only pay out once the asking sum has been reached. In Germany, the best known example of crowdfunding is the 'Stromberg' TV-series, for which fans invested over a million Euros to keep the series alive. On kickstarter, the designers of a digital watch called 'Pebble' that can communicate with a smartphone raised over 4.7 million US dollars to start their firm.

The fields of application for crowdsourcing are diverse, as can be seen from the list of examples discussed above. New ideas are constantly emerging on the internet, making it all the more important for companies to engage with the topic and to learn how to use it effectively. The internet provides an easy link to the masses, and used in the right way, it can supply firms with new ideas and solutions to pressing problems.

The present chapter will provide a brief overview on the managerial challenges that are associated with crowdsourcing. For this purpose, firstly the crowdsourcing process will be introduced and divided into five single steps. Subsequently the opportunities and potential risks of crowdsourcing will be discussed, and the chapter will finish with a short list of the most important aspects managers should keep in mind when engaging in crowdsourcing.

## 2 The Crowdsourcing Process

The use of crowdsourcing is diverse. Whether it is amateur geologists discovering the ideal location for new gold mines, students designing best-selling t-shirts, or retired physicists solving problems that previously depressed dozens of engineers, crowdsourcing has proven to be a ready answer to many problems. To provide a deeper understanding of the method, the entire process has been divided into five distinct steps:
- Preparation
- Initiation
- Implementation
- Evaluation
- Utilization

Each step is associated with particular challenges and needs to be managed so that crowdsourcing can be used successfully as a problem solving tool.

## 2.1 Preparation

If a company starts thinking about using crowdsourcing, usually a problem or a question already exists, and the company believes that the solution or at least a meaningful contribution can be found by someone outside the corporate boundaries:

- *BMW* recognizes that their motorcycle customers aging compared to their competitors' customers and is looking for innovations that appeal to a younger audience.
- The Swiss electricity supplier *EWB* would like to participate in the growing market for electric vehicles, but has no concrete ideas about the best approach.
- An anonymous pharmaceutical company is looking for a biomarker to track the progress of amyotrophic lateral sclerosis, an incurable disease.
- In a project called Crystal Vision D., *Swarovski* is looking for innovative concepts and designs based on Swarovski crystals that can be sold in existing stores.

As can be seen even within a corporate for-profit ambit, the possibilities of crowdsourcing are diverse. However, to make use of crowdsourcing successfully, the company seeking ideas must choose the right approach, and errors can happen in every phase of the project. A few questions help to determine if crowdsourcing makes sense at all, and whether it makes sense to use an existing crowdsourcing platform or build a new one:

**What Is the Desired Result?** Results of crowdsourcing projects can vary considerably. While the search for a biomarker for amyotrophic lateral sclerosis is highly specific and science-driven, the search for new product and service ideas in the field of electric mobility is rather broad. Another difference lies in the expectations towards the innovators or solvers. While in the first case a patentable principle or a working prototype is expected, in the second case the customer is satisfied if he can gather new interesting ideas or confirm internal ones. In the first case, the result is often the product of months of research; in the second case, it can be a statement containing a mere 150 words.

**Which Platform to Choose?** There are numerous intermediaries providing a crowdsourcing community and the corresponding IT platform. Nevertheless, some companies prefer to build their own platform and thus their own community. In these cases, service providers often operate in the background and take care of the technical and methodological questions associated with such projects (e.g., the company *HYVE* that provides software and services for idea competitions). At first glance, it seems to make little sense to renounce the network of an existing community and their skills on purpose. Still, there are reasons why companies opt for their own new platform.

- **An existing specific community:** Although many crowdsourcing platforms advertise the size and quality of their community, whoever wants to use these platforms has to assess whether these are the innovators he or she needs for the

specific challenge. Especially companies with 'high-involvement' products, such as cars or sports equipment, can often build on a strong brand community. BMW built its own innovation platform VIA (Virtual Innovation Agency), where BMW enthusiasts contribute their ideas and input on the car manufacturer's new and existing models.
- **Long-term interest in a company-specific community:** When a company plans to ask the online world for their input regularly, this can justify setting up a separate community. Tchibo ideas, which posts new questions every month, can be cited as an example. When thinking about building a new community, a company's representatives should keep one question in mind: Is it a one-time problem or will the community members be involved in the innovation process regularly in the future?

## 2.2 Initiation

After the decision for crowdsourcing has been taken and the choice for the appropriate platform has been made, the next step is to ask a question. A question or a task on a crowdsourcing platform usually consists of a few sentences. It is therefore most important to ask the right people the right questions and to formulate them accurately.

**What Can Be Asked?** Through the internet, innovators with almost every background and personal interest can be found. In the German speaking world alone, one can find various sites discussing which machine and which setting will produce the optimal espresso (*e.g. coffeeright.de, guter-kaffee.de, espressobar-24.de, coffee-community.de*). Even less familiar topics, such as repairing half-timbered houses (*fachwerk.de*) or growing miniature vegetables (*hausgarten.net*) are actively discussed on the internet. Networking platforms like *Linkedin* or *Facebook* also unite people sharing the same interest and are ideal places to identify people that might contribute to a given specific topic.

If someone has a specific question, he or she can almost be sure someone else on the internet who deals with similar topics. Therefore, there are no limits to the questions to be asked, except legal restrictions and limits imposed by good taste.

**Can the Problem Be Split into Crowdsourceable Parts?** Crowdsourcing can provide the answer to a specific question, but it cannot replace a company's internal R&D or business development work. Therefore, it is important to think carefully about which question the community can contribute to. With the help of *InnoCentive*, *Colgate Palmolive* found a solution to placing fluoride particles in a tube of toothpaste without spilling half of them. The idea came from a retired physicist. InnoCentive helped solve a specific problem, but did not develop the new packaging line. In order to achieve a satisfactory result, complex issues must be split into manageable and crowdsourceable parts.

**What Is an Answer Worth? How Can Innovators Be Motivated?** There are a number of different reasons why innovators sacrifice their free time to help companies (usually without being paid for their efforts). Literature offers a large number of explanations; they include interest in the topic, interest in an improved product, or even pure altruism. Send and Friesike (2013) present an overview on current knowledge on this topic. When starting a crowdsourcing competition, companies searching for the *wisdom of the crowd* have to be aware of these reasons. It is not immoral to make use of the help of volunteers, but especially from a long-term perspective, it has to be fair.

Therefore, it is important to think about how to reward innovators for their ideas. Depending on the question and setting, ideas can be for free or worth $1,000,000, as in some InnoCentive challenges. While one can easily ask for free ideas for a good cause such as climate protection, one should not intend to use freelancers' work in return for a couple of dollars. In general, the more professional results are expected (e.g., for scientific problems in the pharmaceutical industry), the better the innovators know the value of their idea and adequate financial rewarding becomes important. In other cases, monetary compensation can play a secondary role, and a creative compensation may increase the attractiveness of the competition and thus indirectly impact the quality while saving costs for the company. For people interested in technology, for example, a visit to a research center or a test ride on a motorcycle of the latest generation can already provide a sufficient incentive to participate.

**How to Approach the Right Participants?** If the objective is to use the knowledge of the masses, it can be a waste of everybody's time to ask the usual suspects. In a crowdsourcing challenge, the Swiss energy supplier *EWB* asked the following question on the platform *Atizo*: "Which products and services do you want from an innovative energy company?" Although satisfied with the results, the project manager noted that especially tech-savvy men had submitted ideas and therefore the results consisted almost exclusively of technical solutions. Maybe a completely different community would have generated different and much more radical ideas. Deliberately building a heterogeneous community by posting the challenge in a completely different internet forum that attracts different people could be a viable option.

**IP Risks of Crowdsourcing** The basic idea of crowdsourcing is and remains openness and transparency. A good idea should rather be published than never emerge at all. Questions about intellectual property and how to deal with it do have their legitimate place when discussing the topic of crowdsourcing. From an intellectual property perspective, different risks exist and have to be addressed and discussed:
- **Revealing strategic intent**: Revelation of the company's strategy and innovation activities.
- **Risk of followers**: Investing in ideas that have become public in a crowdsourcing challenge and thus might already be known to the competitors.

- **Patent risk**: An idea that has been published in a crowdsourcing challenge is public and thus cannot be patented anymore.

These risks should have an influence on which platform is chosen and how the questions are asked:
- Are the answers visible to anyone?
- Can innovators comment on each other's ideas?
- Is the company name visible or hidden? (For an in-depth analysis and recommendations on patents, see Gassmann and Bader 2010).

## 2.3 Implementation

After the project has been started, the first answers, ideas and comments will arrive. A lot of ideas may be generated in a short period of time, especially with virtual brainstorming platforms like *Atizo*. At this stage, the course of the project is almost set. However, even in this later phase there are still possibilities to influence the project's outcome.

**Facing Resistance Within the Company** Especially in more conservative industries and companies (e.g., B2B suppliers in mechanical construction or federal reserve banks), integrating the entire world into the innovation process might cause resistance among colleagues and reluctance among the management to take the project seriously. However, the acceptance of the crowdsourcing project within the company is decisive for the entire initiative's success. If this has not already been done, this is the point at which all stakeholders in the company should be informed about the project, its objectives, and expected results.

**Moderating the Idea Generation Phase** Even if one has thought about nearly everything in the previous phases, there is still a risk that the community or individual members of it do not understand the question in the same way as the company looking for external input. It can therefore make sense to look at incoming contributions, especially as this will not expend too many resources. Giving explanations, commenting on ideas, and asking further questions is all the more important the more technical the project is, but even with less technical questions it is better to avoid misunderstandings by quickly answering questions as they arise. Also, in a crowdsourcing contest companies can learn a lot through interactions with their customers.

**External Support During the Idea Generation Phase** Depending on the type of contest and question, a large number of contributions can be gathered during a crowdsourcing contest. Especially with online brainstorming platforms, this number can reach into the hundreds or even thousands of contributions. It can thus make sense to rely on external help for the first analysis and bundling of ideas, which makes it easier to arrive at the final evaluation. Such services are often available through the crowdsourcing platform itself, Atizo; for instance, offers them in its extended packages. Furthermore, statistical analysis of the incoming contributions can provide additional valuable information. Different platforms support their customers with this kind of statistical information.

## 2.4 Evaluation

After all proposed solutions have been received, the project appears to be finished. However, this is the moment when most of the work has to be done within the company looking for external ideas. During EWB's above mentioned contest, 428 ideas arrived that had to be evaluated. Even if one third of the ideas were absolutely outlandish and only 5 % contained a certain degree of novelty to the firm, every idea had to be read and assessed.

**Who Assesses the Ideas and What Are the Criteria?** Basically, assessing ideas in crowdsourcing is like correcting an exam at school. All contributions should be assessed on how the task has been fulfilled. But there is one main difference. While students are forced to take the exam, innovators in crowdsourcing competitions do so in their leisure time. If a company does not want to lose its reputation in the community, the assessment has to happen in a fair way. The assessment criteria must be communicated clearly from the beginning, and employees of the company have to assess the ideas carefully against those standards.

Another way to achieve a high level of fairness can be community voting, where the innovators decide which idea is best. Indeed, this method corresponds to the democratic ideal of an internet-based open innovation community, but while community voting might work in some situations, it also has its drawbacks. Especially with a small number of participants, community voting is often influenced by human characteristics, such as personal sympathy, antipathy and alliances. When idea reviews within some communities are analyzed, it becomes apparent that some innovators mutually exchange good reviews. Still, many platforms offer participants the possibility to comment on other ideas. The final evaluation, however, is normally made within the company seeking ideas. The company's internal evaluation committee might take the community vote into consideration, but is not obliged to do so, and can attribute higher importance to other (preferably pre-defined) criteria.

## 2.5 Utilization

After the ideas have been collected and assessed and prizes have been distributed, the next – and often most difficult – step is to make use of these ideas within the company.

**Comparing Ideas Against the Project's Objectives** If a crowdsourcing project aims to search for a long-awaited technical solution and viable ideas have been provided during the project, there is no need to worry whether every possible effort is made to implement them. By contrast, when the objective of the project was to identify potential customer needs, there is a considerable risk that the company will only take cursory notice of them without flagging them for further action or development. It must therefore be clear why the project has been carried out and what its potential consequences are. In the case of EBW cited previously, the ideas with the highest potential were automatically followed up on, as EWB had credited each of them with business potential.

**Fairness Towards the Contributors** If innovators participate in crowdsourcing competitions, they believe in their ideas. Although most innovators might be better off with a lump-sum payment, most of them prefer to participate in the profit generated through their idea. Regardless of the type of compensation, it must be made clear to the participants what will happen with their ideas. Even if the terms and conditions of the respective platform allowed for it, it would be fatal to use ideas without considering the submitters, or even without informing them. If a product or business model is based on the results of a crowdsourcing competition, it is important to reward the contributors accordingly, even if this should not be necessary from a legal perspective.

**Maintaining and Further Developing the Community** Through a crowdsourcing project, the company seeking ideas has found a few to a few 100 people that are willing to work on its problems without guaranteed payment. These people, at least those who have distinguished themselves by competent, creative and interesting posts, have to be retained as potential contributors. Especially when a company has built its own platform, it has to address the community with new challenges every now and then. If the company fails to do so, relying on an existing crowdsourcing intermediary would make more sense (Fig. 25).

## 3    Opportunities and Risks of Crowdsourcing

Crowdsourcing is a phenomenon that will increase in importance and frequency over the coming years. The rise of mobile computing has given rise to a situation in which access to the internet is virtually omnipresent. And an ever increasing number of platforms and service providers make use of this situation and develop new offerings. The service *wheelmap.org*, for instance, enables anyone in Berlin to post whether a place is accessible for wheelchair users. What started as a small community of wheelchair users has become a phenomenon in Berlin that is valuable for other groups, such as young mothers with strollers, too. The amount and the scope of data and insight that can be gathered by a large group of individuals using modern communication tools has yet to be fully understood. However, faced with this enormous potential, many firms do not ask more of the crowd than suggested names for a new product. Firms are urged to engage in crowdsourcing, to study the phenomenon, and to develop a better understanding of how it can be integrated into

# Crowdsourcing as an Innovation Tool

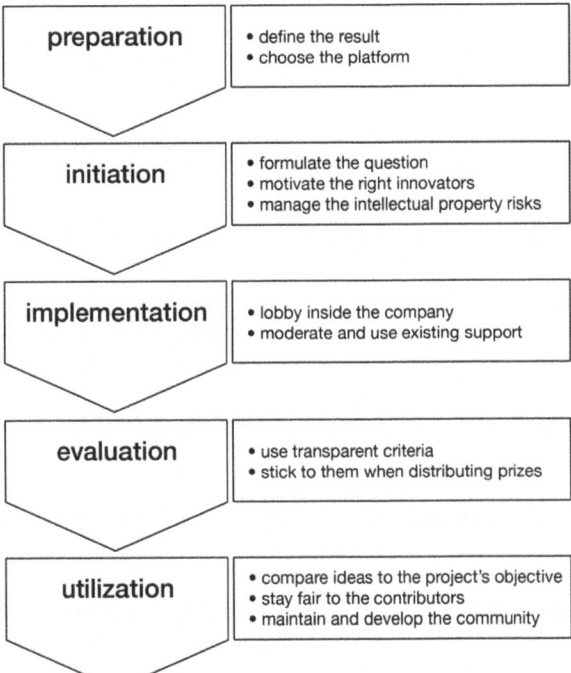

**Fig. 25** Key elements of the crowdsourcing process

the innovation process. Experience shows that the insights gained from crowdsourcing projects are especially valuable in early phases of the innovation process. New ideas and new links to similar problems in other fields often help firms to broaden their solution space. In general, crowdsourcing can offer five diverse opportunities to the firms engaged in this practice:

- **Solving a problem**
  At least in the idea and concept phases of an innovation, problem solution is usually significantly faster, more effective and less costly than internal processes.
- **Reviving an idea discarded internally**
  An additional opportunity lies in the revival of ideas previously discarded internally. Crowdsourcing may show that already rejected ideas are wanted by customers after all and might be worth looking into again.
- **Overcoming organizational blindness**
  Overcoming organizational blindness is an added benefit of crowdsourcing. This has been observed in many cases; in retrospect a solution seemed almost trivial, but because of the mindset shaped by a particular company's industry nobody came to think of it.
- **Understanding needs and desires**
  Another major advantage of crowdsourcing projects lies in recognizing needs and desires and thus getting new market impulses. For example, a Swiss gas and

energy firm learned that customers often associated gas cylinders with massive explosions. A thinking that was probably influenced by Hollywood led the company to increasingly inform on safety issues.

- **Advertising**

    Since crowdsourcing is a comparatively young innovation tool and involves a broad mass of external individuals, it is suitable for indicating the innovation activity of a company. Often 'innovativeness' can only be claimed by a firm, and often such a claim is perceived as a marketing stunt rather than the honest truth. Crowdsourcing, however, is a tool that can show innovation and raise awareness for a firm's engagement in product development.

As with all new technologies, there are also risks involved when crowdsourcing projects are carried out. The good news is that most of these risks can be eliminated by a certain amount of preparation and dedication to the task. However, this chapter would paint an unrealistic picture if it did not report on the potential risks a crowdsourcing project can entail. For instance, jointly with a restaurant chain, the *Pepsi* brand Mountain Dew searched for a new name for one of its soft drinks. The crowdsourcing project had to be taken down after pranksters had taken over the competition. Among the top 10 entries were ideas like 'Soda', 'Diabeetus', 'Moist Nugget', or 'HIV juice'. In another example, the country singer Taylor Swift orchestrated a crowd voting to determine at what university college she would give a free concert. Pranksters all over the internet joined forces and voted for the 'Horace Mann School for the Deaf'. In 2010, *Time magazine* asked the crowd online to determine the 'Person of the Year'. Wikileaks' Julian Assange won by a landslide, the magazine ignored the vote and chose Mark Zuckerberg instead. Zuckerberg had only attracted a fraction of the votes. The magazine faced harsh criticism afterwards. These examples illustrate some of the risks associated with crowdsourcing. Generally, there are four different risks that firms may face when engaging in a crowdsourcing project:

- **Total costs**

    The costs that a company incurs until a project has been completed are far above those caused by the crowdsourcing process itself. The submitted solution must be implemented, completed and embedded in the organizational context. For all these steps, resources have to be calculated. It is important that crowdsourcing projects are not launched as an isolated activity, but in the context of an overall strategy, and that the necessary financial and human resources are made available. Otherwise companies will run the risk of generating even more creative ideas, which will remain unused for everyday business and will only cause the dissatisfaction of everyone involved.

- **Low compensation**

    Participants usually receive little or no compensation for their services. In many cases, only the best solutions are rewarded. Again, this raises the question of whether companies ought to compensate the appropriate participants. Damage to a company's reputation is a possible consequence in extreme cases.

- **Motivation**
  Another risk lies in the contributors' motivation. Given the low compensation and the fact that the participants usually have to give up all their rights, their personal interest in an optimal solution can be compromised.
- **Legal problems**
  The fourth and perhaps most pronounced risk deals with legal problems. How much is an idea worth? Most of the time the appropriate amount cannot be quantified in advance. The premiums paid in advance can lead to resentment. Participants would mostly like to benefit from the market success of their ideas, companies usually try to obtain exclusive rights to these ideas. It is essential to communicate this clearly in advance.

As with other projects, the success of a crowdsourcing campaign depends on its preparation, which is especially challenging as the project itself takes place in a virtual environment. As can be seen from the presented list of opportunities and risks, most of the latter can be managed by intensive and careful preparation. It is important not to jump into a crowdsourcing project blindly, but to make informed decisions on what, when and how to engage in crowdsourcing.

## 4 Using Crowdsourcing Successfully

Given the relatively young history of crowdsourcing, many companies appear to rush into the experience and do not seem to plan the management of the project well in advance. Thus, many crowdsourcing projects fail and sometimes the companies are even scared away from the topic for good. It is unfair to simply blame a process if its execution was at fault. Aside from the process-related factors provided in this chapter, three fundamental cornerstones of successful crowdsourcing projects can be distinguished. Companies should thus make an effort to consider the following:

- **A necessary understanding of the web 2.0 culture**
  Crowdsourcing is an instrument that asks an anonymous crowd for help, understanding how internet users behave in other settings, such as YouTube Videos, Reddits, or on 4Chan, is key to managing a web 2.0 project.
- **Professional preparation**
  Too many crowdsourcing projects feel rushed, and not enough thought is given to the question of what might go wrong. Like in many other settings, proper preparation is key. This is especially true in crowdsourcing projects, as an anonymous group is given a task and might misinterpret it.
- **Goals and responsibilities**
  A clear understanding of what a company wants from a crowdsourcing project and who is responsible for it helps the entire project. It helps to communicate why the project is carried out, helps to level expectations, and helps to make sure a single person feels responsible and acts as spokesperson for the project inside the company.

# Revolutionizing the Business Model

Oliver Gassmann, Karolin Frankenberger, and Michaela Csik

## 1    New Products Are Not Enough

There are many companies with excellent technological products. Especially in Europe, many firms continuously introduce innovations to their products and processes. Yet, many companies will not survive in the long term despite their product innovation capabilities. Why do prominent firms, which have been known for their innovative products for years, suddenly lose their competitive advantage? Strong players such as *AEG, Grundig, Nixdorf Computers, Triumph, Brockhaus, Agfa, Kodak, Quelle, Otto,* and *Schlecker* are vanishing from the business landscape one after the other. They have lost their abilities to market their former innovative strengths. The answer is simple and painful: these companies have failed to adapt their business model to the changing environments. In the future, competition will take place between business models, and not just between products and technologies.

New business models are often based on early weak signals: Trendsetters signal new customer requirements; regulations are discussed broadly before they are eventually approved. New entrants to the industry discuss new alliances at great length; disruptive technology developments are the results of many years of research. Even *Kodak*'s filing for bankruptcy in 2012 is the result of a long chain of events. The first patents for digital cameras had already been acquired by *Texas Instruments* in 1972. Kodak realized the potential of the new technology and in the 90s entered into a digital imaging alliance with *Microsoft* in order to conquer this new field. But – as can as can be frequently observed in other firms as well – the disruptive move was faint-hearted. When the first digital cameras entered the market in 1999, Kodak forecasted that 10 years later digital cameras would account

---

O. Gassmann (✉) • K. Frankenberger • M. Csik
University of St. Gallen, Institute of Technology Management, Dufourstr. 40a,
9000 St. Gallen, Switzerland
e-mail: oliver.gassmann@unisg.ch; karolin.frankenberger@unisg.ch; michaela.csik@unisg.ch

for only 5 % of the market, with analog cameras remaining strong at 95 %. In 2009, the reality was different: Only 5 % of the market remained analog. This miscalculation was so grave with such powerful repercussions that it was too late when Kodak physically blew up its chemical R&D center in Rochester in order to change the corporate-dominant logic of analog imaging. Between 1988 and 2008, Kodak reduced the number of its employees by more than 80 %, in 2012 Kodak filed for bankruptcy protection.

It is often said that existing business models 'don't work anymore'. Still, the typical answers provided by R&D engineers are new products based on new technologies and more functionality. By contrast, the underlying business logic is rarely addressed despite the fact that business model innovators have been found to be more profitable by an average of 6 % compared to pure product or process innovators (BCG 2008). As a consequence, managers consider business model innovation to be more important for achieving competitive advantage than product or service innovation, and over 90 % of the CEOs surveyed in a study by IBM (2012a) plan to innovate their company's business model over the next 3 years. But a plan is not enough.

When it comes to making the phenomenon tangible, people struggle. Very few managers are able to explain their company's business model ad-hoc, and even fewer can define what a business model actually is in general. The number of companies which have established dedicated business model innovation units and processes is even lower. Given the importance of the topic, this lack of corporate institutionalization is surprising; however, considering the complexity and fuzziness of the topic, it is to be expected.

Before discussing how to innovate a business model, it is important to understand what needs innovating. Historically, the business model has its roots in the late 1990s when it emerged as a buzzword in the popular press. Ever since, it has attracted significant attention from both practitioners and scholars and nowadays forms a distinct feature in multiple research streams. In general, the business model can be defined as a unit of analysis to describe how business is transacted. More specifically, the business model is often depicted as an overarching concept that the different components which constitute a business and puts them together as a whole (Demil and Lecocq 2010; Osterwalder and Pigneur 2010). In other words, business models describe how the magic of a business works based on its individual bits and pieces.

Business model literature has not yet reached a common opinion as to exactly which components make up a business model. To describe the business models throughout our study, we employ a conceptualization that consists of four central dimensions: the Who, the What, the How, and the Why. Reducing it to the four dimensions renders the concept much easier to, while, at the same time, exhaustive enough to provide a clear picture of the business model architecture.

**Who:** Every business model serves a certain customer group (Chesbrough and Rosenbloom 2002; Hamel 2000). Thus, it should answer the question 'Who is the customer?' (Magretta 2002). Drawing on the argument from Morris et al. (2005, p. 730) that the "failure to adequately define the market is a key factor

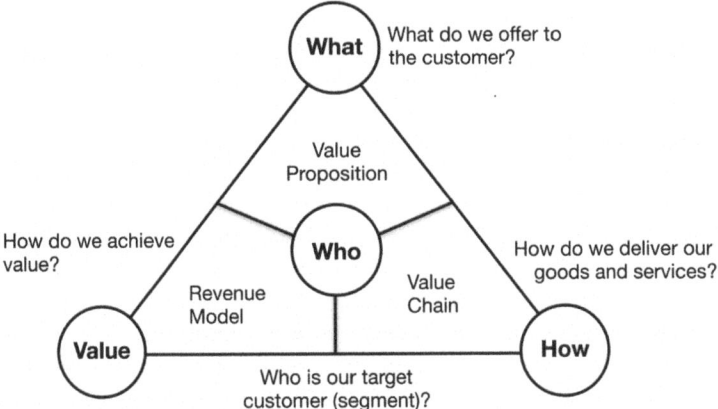

**Fig. 26** Business model definition – the magic triangle

associated with venture failure", we identify the definition of the target customer as one central dimension in designing a new business model.

**What:** The second dimension describes what is offered to the target customer, or put differently, what the customer values. This notion is commonly referred to as the customer value proposition (Johnson et al. 2008), or, more simply, the value proposition (Teece 2010). It can be defined as a holistic view of a company's bundle of products and services that are of value to the customer (Osterwalder 2004).

**How:** To build and distribute the value proposition, a firm has to master several processes and activities. These processes and activities, along with the relevant resources (Hedman and Kalling 2003) and capabilities (Morris et al. 2005), plus their orchestration in the focal firm's internal value chain form the third dimension within the design of a new business model.

**Value:** The fourth dimension explains why the business model is financially viable; thus it relates to the revenue model. In essence, it unifies aspects such as, for example, the cost structure and the applied revenue mechanisms, and points to the elementary question of any firm, namely how to generate value (see Fig. 26).

By answering the four associated questions and identifying (1) the target customer, (2) the value proposition towards the customer, (3) the value chain behind the creation of this value, and (4) the revenue model that captures the value, the business model of a company becomes tangible and a common ground for its rethinking is achieved. A central virtue of the business model is that it allows for a holistic picture of the business by combining factors located both inside and outside the firm (Teece 2010; Zott et al. 2011). For this reason, it is often referred to as a boundary-spanning concept that explains how the focal firm is embedded in, and interacts with, its surrounding ecosystem (Shafer et al. 2005; Zott and Amit 2008). The task most commonly attributed to the business model is that of explaining how the focal firm creates and captures value for itself and its various stakeholders within this ecosystem.

Considering the vast scope that is subsumed by the business model umbrella, it becomes clear that, in the real world, a firm's business model is a complex system full of interdependencies and secondary effects. We can assume then that changing – or innovating – the business model can be a major undertaking that can quickly become more complex.

Generations of managers have been trained in Porter's five forces of industry analysis. Michael Porter taught us to analyze the industry and try to gain comparative competitive advantage due to better positioning. Kim and Mauborgne (2005) built further on this. 'Beat your competitor without trying to beat your competitor' is the credo that obliges companies to leave their own highly competitive industry and create new uncontested markets in which they can prosper. It is a mantra for business innovators as we have seen in our own research and company coaching over the last decade. *IKEA* revolutionized the furniture business, *Apple* successfully re-defined industry boundaries, and *Zara* reinvented the European fashion industry with high-speed cycles. Many others revolutionized their industries in a very radical way: *Mobility car sharing, Car2go, TomTom, Wikipedia, Microinsurance, Better Place, Verizon,* and *Bombardier Flexjet* are only few examples of companies which escaped the traditional industry logic and redefined their respective industries.

So, why do more companies not just come up with a new business model and move into a 'blue ocean'? It is because thinking outside the box is hard to do – mental barriers block the road towards innovative ideas. Managers struggle to counteract the predominant logic of 'their' industry, which they have spent their entire careers understanding. First, many managers do not see why they should leave the comfort zone as long as they are still profitable. Second, it is common knowledge that the harder you try to get away from something, the closer you get to it. Bringing in outside ideas might seem promising in this case. However, the not invented here (NIH) syndrome is well known and will soon quash any outside idea before it can take off in a company.

In view of these barriers, a successful approach that leads to innovative business model ideas must master the balancing act of bringing in stimuli external to an industry to achieve novelty while, at the same time, enabling those within an industry to develop their own innovative business model ideas.

**Research Methodology.** As business innovation research is still a young phenomenon, we used a two-step approach to analyze the basic patterns of business models.

In phase 1, we analyzed 250 business models that had been applied in different industries within the last 25 years. As a result we identified 55 patterns of business models which served as the basis for new business models in the past. More than 5 years of research and practice in the area of business model innovation have culminated in a methodology that helps firms structure and navigate the process: the *Business Model Innovation Map*, which guides the innovator through the many opportunities a company faces (see also Gassmann et al. 2013).

In phase 2, we used that knowledge and, together with selected companies, developed a construction methodology which is based on two basic principles: First, we found in our research group that 90 % of all new business models have recombined previously existing ideas, concepts and technologies. Consequently this fact has to be used for developing new business models. Second, we applied the iterative process of design thinking, which was developed at the Institute of Design at Stanford University. This action-based research approach helped us to learn more about the practical use of the design of new business models.

We applied the methodology with teams in the following companies: *BASF* (chemicals), *Bühler* (machinery), *Hilti* (construction tools), *Holcim* (cement), *Landis+Gyr* (electricity metering), *MTU* (turbines), *SAP* (software), *Sennheiser* (audio technology), *Siemens* (health care), *Swisscom* (telecom). In all companies, investments have been initiated as a result of the business model project, in some companies up to double-digit million amounts are invested. In addition we used the approach during 3 years of teaching Executive MBA students at the Executive School in St. Gallen and applied it in a 1-day workshop for more than 50 companies. This experience has been built into the methodology as well.

## 2  Creative Imitation and the Power of Recombination

The phrase 'reinventing the wheel' describes the fact that, at a closer look, only few phenomena are ever really new. Often, innovations are slight variations of something that has existed elsewhere, in other industries, or in other geographical areas. We have looked at several hundreds of business model innovators and were not surprised to find that about 90 % of the innovations turned out to be such re-combinations of previously existing concepts. We identified 55 repetitive patterns that form the core of many new business models (see Gassmann et al. 2012, 2013). The business model innovation map (see Fig. 27) depicts the 20 most popular patterns as lines, along with the companies which applied them in their new business models.

The 'razor and blade' pattern, for example, goes back to *Gillette*'s 1904 move to give the base product (the razor) away for a low price and earn money through higher-priced consumables (the blades). The pattern, which defines the value proposition and revenue logic of a business model, has spread across many industries since then. Examples include inkjet printers and cartridges, blood glucose meters and test stripes, or *Nespresso*'s coffee machines and capsules. In the world of business models, there is really not much that is actually new, but many powerful adaptations and applications contexts and industries can be found.

What can we learn from this observation? Clearly, the patterns of business models identified can serve as an inspiration when innovations of business models are considered. If they could be adopted elsewhere, why not apply them to one's own company? This approach both brings in external stimuli enough room to prevent the NIH syndrome. Over time, we developed the 55 business model patterns identified into the central ideation tool of our *Business Model Innovation Navigator* methodology.

The BMI Navigator transforms the main concept – creating business model ideas by utilizing the power of re-combination – into a ready-to-use methodology, which

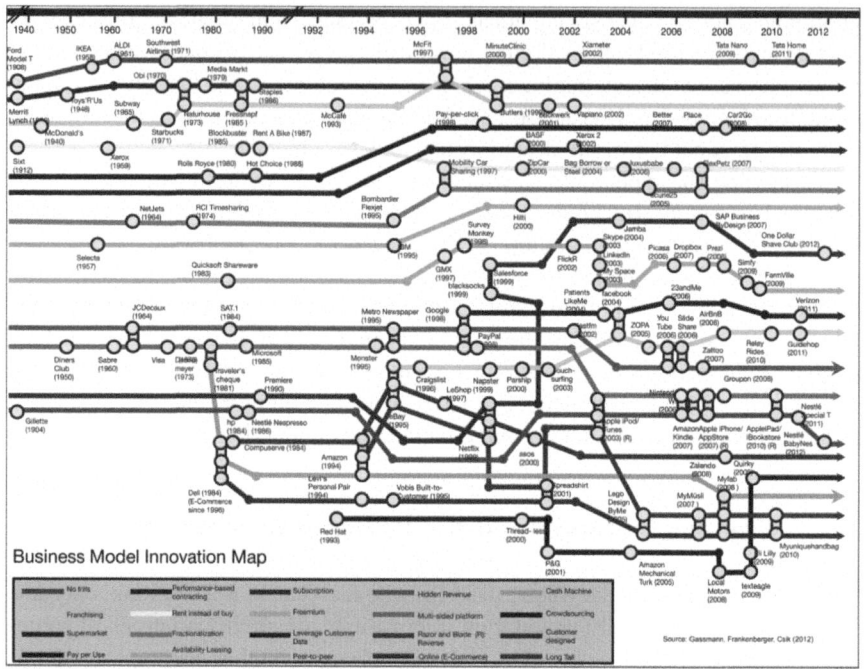

**Fig. 27** The business model innovation map: every node represents a revolution in an industry

has proven its usefulness in countless workshops and other formats. Three steps pave the road to a new business model:

**Step 1: Initiation – Preparing the Journey.** Before embarking on the journey towards new business models, it is important to define a starting point and rough direction. Describing the current business model, its value logic, and its interactions with the outside world is a good exercise for getting into the logic of business model thinking. It also builds a common understanding of why the current business model will need an overhaul, which factors endanger its future, or which opportunities cannot be exploited in the current business model. Investigating these woes and the predominant industry logic provides a rough direction according to which the generic business model patterns should be interpreted in step 2.

**Success Factors**
- Involve open-minded team members from different functions; the involvement of industry outsiders supports thinking outside the box.
- Overcome the dominant industry logic: sentences like 'this has always worked like that in our industry' are taboo. Instead, a eulogy for one's firm helps to overcome the past. Why did the company die? This is a fascinating exercise which *McKinsey* has often used successfully in change projects when individuals needed to overcome mental barriers.

**Fig. 28** Pattern card set

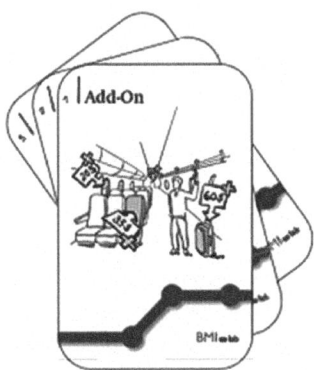

- Use methodological support, e.g. card sets, e-learning based innovation, an app (see www.bmi-lab.ch for our methodological approach and background information).

**Step 2: Ideation – Moving into New Directions.** Re-combining existing concepts is a powerful tool to break out of the box and generate ideas for new business models. To ease this process, we have condensed the 55 patterns of successful business models into a handy set of pattern cards. Each pattern card (see Fig. 28) contains the essential information that is needed to understand the concept behind the pattern: a title, a description of the general logic, and a concrete example of a company implementing the pattern in its business model. During the ideation stage, the level of information on the card is just right to trigger the creation of innovative ideas.

The way in which we apply the cards is termed *pattern confrontation* to describe the process of adapting the pattern to one's own initial situation. Participants, typically divided into groups of three to five, ask themselves how the pattern would change their business model if applied to their particular situation.

At first glance, the cards might seem unrelated to the problem; however, the results are quite surprising. Often the stimuli, in the form of pattern cards, cause innovative ideas to emerge, which inspire discussions among the group members. In one instance, for example, the task of fitting the 'subscription' pattern to the business model of a machine manufacturer led to the idea of training sought-after plant operators and leasing them to customers. The concept was implemented and now contributes to the company's turnover while at the same time strengthening ties with customers – which had been the original reason for thinking about a new business model.

**Success Factors**
- Try not only the close patterns, but also confront more distant patterns. We had very surprising results when a first tier automotive supplier applied the question:

"How would *McDonald*'s conduct your business?". For example, McDonald's front desk employees are fully productive after a 30 min introduction. The automotive supplier had to learn that reducing complexity would lead to totally new business models and would also stimulate quick learning.
- Keep on trying. At first, it seems impossible to learn something from industry outsiders. Especially individuals with a profound background in the existing industry have difficulties in overcoming the dominant industry logic.

**Step 3: Integration – Completing the Picture.** There is no idea that is clear enough to be immediately implemented in a company. On the contrary, promising ideas need to be gradually elaborated into full-blown business models that describe all four dimensions (who? what? how? why?) and also consider stakeholders, new partners, and consequences for the market. A set of checklists and tools, such as the value network methodology, are available in the BMI Navigator to ease the step of quickly developing the business model around a promising idea. The list of example companies on each pattern card makes it possible to draw inspirations from other companies which implemented the same pattern.

**Success Factors**
- Be consistent. Consistency between the internal and the external world is necessary. There has to be a fit between the internal core competencies, the competitor's perspective, and the perceived customer value.
- Try hard. Developing a business model and implementing the idea in one's own company requires a lot of work.

### Conclusions

With the BMI Navigator a new methodology has been developed that structures the process of innovation of a company's business model and encourages out-the-box thinking, which is a key prerequisite for successful business models. Wellgrounded in theory, it has proven its applicability in practical settings many times over.

In order to achieve successful business model innovations within a company it is important to not only acknowledge the importance of business model innovation, but to implement an effective business model innovation process within the firm. This is the most difficult, but also the most important step. Various tools have been developed to support managers during the business model innovation process:

# Revolutionizing the Business Model

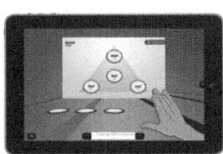

**Business model navigator**
Innovative software tool, which allows exploring the 55 business model patterns and the map interactively. The software tool supports the construction of a new business model based on the St. Gallen Business Model Innovation Navigator throughout the company on a worldwide scale

**Online learning**
The online learning course is aimed at employees and in an interactive way explains the logic and importance of business model innovation and the power of recombining existing business model elements

**55 business model cards**
A set of the 55 business model cards supports the creative ideation process during workshops. For more details check www.bmi-lab.ch

For practitioners using this new approach to revolutionize their business following managerial implications can be given:
- Challenge the dominant logic by using confrontation techniques. The 55 patterns of business models identified support this challenging task.
- Use an iterative approach with many loops.
- Use haptic cards or other devices to stimulate the creative thinking process.
- Carefully assign the role of a pivot thinker for changing the direction between divergent and convergent thinking.
- Create a culture of openness: there are no sacred cows in the room.

Given the overwhelming demand for that new business model innovation methodology, the journey of the BMI Navigator will continue. The future race for comparative competitive advantages has shifted from pure products and services to business models. Firms need to get ready for that race. Identifying the opportunity is not enough, innovators and entrepreneurs have to capture the opportunity and start moving. Knowing the past helps in creating the future.

# Managing the Intellectual Property Portfolio

Martin A. Bader, Oliver Gassmann, and Nicole Ziegler

## 1  Introduction

In today's society driven by knowledge, *intellectual property* (IP) has moved from a legal matter to a strategic issue (Smith and Hansen 2002). In research and practice the management of intellectual property has gained in recognition as a powerful instrument of corporate strategy and a main source of competitive advantage. Patents have become increasingly important for firms in order to protect technologies from imitation, achieve a stronger position in global markets, strengthen the firm's technological leadership, and enable the trading of intellectual assets (Davis and Harrison 2001). This trend is underlined by the growing number of patent applications over the last century. Since 1985, the worldwide yearly patent filings have more than doubled. The World Intellectual Property Organization (WIPO) reports that 1.98 million patent applications were filed worldwide in 2010 (WIPO 2011). The countries filing the most patent applications are the United States with 490,226 patent applications, followed by China with 391,177, and Japan with 344,598 patent applications. At the European Patent Office, 150,961 patents were filed in 2010. These numbers show that firms are rapidly accumulating patent rights and therefore the firms are challenged to effectively manage their growing *patent portfolios*.

---

This chapter is based on an earlier publication in *Drug Discovery Today*.

M.A. Bader (✉)
BGW AG Management Advisory Group, Varnbuelstr. 13, 9000 St. Gallen, Switzerland
e-mail: martin.bader@unisg.ch

O. Gassmann • N. Ziegler
University of St. Gallen, Institute of Technology Management, Dufourstr. 40a, 9000 St. Gallen, Switzerland
e-mail: oliver.gassmann@unisg.ch; nicole.ziegler@unisg.ch

In addition to the increasing number of intellectual property rights, and apart from their intellectual value, the financial value of intellectual property has also increased tremendously: in 2006, for example, *Research-in-Motion*, the developer of the BlackBerry smartphone, was forced to pay US $612.5 million to NTP, a firm focusing on the acquisition and deployment of patent rights, due to technology infringing on NTP's patent portfolio. More recently, in 2011, in the litigation case between *Intel* and *Nvidia* (a computer graphics specialist), Intel paid the record sum of US $1.5 billion to Nvidia for a license for Nvidia's entire technology portfolio.

Overall, patents are an important instrument for maintaining competitive advantage through temporary monopolies and are an important means of appropriating increased returns on investment (Thumm 2001). Many successful firms, such as *IBM, Philips, Dow Chemical, Roche, Novartis,* and *BMW*, have established well-structured IP management processes and organizational structures and consider intellectual property a major corporate asset. Despite the increasing importance of IP management in literature and practice, insights into how firms manage their patents from a holistic, strategic perspective, and how the portfolio value of patents can be optimized, are scarce. Existing intellectual property management frameworks end with the application for the patent rights, and lack a more comprehensive patent management model. Based on the findings of our interviews with leading research- and technology-based companies, we argue that the management of patents should not stop at their filing of the patents, but that it should be linked to the innovation process of a technology.

## 2 Managing Patents Along Their Life Cycle: A Management Model

Patents are the result of a firm's research and development activities and thus are directly related to the technology and product strategy of the firm. Hence, the patent strategy should be derived from corporate strategy and intend to both help generate new business potential and secure existing and realized potential. To this end, successful patent management follows technology management, i.e., patent management is strongly linked to the *life cycle of technologies* starting with the discovery of ideas and continuing until a product is discarded from the firm's portfolio. Based on this, five distinctive phases reflect the *patent life cycle* management activities:
- Explore
- Generate
- Protect
- Optimize
- Decline

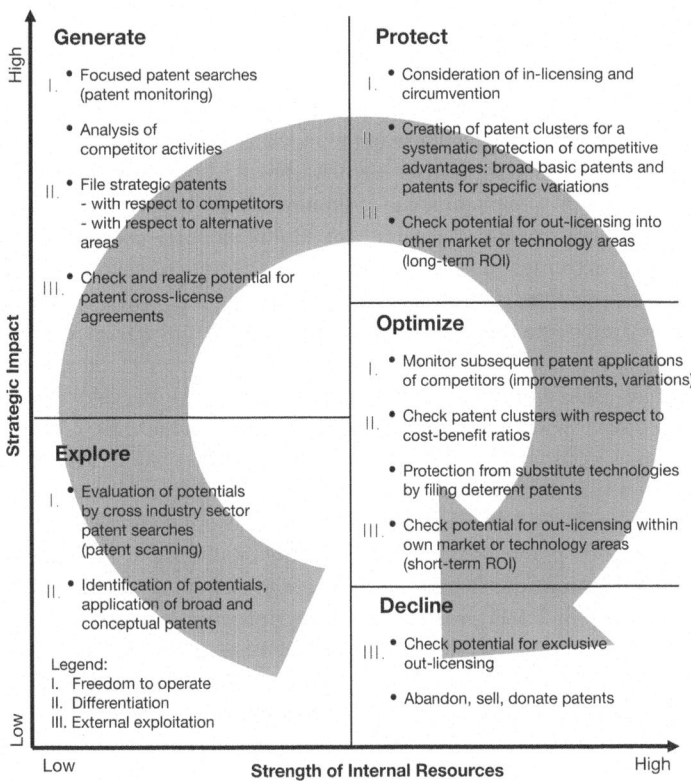

**Fig. 29** The patent life cycle management model

Furthermore, the way patents are managed largely depends on the patent's strategic value and the firm's internal resources. The patent's strategic value refers to the strategic value of the technology or patent relative to existing markets, competitors, and substitution technologies. The strength of internal resources refers to the firm's assets, such as employees, know-how, and experience regarding a certain technology. While in the first two phases, i.e., explore and generate, the firm accumulates new competences about a new technology, these competences remain at a high level in the protect and optimize phases. This is true even for the decline phase, although here, the firm may decide to discard the patent. Thus, our management model displays patent life cycle management as a function of the strategic impact and internal resources available in which each phase addresses three core dimensions of patenting: freedom to operate, differentiation from competitors, and external patent exploitation (Fig. 29).

In the following sections, the phases of the patent life cycle management model are described in detail and illustrated through examples.

## 2.1 Explore

In the first phase, i.e., explore, the firms collect ideas for new inventions. The strategic impact is still low or unpredictable and technological trends are explored through broad cross industry patent searches, such as, for example, patent scanning. Researchers or marketing specialists identify specific keywords and define search profiles to narrow down relevant areas of interest. Also the patentability of existing technologies and the freedom to operate are simultaneously checked during these patent scanning activities. The car manufacturer *DaimlerChrysler*, for instance, starts its development projects with searches regarding the relevant state-of-the-art and third-party intellectual property rights. Based on this, the relevant situation of intellectual property rights is recorded and assessed. Also *Prionics*, a Swiss life science company and world leader in farm animal diagnostics, follows the strategy that a new project always starts with a comprehensive patent search. For each new project, Prionics compiles an individual search profile – often with the help of external experts like the national patent and trademark office. Patent monitoring is conducted on a monthly basis with the databases of Medline and Derwent. During this search, 400–500 potentially relevant literature citations and 75–120 potentially relevant patent citations are identified. In a second screening, an internal group of experts, consisting of R&D project leaders and product managers, evaluates the search results and filters 30–50 relevant literature citations and 10–25 relevant patent citations. This kind of search process is very successful at Prionics and has been established as an integral part of new product development.

## 2.2 Generate

The exploration phase is succeeded by the generate phase where ideas are realized through the development of new products. If the developed solution incorporates patentable technologies, the right patenting approach has to be chosen. For example *Henkel*, a German consumer goods manufacturer, pursues the strategy of striving for the exclusive protection of its products, technologies, packaging, and substances for its core competence areas. Inventions in non-core competence areas often are not protected through patents, but are published, e.g., in professional journals to prevent potential patenting by competitors. At *Cytos Biotechnology*, a Swiss high-tech biotechnology SME specializing in therapeutic vaccines, the strategy is to identify and patent new specific substances as early as possible and to partner with large pharmaceutical companies for further clinical development.

In addition, selective patent searches should be carried out as soon as relevant areas with increasing strategic importance have been identified. The aim is to selectively monitor further developments in specific technological fields and for specific competitors. *Erbe Elektromedizin*, a medium-sized German medical engineering company, systematically monitors its competitors:

- Every month the patents department receives the new publications from the previous month generated by property rights monitoring. Property rights are mostly monitored through the search department of an external patent attorney who defines the search parameters on the basis of a specified filter. In urgent cases the department can also carry out searches itself.
- The patent department examines and preselects the documents, which are then forwarded to the relevant technical experts in R&D. Thus, an engineer receives precisely those documents that relate to his/her technical fields.
- The technical experts prepare synopses of the documents presented to them. They are allowed 3 min for their report.
- The technical experts' brief reports are presented in the course of a monthly patents round, for example every first Tuesday of the month at a specified time. Each report is followed by a brief discussion, and the next steps are agreed on, for example the decision to file an opposition or to include a document in document monitoring. Since the patents round meets once a month, there is the possibility of filing an opposition to any of the documents discussed there.
- Finally, the patent department prepares minutes which are distributed to those involved in the patents round.

This procedure has several advantages: the fixed date for the patents round guarantees considerable regularity, which, in turn, ensures that the engineering specialists are always fully aware of the property rights situation. The obligation to provide synopses ensures that the engineers analyze the patent documents at the appropriate time (period for opposition) and report back directly to the patents department and colleagues; moreover, the procedure ensures that there is a lively discussion of the documents, that specific suggestions are submitted to each R&D work group, and also that duplicate developments and duplicate applications are avoided.

Furthermore, opening up research processes, acquiring external technologies to complement one's own technology portfolios, and engaging in collaboration is becoming increasingly important in this phase. *Roche*'s research and development network, *Bayer*'s engagement in cross-licensing, and *Biotronik*'s strategy to in-license technology are examples of ways to complement internal know-how. Although *Henkel* is rather reluctant to open its innovation process, this company also uses cross-licensing agreements with cooperation partners for specific parts of the portfolio. In these cooperations, Henkel aims to avoid financial compensation, but tries to agree on a patent exchange based on the quality of the patents. Another example of a collaboration is furnished by *CeramTec*, a subsidiary of the Swedish *DynamitNobel Group*. The company collaborated with a supplier of the automotive industry to develop a cylinder head for engines. Negotiations regarding the use of the ensuing intellectual property rights were carefully conducted. While a joint use was agreed on for engine applications, CeramTech obtained the exclusive rights for the ceramic markets.

## 2.3 Protect

The high strategic impact of the technology and strength of resources are characteristic of the *protect* phase. The firms have accumulated comprehensive know-how in a field of competence with a high level of strategic importance. The potential for filing broad basic patents is declining since public knowledge in these fields has already greatly increased. The patent applications now focus increasingly on more detailed, very specific embodiments, often with the motivation to build patent fences around a core invention to preclude patenting of substitutes by rivals. Therefore, firms should increasingly seek to create patent clusters in strategically important fields of technology. This involves generating patent portfolios which have a broad scope (growing), but which later, when it is easier to estimate which ideas are technically and commercially viable, are thinned out again (pruning). Henkel successfully uses the growing and pruning method to protect as many variants as possible at an early stage of development and to prevent the patent portfolio from later incurring excessively high costs. At this stage, searches often no longer produce the desired up-to-date information since – owing to the 18-month waiting period for publication – it is impossible to say which variants competitors are continuing to develop or which technical means have been selected to solve a problem. Especially in the case of competences generated with external collaboration partners, consideration should be given to the extent to which it is possible to out-license in other technical fields or market segments in order to be able to generate revenue from licensing agreements in the long term. The Swiss pharmaceutical company Roche continuously builds up patent clusters by filing patents for back-up compounds and follow-on patents to enhance the protection of its products. About 1–10 basic patent applications and patents for back-up compounds are filed per project. During the clinical development and early commercialization phases, about 3–30 follow-on patents are filed. Follow-on patent applications include, for example, patents for polymorphs, salts, or alternative formulations.

Also in-licensing opportunities or collaborations to access third party know-how should be considered in this phase. For the development of the central multifunctional *idrive* control element, the car manufacturer *BMW* collaborated with the small Californian software company *Immersion*. This company had already developed relevant competences in the field of force feedback technology, which is used in joysticks, controllers in the field of design engineering, and in medical technology. It was agreed that BMW would acquire exclusive rights, for a limited time, to the development results in the automotive field, while Immersion would be entitled to engage in independent exploitation and marketing outside the automotive sector.

## 2.4 Optimize

In the *optimization* phase, the firm has a high level of competence in the respective technological field, but the strategic importance with regard to customers, markets, competition, or technology is declining. The firms monitor competitors' patenting activities and review their own patent clusters thoroughly in terms of cost-benefit considerations. For example, a potential reduction of the territorial coverage of the patents should be checked regularly. Also the risk of substitute technologies must be analyzed. If there is a risk of competences being replaced by substitute technologies, the firm's own patents in these fields can be used to block property rights to prevent a decline in value of the existing core technologies. The German sports car manufacturer *Porsche*, for example, utilizes intellectual property rights relating to substitute technologies specifically to prevent the premature decline in value and dilution of existing technologies. Where appropriate, exclusive licenses are even acquired for this purpose and kept on a shelf.

For research-based pharmaceutical firms, the risk of infringement especially emanates from generic drug companies. Generic drug companies become competitors when the relevant patent's expiration date approaches. About 5 years before expiration of a patent, the generic companies are able to legally enter the market and use the specific agent. The Swiss-based company *Novartis*, for example, performs active competitive intelligence to identify potential infringement. As generic companies can start regulatory readiness before the innovator's exclusivity period has expired, Novartis keeps an active eye on all developments of 'their' products. The first assessment is undertaken in the preclinical phase; further assessments take place at the beginning of each development phase.

A review also needs to be carried out on out-licensing opportunities, which, unlike when securing potential, also include the company's own technical fields or market segments so that revenue can also be generated from licensing agreements in the short term. Sometimes it is even possible to stimulate a market segment by opening it up so that the likelihood of substitution can be delayed further by greater standardization and price reductions. For example, after a patent dispute, the Danish hearing aid manufacturer *ReSound* was able to buy a strong patent portfolio from *3M*, which ReSound contributed to the hearing instrument patent pool HIMPP (Hearing Instrument Manufacturers Patent Partnership). Companies can join this pool, which was set up by *Danavox, Oticon, Phonak, Starkley* and *Widex*, on payment of a membership fee. In practical terms, this creates market entry barriers for potential new competitors.

## 2.5  Decline

In the *decline* phase, when the strategic importance of a technology or competence has greatly decreased, the corresponding patents are reviewed to determine whether they still add value to the firm and to define the divestment strategy. Before the patents are abandoned, the firms should check the patents for out-licensing, selling, or donation opportunities. The industrial process and measurement engineering company *Endress+Hauser*, for example, selects or sells any patents whose subject areas do not involve its own products or production processes within a period of approximately 7 years. In the early 1990s, *Dow Chemical* conducted a full review of its entire intellectual property right holdings in order to sell, out-license, donate, or abandon obsolete intellectual property rights. By doing this, abandoned or donated intellectual property rights enabled Dow to realize savings of US $50 million in unnecessary renewal fees and tax advantages.

*Biotronik*, a German-based multinational biomedical company, evaluates its patent portfolio in a yearly review and decides on how to proceed with obsolete patents. The intellectual property review board, supported by R&D and management, is responsible for assessing obsolete patents with regard to external exploitation through out-licensing. The board checks if the patent protects one of Biotronik's products or if its internal exploitation is planned in the future. Furthermore, it analyzes whether competitors could potentially use Biotronik's patents for their technologies and also whether the patents could be enforced in case of infringement. Finally, the overall costs and efforts are estimated before Biotronik decides on licensing or abandoning the patents.

## Conclusions

Effectively managing and optimizing the value of the patent portfolio is a major challenge for many firms. We adopt a strategic patent management perspective and suggest a holistic patent life cycle management model for an efficient management of intellectual property. The key message is that patent management should not be regarded as an isolated function, but as an integrated activity that considerably contributes to a firm's success because intellectual assets have become critical firm resources. Firms should therefore take a holistic view on their patent management and consider the following checklist:

- The basic recommendation is that patent strategy must be aligned with the firm's overall and R&D strategies to optimize the return on investment in new technologies. This allows firms to move their patent management in the same direction as their R&D management and – more broadly – as corporate management.
- Identifying new technological challenges is an important factor in creating innovation. Thus, firms should establish active technology scouting and patent scanning processes. The earlier technological trends are identified, the better the firm can react and obtain a first-mover advantage. It is especially important that these technology scouting and patent scanning activities are conducted and updated on a regular basis. Also, firms should ensure that the respective employees, e.g., R&D employees, patent managers, and business developers, are given access to the results.
- During the development of new technologies and products, it is important to keep an eye on competitor and market activities. Firms should therefore establish a patent monitoring system which regularly observes their environment. Special attention should be paid to identifying substitute technologies because these might weaken the firm's temporary monopoly gained through patent protection.
- External leveraging of patents through, e.g., out-licensing, cross-licensing, sale, strategic alliances, and joint ventures may enable firms to generate

additional returns on investment and to reap strategic benefits. Therefore, considering external exploitation opportunities at all stages of the patent life cycle should be a standard activity of any firm's patent management.
- Firms are advised to conduct regular (e.g., yearly) patent audits where the value of the patent portfolio is assessed. These audits should also be used to keep an overview on how each patent is exploited, i.e., which patents protect which products or technologies, which patents have a blocking function, which patents are out-licensed and to whom, and also which patents are currently not used for any competitive advantage or financial benefit. Based on the audit, decisions on when and where new patents should be filed and which patents could be out-licensed to generate additional income can be made and further steps for implementing these activities can be defined.
- Issued, but unused patents cause unnecessary maintenance fees. Hence, firms should make proactive patent divestment decisions to avoid the accumulation of unused patents. First, potentially obsolete patents should be evaluated and balanced with regard to the benefits and costs for the firm. If the patent reveals potential attractiveness for other firms, out-licensing, sale, or donation should be considered.
- Finally, patent life cycle management should be seen as a holistic and interdisciplinary task. Thus, the aforementioned recommendations should be implemented by a small group of senior executives, consisting of heads of intellectual property, R&D, business development, product development, and marketing.

# Applying Cross-Industry Networks in the Early Innovation Phase

Ellen Enkel and Sebastian Heil

## 1 Introduction

Rapidly changing and hyper-competitive global business environments, ever-increasing complexity of technology, and the growing mobility of highly experienced people which leads to a fragmentation of knowledge are undisputed challenges of modern times. Consequently, firms have realized that going it alone when sourcing new knowledge and technologies to develop new products and services is no longer a promising option (Chesbrough 2003). External knowledge provides greater prospects for the combination and recombination of knowledge for innovation purposes (Laursen and Salter 2006). For example, single R&D cooperations with existing suppliers and customers provide flexible and efficient solutions to knowledge and technology sourcing in the early phase of innovation; however, these are not enough to keep track of all technological and market developments that might be important in the future.

Mergers, especially those brought about by technological convergence, have gained increasing managerial attention in recent years. In this special form of technological and market change, the confluence of previously distinct knowledge bases from different industries gives rise to the creation of new applications and business models (Hacklin 2008). Historically separate industries, such as the communications, consumer electronics, and media industries, are increasingly merging into a single digital industry, which, from a business perspective, creates opportunities in new, previously untapped areas. Similarly, the move towards an electrically mobile future requires a new system approach on the part of different industries, such as the automotive, mechanical engineering, electronics, information and communication technology (ICT), and energy industries. This convergence

E. Enkel (✉) • S. Heil
Zeppelin University, Dr. Manfred Bischoff Institute for Innovation Management of EDAS,
Am Seemoser Horn 20, 88045 Friedrichshafen, Germany
e-mail: ellen.enkel@zu.de; sebastian.heil@zu.de

implies significant disruptions to the established automotive sector and promises the creation of new value for all the major motor vehicle manufacturers, but also for suppliers and other firms from various industries. A prerequisite for recognizing the pervasiveness of convergence and introducing new ways of solving problems are multiple inter-organizational relationships beyond established industry boundaries.

In this chapter, the term *cross-industry network* is used to describe the situation when a focal firm uses a systematic approach of managing diverse *bilateral relationships* simultaneously and across industries to access additional technological and market knowledge. Inside the bilateral network structure, the relationships can take any form between informal ties and equity-based joint ventures. More specifically, cross-industry networks comprise an intensive exchange and development of knowledge within a process of mutual learning and adaptation through collaborative arrangements with specific innovation partners. These partners may stem from the focal firm's own industry (e.g., suppliers, customers, and competitors), foreign industries (e.g., other firms), as well as related and unrelated fields of expertise (e.g., individual experts, universities, and research institutes).

Cross-industry network relationships can focus on the fuzzy front end (FFE) of the innovation process when they enhance knowledge sharing and innovation. These relationships are typically characterized by a profound interaction between parties and tend to result in context-specific solutions which have originated in areas other than the established ones (Hamel 1991; Lane and Lubatkin 1998). In particular, when innovations emerge at the intersection of clearly defined industry boundaries (cross-industry innovation), they allow for highly novel product and service solutions (Enkel and Gassmann 2010). This is because the combination of more distant knowledge can possess a higher innovation potential. The combination of different technological trajectories may also yield outcomes which in their innovation performance exceed the sum of their parts, and thereby create new business models (Enkel and Mezger 2013). Furthermore, the benefits of inter-organizational learning in cross-industry networks and of the integration of the analogical knowledge of partners from foreign industries can be observed in a reduction of development time, cost, and risk (Kalogerakis et al. 2010). For example, viable technologies for efficient and powerful electric vehicles need to be developed quickly, applying already existing technologies from other industries to avoid time-consuming and costly developments. Moreover, the automotive industry is known for its distinctive vertical R&D alliances between manufacturers and suppliers and the focal position automobile manufacturers – Original Equipment Manufacturer (OEMs) – hold in these alliances. However, vertical R&D OEM/supplier alliances are trapping OEMs in a setting where they are increasingly running the risk of failing to seek out alternative technologies. Ultimately, auto OEMs may miss opportunities for innovation, and consequently their innovativeness may suffer substantially (Gassmann et al. 2010b). For example, in the field of electric mobility, links between vehicle manufacturing and renewable energy industries will help to utilize spillovers from mutual learning processes.

Thus, cross-industry network relationships prevent firms from becoming locked in a specific field through a combination of distant pieces of knowledge and interaction with 'idea suppliers' from various technological domains. Furthermore, these network activities are beneficial for the overall performance of firms in terms of survival, growth, and innovativeness, although establishing network relations should not be seen as a panacea. Firms engaging in innovation networks across industries experience challenges in balancing the costs of learning and the benefits in terms of enhanced innovation performance. Obviously, applying innovation networks focused on the FFE is complex and difficult, but has received only scant attention in open innovation research (Enkel et al. 2009). Mechanisms and guidelines with which to enhance firms' full potential for harnessing ideas from cross-industry networks have not been sufficiently acknowledged as yet. This chapter provides insights into how the right network partners can be effectively found and how network competence can be developed beyond established industry boundaries.

## 2 How to Prepare for Distant Collaboration in Cross-Industry Networks

Whether the partners in a bilateral cross-industry network relationship gain incremental or radical results depends on the heterogeneity of knowledge between the collaborating firms, referred to as the *organizational cognitive distance*. The relation between differences in cognitive distance and innovation performance has been found to follow the shape of an inverted U (Nooteboom et al. 2007). This signifies that up to a certain point innovation potential increases with rising cognitive distance. After this point has been reached, novelty reduces performance, because too much diversity hinders efficient absorption (Cohen and Levinthal 1990). The crucial implication of these opposite effects is that firms engaging in distant collaboration have to perform the dual task of assessing cognitive distance with potential partners from other industries and of ensuring that distant knowledge can be adequately absorbed.

### 2.1 Finding the Right Network Partners

Enkel and Heil (2012) developed a new, empirically grounded measure that can be used ex ante in order to identify the industry of potential partners according to the intended outcome of the collaboration. Survey data on 215 bilateral cross-industry collaborations between firms was used to conduct a *network analysis* and capture cognitive proximity in terms of knowledge redundancy between firms based on an industry level analysis of structural equivalence. Two industries are structurally equivalent to one another when the aggregate firms from these industries are connected to firms from the same other industries in the cross-industry network. In other words, structurally equivalent industries have the same relationships to every other industry in the network, share the same external sources of information, and therefore are characterized by a high degree of knowledge redundancy (Burt

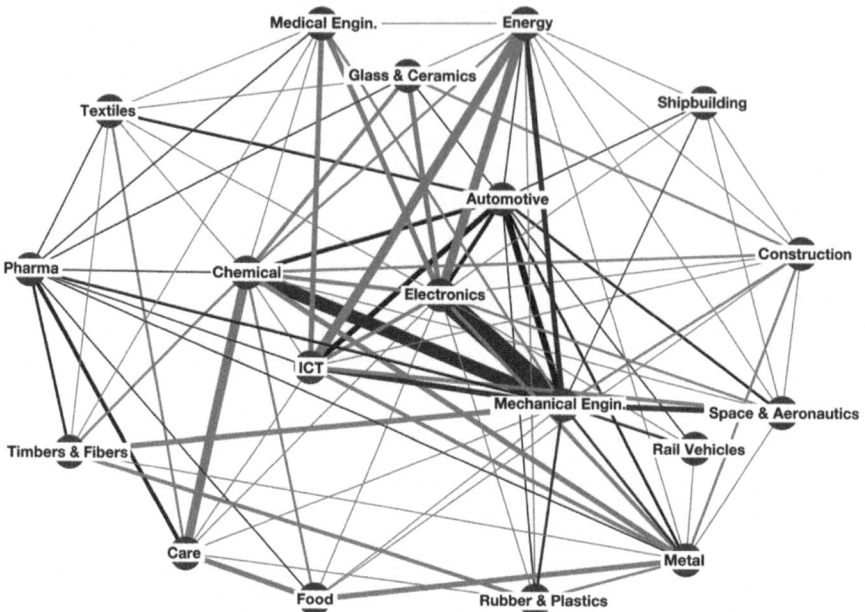

**Fig. 30** Cross-industry network graph; *ICT* information and communication technology

1992). Based on the aggregate firm level data, structural equivalence among industries was measured by computing the correlation coefficient of every pair of industry collaboration profiles. The values for this variable range from +1 (high redundancy) to −1 (non-redundancy).

The network graph shown in Fig. 30 illustrates the collaboration patterns among industries based on the network analysis. The graph represents the network as a series of nodes, which denote industries, connected by lines indicating the presence and strength of a relationship. Furthermore, the network structure indicates that the nodes automotive and mechanical engineering, for example, have similar ties to other industries (cognitive proximity: their knowledge overlaps). By contrast, there is a strong tendency for automotive to have ties to industries that pharmaceutical does not have, and vice versa (cognitive distance: their knowledge does not overlap). This network analysis may guide partner selection when determining the architecture of the cross-industry network. Firms in search for new knowledge will benefit from their network activities when they develop ties to partners that are themselves not connected to a firm's current group of partners. Such a tie provides access to new information by allowing firms to draw on new and different areas of knowledge and on new settings for product applications rather than focusing on current uses and markets.

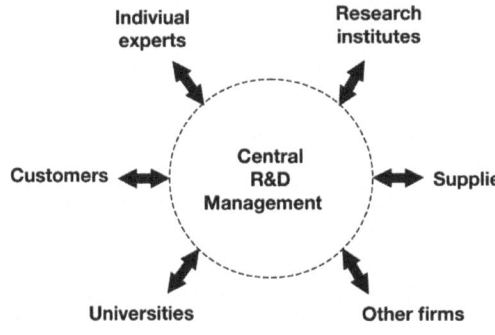

**Fig. 31** Beiersdorf cross-industry network Pearlfinder

## 2.2 Beiersdorf Cross-Industry Network Pearlfinder

The purpose of *Beiersdorf's Pearlfinder network* is to integrate ideas and functionalities from cross-industry partners to solve an existing problem (see Fig. 31). *Beiersdorf*, a global skin care company, with its 450 scientists and packaging engineers launched Pearlfinder as a trusted network. Whenever R&D managers face a question or task in the early phase of innovation and internal ideas cannot provide a satisfactory answer, they may ask external partners to solve the problem. More specifically, through the network, Beiersdorf integrates potentially valuable external partners, such as other firms, research institutes, universities, and individual experts from a variety of industries, in innovation projects at an early stage. Pearlfinder allows external partners in all regions of the world online access to the firm's confidential scientific and technological challenges in order to propose ideas and solutions. Additionally, partners can offer finished formulations, products and packaging or new technologies and also act as contact mediators through their own networks. In return, they will receive qualified feedback from Beiersdorf within 8 weeks. The exchange of ideas can lead to longer-term collaborations and business with Beiersdorf.

Pearlfinder is secured both internally and externally, so that external partners have a guarantee that their ideas are safe. Beiersdorf ensures that no other competitors will be able to look at proposals and that briefings will be viewed only by a pre-selected group of employees. Conversely, by registering on Pearlfinder, partners become community members and accept the agreements on confidentiality and legal terms and conditions. After registration, Pearlfinder offers its community members insights into Beiersdorf's knowledge of skin and beauty care as well as into consumer needs. This information provides a deeper understanding of the firm's needs and requirements. Overall, Beiersdorf's network is largely structured into separate bilateral cooperations with different partners from various industries.

Consequently, the focus of the instruments measuring performance for this type of network is directed towards two central aspects. First, inside the network there are predominantly bilateral contacts that provide requested input. The management is therefore oriented towards bilateral relationships, and sophisticated analyses of potential partners' expertise are conducted. The proposals need to convince the network committee that the partner provides the abilities needed in the specific case. Having selected a partner, the next step will be to clarify and discuss the details and conditions for further cooperation together with the prospective partner. Second, since a specific outcome is expected, the progress towards the objective is controlled using joint project management frameworks together with the partner.

## 2.3 Developing Networking Competence

Existing skills and knowledge are very important in the knowledge sharing/transfer process. Especially in cross-industry networks, the ability to understand and adopt distant knowledge depends strongly on the correlation of transferred knowledge to the existing knowledge base. If the relation of transferred to existing knowledge is very small and knowledge gaps are too large, knowledge transfer is more difficult and its success can be hindered (Mowery et al. 1996, 1998; Stuart 1998). Hence, cross-industry network competence denotes the capability of a firm to benefit from new knowledge that is shared in networks, particularly when transferred beyond industry boundaries. Cross-industry network competence is even more important if the networks are focused on the FFE of the innovation process due to the inherent complexity and dynamics of this early innovation phase. Thus, firms have to exhibit a highly developed *potential absorptive capacity* and possess adequate internal knowledge and capabilities to get access to and gain from externally generated knowledge (Zahra and George 2002; Cohen and Levinthal 1990). Consequently, it can be expected that a highly developed potential absorptive capacity will allow a firm to collaborate with external partners in cross-industry networks with a high cognitive distance (see Fig. 32, adapted from Nooteboom 1999).

Potential absorptive capacity makes firms receptive to understanding and evaluating external knowledge. It prevents them from becoming locked in a specific field, and thus running the risk of failing to seek out alternative technologies, by providing them with the strategic flexibility to adapt to various industry contexts. In the context of cross-industry innovation, potential absorptive capacity comprises the process stages of
- Recognizing potentially valuable external knowledge from other industries,
- Assimilating valuable new knowledge, and
- Maintaining it over time

to set the stage for future knowledge transfer (Zahra and George 2002; Lane et al. 2006). In this context, Zahra and George (2002) propose a positive relationship between a firm's exposure to various and complementary external areas of expertise and its potential absorptive capacity. Once developed, potential absorptive capacity

**Fig. 32** Implications of increased potential absorptive capacity on learning and cognitive distance (Adapted from Nooteboom 1999)

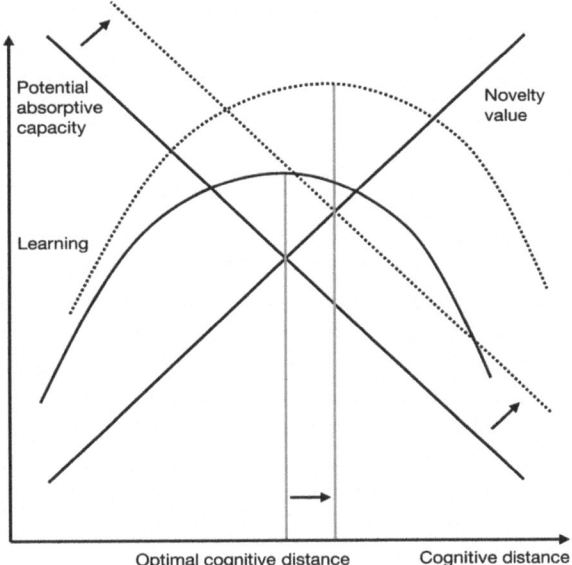

can be redeployed to leverage more resources from other, similar network initiatives.

In recent case study research, three alternative approaches have been identified which allow recognizing, assimilating and maintaining externally generated knowledge to enhance a firm's potential absorptive capacity. Based on the measures of structural equivalence described in the previous section, the average distance or proximity in past cross-industry collaborations of 90 firms was measured. Fifteen leading firms, ranked by market share, from different industries were identified as possible candidates for in-depth case studies, as they displayed a large portfolio of cross-industry collaborations at moderate and high distance to accomplish radical innovation. After all of these firms had been contacted, seven were willing and able to provide detailed information about three to five distant collaborations during the five previous years. The firms range from established medium-sized to large firms, and collaborations were selected with a view to variance across industries. The primary data source for studying the cases were semi-structured interviews with executives from the areas of innovation, R&D, and new businesses that had all held important project responsibilities. Moreover, the interview data was supplemented with secondary data, both publicly available and provided by the respective companies. Furthermore, while the interviews were mainly conducted by telephone, personal interaction through site visits provided further insights and substantiated the findings learned from the interview data. The resulting three approaches to developing potential absorptive capacity are presented in the following (Enkel and Heil 2012).

### 2.3.1 Innovation Flexibility Approach

Firms with decentralized, well-funded technology sourcing apply a wide search scope, make regular use of a broad range of mechanisms for recognizing and assimilating distant knowledge, and foster knowledge sharing across the organization. This *innovation flexibility approach* implies an increase in the variability of distant knowledge that can result in both more exploratory innovation and increased failure (Katila and Ahuja 2002). The larger the scope of external search mechanisms, the more likely the firm is to identify valuable external knowledge that can be combined with internal knowledge in a novel way. A wide scope applied to searching for novelty through many decentralized activities increases the proportion of new knowledge to be integrated into a firm's knowledge base (Fleming and Sorenson 2001). Additionally, firms adopting an innovation flexibility approach may benefit from decentralized activities of sourcing technology, as loose social relations between unit members increase deviant behavior and the search scope (Rowley et al. 2000).

For example, at *Procter & Gamble*, a team of over 70 specialists, who know about and are involved with the technological needs of one or more business units, makes regular use of advanced data mining tools to systematically monitor and scan patent applications across industries and to systemically search for new solutions P&G is in need of. They also search web pages and scientific literature as a source of information on technological developments in fields which appear unrelated at first sight. These seemingly unrelated fields are often chosen with the help of modeling and simulation (M&S) tools that help to generalize specific problem statements. Furthermore, these specialists attend conferences, seminars, and fairs outside the respective fields of expertise and relay new knowledge back into the firm. They also regularly and informally interact with outside organizations from distant industries and areas of expertise to learn about new sources of alternative technologies and future technological trends.

### 2.3.2 Resource Efficiency Approach

By contrast, firms with centralized technology sourcing and relatively low resource investments show a much smaller number of mechanisms and apply a problem-oriented, narrow search. When this *resource efficiency approach* is adopted, the costs of search and integration of distant knowledge decrease (Henderson and Clark 1990). However, there is a limit to the number of new insights that can be found by applying a narrow search scope (Katila and Ahuja 2002). In other words, firms with a resource efficiency approach run the risk of missing opportunities for innovation, but are more resource efficient than companies following the innovation flexibility approach.

For example, *Binder+Co* shows a smaller number of mechanisms for recognizing and assimilating distant knowledge and uses a narrow search scope in response to certain technology needs. Representatives of the senior management and R&D department periodically approach a certain research institution to identify alternative domains of technological knowledge on specific issues. The appointments with the research institution are an effective way of eliciting knowledge that resides in individual experts who have a broadened perspective of similar issues in different industries. Thus, at Binder+Co, only a few qualified specialists, who are familiar with the firm's prevailing technology needs, investigate potentially valuable, distant knowledge.

### 2.3.3 Combinatorial Approach

Firms with centralized technology sourcing and reasonable resource investments are also limited in search scope, but stay focused on certain key industries. They use intelligent mechanisms for recognizing and assimilating distant knowledge. In so doing, they gain a combinatorial advantage by applying a targeted identification and assimilation of distant knowledge while simultaneously complementing their external search activities. This *combinatorial approach* allows for a proper balance between the costs of searching and the benefits of acquiring a variability of distant knowledge. On the one hand, given the infinite size of the technological search space outside the industry's boundaries, concentration allows firms to spot opportunities that are really valuable. On the other hand, mechanisms that efficiently leverage external scope in recognizing and assimilating distant knowledge deepen the awareness of what type of external knowledge the firm may need. In turn, they help to develop a more refined filter when actively searching the technological environment for valuable knowledge.

For example, *Dräger*, a leading international company in the fields of medical and safety technology, makes regular use of the prevalent scouting and screening mechanisms similar to those of Procter & Gamble, but performs its activities on domains predefined according to the most advanced technology needs. Dräger conducts broad observations of technological developments in certain key industries with high cognitive distance, such as consumer electronics. This industry is expected to induce technological change which will influence the development of medical devices. However, in order not to miss opportunities for innovation because of the search scope being defined too narrowly, Dräger employs disparate teams with experts of different industry backgrounds to enhance the heterogeneity of the team's knowledge base. Dräger has hired new employees from the computer entertainment industry, for example, on the account of their potential to more effectively absorb and contribute distant knowledge relevant for future innovation. In so doing, Dräger gains a combinatorial advantage by applying a targeted identification and assimilation of distant knowledge while simultaneously complementing its external search efforts by putting together teams of experts from different industries. This leverages the scope of the resources available in recognizing and assimilating distant knowledge without necessitating significant investment in infrastructure or people.

In conclusion, all approaches have yielded promising results, enabling different types of innovative firms to prepare for distant collaboration in cross-industry networks. Although small and medium-sized firms might utilize fewer activities of sourcing technology owing to resource constraints (Nooteboom 2004), they can still achieve a comparable quality of distant knowledge by adopting the combinatorial approach. The identification of synergetic effects across the process stages of potential absorptive capacity helps technology managers to enrich their understanding of how intelligent mechanisms can complement directed search efforts.

## 2.4 Fostering Network Managers' Capabilities

Regarding all these approaches towards developing potential absorptive capacity, the role of individual network managers with regard to the interface between the firm and its external environment needs to be highlighted, as they serve as initiators of new network relationships (Reid and De Brentani 2004). Operating on their own or in groups, network managers recognize technological and market discontinuities and facilitate the information flow between the external environment and the organization. Key outcomes of knowledge search which impact the FFE innovation movement include the depth (degree of detail of the information), breadth (amount of information collected), and filtering level (relation of information retrieved/ discarded) of knowledge search (De Brentani and Reid 2012). However, network managers face a number of *challenges*:

- Identifying useful external knowledge is costly in terms of resources needed to keep track of changing technological opportunities and market demands (Laursen and Salter 2006).
- Assimilating new external knowledge to existing knowledge is exacerbated, as it entails cognitive distance (Nooteboom et al. 2007) and cannot be easily aligned with existing organizational categories (Lane and Lubatkin 1998).
- Utilizing an external idea internally is particularly difficult as a result of the not invented here (NIH) syndrome (Katz and Allen 1982).
- Successfully integrating external ideas into the firm's activities also demands familiarity with the idiosyncratic needs of, and profound competences in, the firm's or business unit's established fields (Cohen and Levinthal 1990).

Thus, when moving beyond local search, network managers need to strengthen their capabilities to effectively locate and capture novelty value in their technological and market environments. The way in which a network managers' information search plays out in the FFE is related to their *individual-level absorptive capacity* (ter Wal et al. 2011; Cohen and Levinthal 1990). This denotes the capability of network managers to identify external knowledge, assimilate it, and utilize it to commercial ends. Greater individual-level absorptive capacity leads to greater depth and breadth of external search effort as well as better information filtration; in turn, this induces better quality of information. In addition, abstracting the problem before engaging in an in-depth technological and market search is pivotal when searching for providers of novel solutions (Gassmann and Zeschky 2008). The chances of identifying potentially valuable innovation partners are increased if the problem is abstracted to the level of its structural similarities to other contexts. The identification of structural similarities is also supported when firms not only rely on the cognitive abilities of the individual, but also employ a systematic search based on abstract search terms.

Moreover, *knowledge integration capability* underlies all process stages of individual-level absorptive capacity (De Brentani and Reid 2012). This capability determines the extent to which network managers see knowledge integration as difficult when combining new and existing knowledge across the different process stages of individual-level absorptive capacity. It is likely to impact knowledge flow, both in terms of how quickly and how well new knowledge is integrated. Knowledge integration capability requires a solid knowledge base in all the fields in which the network manager's firm or business unit wishes to innovate. Hence, network managers not only have to scout their external environment, but also have to remain on the cutting-edge of their fields by reading journals, visiting fairs, and meeting with other experts of the firm, for example. In this process, the eventual outcome is not expected to be in a specific direction; rather, it concerns an undirected gathering of fundamental knowledge necessary for successful integration of external knowledge into the firm. A network manager's knowledge integration capability also depends on his ability to use networks within the firm and to carry external ideas into and across the firm (Bartol and Srivastava 2002; Mowery 2009). The internal networks of network managers are important, because good relationships and trust can enhance knowledge flow through knowledge sharing. Therefore, network managers need to employ a loose network of internal experts whom they can refer to and draw on to shepherd external knowledge through internal procedures of decision making. Furthermore, the informal roles played by network managers can impact knowledge integration. Especially championing can have a profound effect on the speed and quality of knowledge sharing and is of particular importance when the origins of the idea lie outside the firm in the presence of NIH attitudes (Howell and Shea 2001; ter Wal et al. 2011). Figure 33 represents the different process stages of individual-level absorptive capacity and underlying capability of knowledge integration.

## 3 Creating a Multilateral Cross-Industry Network

Following the discussion of preparing for distant collaboration in cross-industry networks conducted in the previous section, this section will briefly describe how to set up a multilateral cross-industry network. In contrast to the aforementioned bilateral network, which is structured into separate single relationships, network members in a *multilateral network* are integrated into one powerful working relationship. In other words, a multilateral network structure is typically characterized by the profound interaction and joint development of knowledge between multiple members over a longer period of time. The initial framework of multilateral cross-industry networks encompasses the following elements:

- Actors as individuals, groups, or organizations.
- Relationships between all actors which can be categorized by form, content, and intensity.
- Resources which may be used by actors to network with other individuals, groups or organizations.
- Organizational properties, including structural and cultural dimensions.

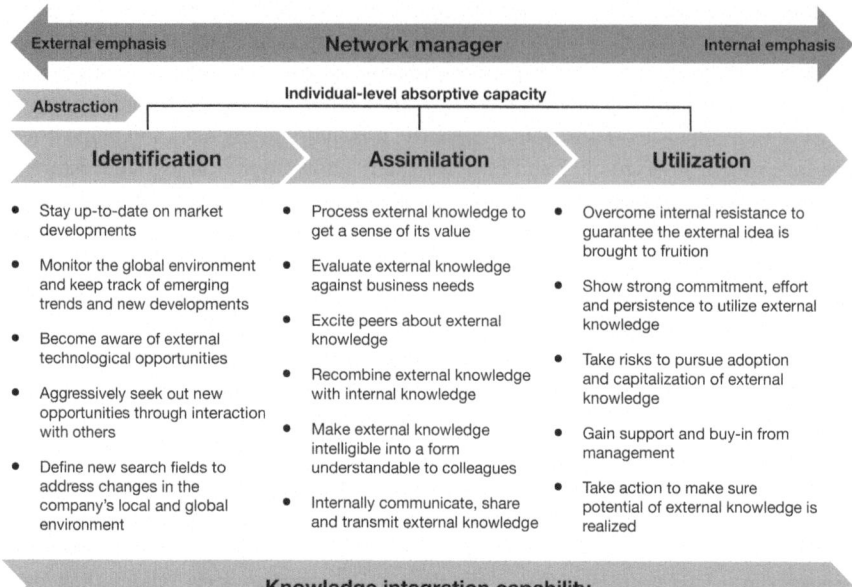

**Fig. 33** Network manager capabilities (Adapted from ter Wal et al. 2011; Enkel and Heil 2012; Jansen et al. 2005; Danneels 2008; Teece 2007; De Brentani and Reid 2012)

When an integrated perspective of these elements is employed, multilateral cross-industry networks are conceptualized on the following three pillars (Back et al. 2005): cultural and structural facilitating conditions, knowledge work processes (i.e., locating/capturing, sharing/transferring, creation), and cross-industry network architecture (i.e., information and communication tools). In the following, the discussion will focus on cultural facilitating conditions (Ritter and Gemünden 2003; Koschatzky 2001), which are of particular importance for setting up multilateral cross-industry networks. In order to implement multilateral cross-industry networks, cultural issues must be considered as knowledge work processes beyond organizational boundaries entail organizational cognitive distance. An established *network culture* facilitates the sharing/transfer of knowledge beyond industry boundaries. Hence, understanding the different components of network culture is one of the most important steps in establishing successful multilateral cross-industry network activities.

## 3.1 Trust and Openness

Poor relationships within multilateral cross-industry networks hinder or complicate knowledge sharing/transfer activities, whereas positive relationships – which are built on trust – favor these activities. Trust can be defined as an optimistic expectation during a common task in which the trusting party has something at stake and has no control over the other party. This definition can be adapted to the situation of knowledge sharing/transfer between two people or organizations. If there is a lack of trust, it will lead to negative effects on the openness of network members, i.e., their willingness to openly communicate and not hide information, motivations, etc. The permission to make mistakes and to ask for help without negative consequences also supports openness. If network members are afraid of committing an error, it is very likely that they will not share their knowledge until they rate their work absolutely acceptable. Besides, if asking for help is undesirable, this creates an obstacle to integrating transferred knowledge. Additionally, the personal openness of individual network members and their possibility to contribute both influence the value provided by the network in terms of increased innovativeness, reduced costs, and a better fulfillment of tasks in the home organization (Enkel 2010). Therefore, sufficient time to contribute to networks, the selection of experienced and open-minded members, and a time frame long enough to permit building trust in the network relationships are preconditions for benefiting from multilateral cross-industry networks.

## 3.2  Shared Norms, Values, and Language

Processes of knowledge work within a multilateral cross-industry network are additionally supported by similar views, similar norms and values, and similar communication patterns, i.e., network language. Knowledge sharing/transfer within networks can become more complicated when norms, values, etc. vary considerably. If there is a lack of understanding of norms and values, and/or these do not complement each other, miscommunication and conflict within the network will arise. In particular, the values and norms of a network culture often go against established hierarchy and control within network members' firms. Therefore, anchoring the network culture and its values and norms can facilitate knowledge flow between multiple network members. These values and norms can be transmitted through education and training, for example. Also, if communication styles and tools of different network members vary too much and if adaptation to other communication styles and tools is too difficult, knowledge sharing/transfer within cross-industry networks may prove more difficult. In such cases, it is absolutely necessary to find out if the misunderstanding is a result of different communication patterns. To transfer knowledge effectively, a context of mutual understanding must be created.

## 3.3  Shared Objectives, Aims, and Interests

When shared objectives, aims, and interests within multilateral cross-industry networks are in tandem with knowledge sharing/transfer objectives and activities, these shared objectives will support knowledge processes within networks.

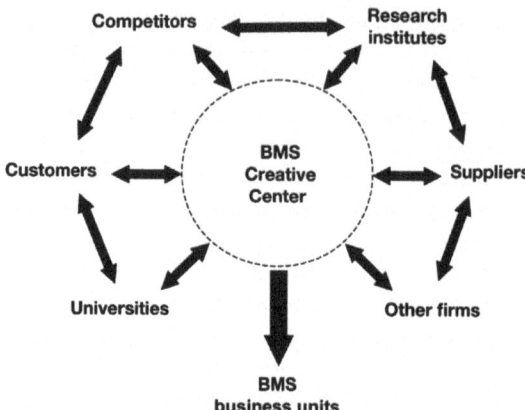

**Fig. 34** Bayer MaterialScience cross-industry network future_bizz

However, when an innovation project is introduced in which network members perceive a conflict of objectives, the motivation for knowledge sharing/transfer might dwindle, and the innovation project may be unsuccessful or not carried out at all. The motivation and commitment of network members towards their 'own' network support knowledge sharing/transfer. The impact of intrinsic motivation on work processes in a knowledge network can be both negative (i.e., 'knowledge is power') and positive (i.e., 'knowledge sharing/transfer is power'). If the network members perceive knowledge sharing/transfer within cross-industry networks as worthwhile and attractive, they are intrinsically motivated in the right direction. In addition, extrinsic motivation, e.g. motivation affected by incentives, can also have a positive impact on knowledge sharing/transfer. To motivate individual network members, incentive systems on a firm or network level have to reward network activities.

### 3.4 Bayer MaterialScience Cross-Industry Network Future_Bizz

The purpose of Bayer MaterialScience's cross-industry network *future_bizz* is to jointly take a look into the future and develop business ideas for all areas of life (see Fig. 34). *Bayer MaterialScience AG*, a world-leading materials provider, formed the network together with firms from different industries and of varying sizes as well as other external partners. Usually getting together in moderated workshops, the participants try to assess future trends and envisage future scenarios to create market opportunities for the future. Another advantage of the cross-industry network is that potential development partners can collaborate early to jointly estimate the opportunities and risks of potential applications, and then successfully shape future markets with appropriate products. For example, a development project of innovative material solutions for the commodity flow and recycling stream in logistics concepts of the year 2020 forms part of the future_bizz network activities.

Bayer MaterialScience AG has institutionalized this multilateral, cross-industry network by founding a department responsible for opportunity generation through networks, its so-called 'Creative Center'. This department has successfully conducted workshops on different topics with a broad variety of partners which had actively been searched for and asked to participate in the network. This enables the participating firms' business units to take note of future demands in other industries and markets that might influence their own innovation activities.

The network is structured as multilateral relationships of equal partners. This equality helps to avoid the pitfall of partners' lack of motivation and commitment and guarantees everyone's equal contribution. Bayer MaterialScience's joint agreements include every participating firm (including Bayer MaterialScience) having to pay a participation fee. This guarantees adequate motivation and finances the services of a professional moderator to manage meetings. In addition, the resultant intellectual property is owned by all the participants. Although intellectual property is rarely an outcome, it gives the companies the guarantee that they would be able to use the results themselves. From the jointly developed future scenarios, each participating firm usually develops its own innovation opportunities without being controlled by the network.

In the case of Bayer MaterialScience's cross-industry network, management assumes the difficult task of balancing two contrary aspects. On the one hand, opportunity generation requires facilitating conditions which foster a free environment to promote creativity and thinking outside the box. On the other hand, the establishment and operation of a network for opportunity generation is expensive and requires properly measuring the results. Bayer MaterialScience measures the outcome of the network efforts by evaluating the ideas in terms of their innovation potential for Bayer MaterialScience's business units. The attractiveness of ideas is measured by how fast the business units 'buy' them from the internal idea market where they are presented. The evaluation of the commercial benefits inside the business units provides the management with an additional retrospective evaluation of the opportunity generation efforts. Due to the nature of opportunity generation efforts, ongoing performance measurement is hard to establish. Bayer MaterialScience therefore balances the allocated budget with the retrospective evaluation of the opportunities to control its department's efficiency. In addition, the network participants' feedback evaluation is used to assess their contribution to the opportunity generating workshops. Based on this assessment, Bayer MaterialScience attempts to steer the composition of its network by only inviting network participants to its workshops that have previously been assessed positively.

## 4 Process Steps to Set Up Cross-Industry Networks

To sum up, the following process steps may serve as a guideline for successfully preparing for distant collaboration in cross-industry networks as well as for creating multilateral cross-industry networks.

## 4.1 Preparing for Distant Collaboration in Cross-Industry Networks

- Define the problem to be solved in the most general and abstract way possible, and think about partners that deal with structurally similar problems.
- Assess cognitive distance with potential partners from different industries and select network partners according to the intended outcome of the collaboration.
- Provide sufficient resources in recognizing, assimilating, and maintaining external knowledge beyond established industry boundaries to enhance potential absorptive capacity and set the stage for future knowledge transfer.
- Adopt a certain approach to developing potential absorptive capacity to be able to benefit from future network activities.
- Gain synergistic benefits by staying focused on certain lead industries while simultaneously leveraging external search efforts through intelligent mechanisms, such as disparate teams with experts from different industry backgrounds.
- Foster network managers' absorptive capacity to effectively locate and capture potentially valuable external knowledge in the global technological and market environment.
- Foster network managers' knowledge integration capability to connect external knowledge with what the firm already knows and to support its application within the firm.
- Carefully prescreen new hires based on individual-level potential absorptive capacity and knowledge integration capability.

## 4.2 Creating a Multilateral Cross-Industry Network

- Specify what skills are needed in the cross-industry network and decide what you want to work on and with whom.
- Form multilateral relationships with relevant external experts or organizations to exchange ideas and build expertise for the early innovation phase.
- Develop trust and a climate of openness within a network environment so that individuals can experiment, reflect, and learn as these activities constitute the foundations of joint network activities.
- Agree on core values and norms, and create a common language that is widely understood within the network.
- Articulate your objectives, aims, and interests, and create a shared understanding of the purpose and function of the network.
- Identify common ground and convene meetings with network members to agree on the overall purpose and function of the network.
- Give rewards and provide incentives in line with the network culture.
- Continually review network targets and assess the value of external partners' contributions.

# Accelerating Learning by Experimentation

Stefan Thomke

## 1  Introduction

At the heart of every company's ability to innovate lies a process of *experimentation* that enables the organization to create and refine its products and services. In fact, no product can be a product without it first having been an idea subsequently shaped through experimentation. Today, a major development project involves literally thousands of experiments, all with the same objective: to learn, through rounds of organized testing, whether the product concept or proposed technical solution holds promise for addressing a need or problem. The information derived from each round is then incorporated into the next set of experiments, until the final product is launched. In short, innovations do not arrive fully-fledged but are nurtured – through an experimentation process that takes place in laboratories and development organizations.

But experimentation has often been expensive in terms of the time involved and the labor expended, even as it has been essential in terms of innovation. What has changed, particularly given the new technologies now available, is that it is now possible to perform more experiments in an economically viable way while accelerating the drive towards innovation. Not only can more experiments be run today, the kinds of experiments possible are expanding. Never before has it been so economically feasible to ask 'what-if' questions and generate preliminary answers. New technologies enable organizations to both challenge presumed answers and pose more questions. They amplify how innovators learn from experiments, creating the potential for higher R&D performance and new ways of creating value for

---

This chapter is based on an earlier publication by the author in *Harvard Business School Press*.

S. Thomke (✉)
Harvard Business School, Soldiers Field Morgan Hall, Room 489, 02163 Boston, MA, USA
e-mail: sthomke@hbs.edu

firms and their customers. At the same time, many companies do not fully unlock that potential because of *how* they learn from experimentation. That is, even deploying new technology for experimentation, these organizations are not organized to capture its potential value – in experimentation, in innovation.

## 2   Learning by Experimentation

All experiments, by definition, generate information, which, at a minimum, becomes an input to additional experiments or is applied to the end result – the intent of the experiment itself – or both. An experimentation process, however, can do more than generate information useful to the process itself. When well structured and integrated into an organization, experimentation generates *learning* that has implications far beyond the 'laboratory'. For example, changes in learning from experiments in custom chips ended up transforming an industry, indeed creating a new multi-billion dollar segment – programmable logic technologies.

At the same time, the *rate* of learning possible is influenced by a number of factors, some affecting the process and others how it is managed. Both sets of factors are equally important. What constitutes good experimentation has been known for a long time. For more than a 100 years, experimentation organized as a group activity has also been codified. A pioneer in managing learning was Thomas Edison. He may be popularly known as 'The Wizard of Menlo Park', but it was his West Orange (New Jersey) industrial laboratory – built in 1887 on 14 acres and subsequently extended well beyond that – that showed how 'group experimentation' could work. Hundreds and eventually thousands of people were employed at this self-styled invention factory, then the largest in the world, whose organization, and the thinking behind it, remains salient today. Edison stressed learning as critical for practical and scientific endeavors:

> Edison's invention factories were the pioneers of industrial research because they carried out organized, systematic research directed toward practical goals. Their work encompassed a broad range of activities.... The laboratory notebooks kept at West Orange provide evidence of Edison and his leading experimenters theorizing about fundamental principles, making deductions from these principles, and testing the results by experimentation. (Millard 1990, p. 19)

In this chapter we will look at what factors drive the learning by experimentation process. We will also look at what impedes that yield – the managerial and organizational factors that inhibit not only the clarity of information but potential learning. Woven through this discussion is the 'case' of Black Magic, Team New Zealand's stunning winner of the 1995 America's Cup. By integrating new experimentation technologies with tried-and-true methods and capturing the results in its organization, Team New Zealand shows how learning by experimentation works. As exciting and recent as these cases are, we should not forget the Wizard. A hallmark of Team New Zealand's approach was the rapid iteration of experimentation 'steps'. Consider what Edison said more than 100 years ago: "The real measure of success is the number of experiments that can be crowded into 24 hours" (Millard 1990, p. 40).

## 3 Types of Experiments

When managers want to go beyond the passive learning through observation or exploration, they may choose to carry out experiments. Such experiments require a directed effort to manipulate or change variables of interest. In an ideal experiment, managers or engineers separate an independent (the 'cause') and dependent (the 'effect') variable and then manipulate the former to observe changes in the latter. The manipulation, followed by careful observation and analysis, then gives rise to learning about relationships between cause and effect, which, ideally, can be applied to or tested in other settings. In the real world, however, things are much more complex. Environments are constantly changing, linkages between variables are complex and poorly understood and often the variables are uncertain or unknown themselves. We must therefore not only move between observation, exploration and experimentation but also iterate between experiments.

When all relevant variables are known, formal statistical techniques and protocols allow for the most efficient design and analysis of experiments. These techniques are used widely in many fields of process and product optimization today and can be traced to the first half of the twentieth century when the statistician and geneticist Sir Ronald Aylmer Fisher first applied them to agricultural and biological science.[1] Today, these *structured experiments* are being used both for incremental process optimization as well as studies where large solution spaces are investigated to find an optimal response of a process. In more recent years, these techniques have also formed the basis for improving the robustness of production processes and new products.

However, when independent and dependent variables themselves are uncertain, unknown or difficult to measure, experimentation itself is much more informal or tentative. A manager may be interested in whether manipulating the incentives of an employee improves her productivity or a software designer wants to know if changing a line of code removes a software error. These *trial-and-error* types of experiments go on all the time and are so much an integral part of innovation processes that they become like breathing – we do them but are not fully aware of the fact that they are experiments. Moreover, good experimentation goes well beyond the individual or the experimental protocols but has implications for firms

---

[1] Over the years, many books have been written on experimental design. Montgomery's (1991) textbook provides a very accessible overview and is used widely by students and practitioners. Box et al. (1978) gets much deeper into the underlying statistics of experimental design. Readers that are interested in the original works of Ronald Fisher may either go to his classic papers on agricultural science (Fisher 1921, 1923) or his classic text on the design of experiments (Fisher 1966).

in the way they manage, organize and structure innovation processes. It is not just about generating information by itself but about how firms can learn from trial and error and structured experimentation.

## 4  An Experimentation Framework

All experimentation consists of iterating attempts to find the direction in which a solution might lie (Allen 1966; Leonard-Barton 1995; Marple 1961; Thomke 1998; von Hippel and Tyre 1995). The process of experimentation typically begins by selecting or creating one or more possible solution concepts, which may or may not include the 'best possible' solutions; indeed, no one knows what these are in advance. Solution concepts are then tested against an array of requirements and constraints. These efforts (the *trials*) yield new information and learning, in particular, about aspects of the outcome the experimenter did not (or was not able to) know or foresee: the *errors*. Test outcomes are used to revise and refine the solutions under development, and progress is made in this way towards an acceptable result.

When Team Zealand developed their winning racing yacht, the design team began with different concepts that were based on prior experience, expertise and creativity. These solutions were tested with the aid of one-quarter scale models in wind tunnels and towing tanks. Team New Zealand's design team was headed by Doug Peterson, an American whose experience spanned more than 30 years and thousands of boats, including the winning boat of the 1992 America's Cup race where he ran over 65 prototype tests and iterations alone. However, in 1995, Peterson planned to tap into the power of computer-aided design, modeling and simulation tools which required him to hire experts in these areas as well. Under Peterson's and Blake's leadership, the team followed a disciplined process of experimentation that emphasized rapid learning.

Specifically, such experimentation comprises four-step iterative cycles (Fig. 35)[2]:

**Step 1: Design.** In this step, individuals or teams define what they expect to learn from the experiment. Existing data, observations, and prior experiments are reviewed, new ideas are generated through brainstorming, and hypotheses are formulated based on prior knowledge. The team then selects a set of experiments to be carried out in parallel and analyzed.

---

[2] Similar building blocks to analyze the design and development process were used by other researchers. Simon (1969, Chap. 5) examined design as series of 'generator-test cycles'. Clark and Fujimoto (1991) and Wheelwright and Clark (1992, Chaps. 9 & 10) used 'design-build-test' cycles as a framework for problem-solving in product development. I modified the blocks to include 'run' and 'analyze' as two explicit steps that conceptually separate the execution of an experiment and the learning that takes place during analysis.

# Accelerating Learning by Experimentation

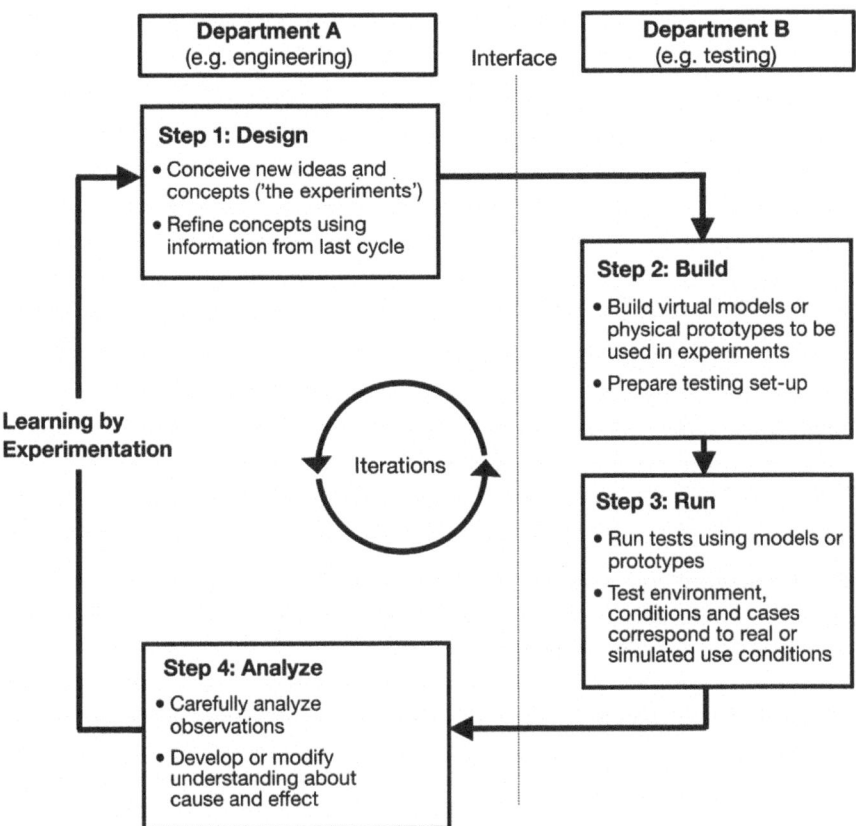

**Fig. 35** Experimentation as four-step iterative cycles

In Team New Zealand's case, the team had to design a light boat with as low a drag in the water as possible. At the same time, the structure had to be strong and flexible enough to withstand the harshest conditions: strong winds and a highly variable sea. While mast and sails were important elements of a boat, most of the team focused on the shape of the hull and the keel. The hull would define a boat's architecture and thus had the potential for significant jumps in performance but also catastrophic structural failures. (In fact, the Australian team did sink one of its boats when it competed against Team New Zealand in an early race.) In contrast, the keel sitting below the hull could be optimized carefully and a gradual optimization could still lead to big gains that were sufficient to win a race.

During this initial step, the team thus brainstorms on different design alternatives that could enhance the performance of the boat. At the start of their development process, these alternatives tended to be more radical departures from known designs (such as new hull concepts), but as time passed by and deadlines loomed, the focus shifted to more incremental improvements on prior experimental iterations (such as tweaking the wing of a keel).

**Step 2: Build.** At this point, one builds (physical or virtual) prototypes and testing apparatus – models – that are needed to conduct an experiment. In yacht design, teams would build a one-quarter scale (20 ft) version of the boat at an expense of about $50,000 and several months of construction time. It was not unusual to build five to six boats in parallel per iteration, and repeat this process 3–4 times.

**Step 3: Run.** The experiment is then conducted in either laboratory conditions or a real setting. In yacht design, wind tunnels and towing tanks simulate the varying conditions of the sea, with the advantage that designers have control over the settings. Storms and high waves can be created without having to wait for the real weather to change. Of course, the trade-off is that laboratory conditions are not real and a test apparatus is often designed for certain purposes. True errors may go undetected or false errors show up because of unique conditions under which the experiment is carried out. For example, the apparatus designed to measure the speed of an airbag deployment in the design of a car is unlikely to be able to detect unanticipated toxicity in the gas used to inflate the airbag, even though information regarding this error would presumably be of great interest to a car company.

**Step 4: Analyze.** The experimenter analyzes the result, compares it against the expected outcome and adjusts his or her understanding of what is under investigation. It is during this step where most of the learning can happen and forms the basis of experiments in the next cycle. At a minimum, the developer will be able to disqualify failed experiments from the potential solution space and continue the search by going to step 1 of another cycle. In many cases, however, an error or a failed experiment can help someone to adjust mental, computer or physical models to reflect what has been observed. The result will be a deeper understanding and less uncertainty about cause and effect.

If the results of a first experimental cycle (steps 1–4) are satisfactory or addresses the hypothesis in question, one stops.[3] However, if analysis shows, as is usually the case, that the results of the initial experiment are not satisfactory, one may elect to modify one's experiment and iterate – try again. Modifications may involve the experimental design, the experimental conditions, or even the nature of the desired solution. For example, a researcher may design an experiment with the goal of identifying a new cardiovascular drug. However, experimental results obtained on a given compound might suggest a different therapeutic use, and cause researchers to change their view of an acceptable or desirable solution accordingly.

---

[3] Simon (1981) notes that traditional engineering methods tend to employ more inequalities (specifications of satisfactory performance) rather than maxima and minima. These figures of merit permit comparisons between better or worse designs but they do not provide an objective method to determine best designs. Since this usually happens in real-world design, Simon introduces the term 'satisfice', implying that a solution satisfies rather than optimizes performance measures.

*Experimentation iterations* like those noted above are performed by individuals and teams that are often divided across different functional departments; in large development projects such as in automotive development, there can be tens of thousands of such cycles – even small projects can involve many iterations. How firms link experimentation activities to major process phases, system stages and development tasks, therefore, is an essential part of effective management practice.

As projects progress and designs mature, cycles tend to include models of increasing fidelity, or representativeness, gradually moving towards functional prototypes and pilot vehicles. These models are used to test decisions affecting design appearance, function, structure, and manufacturability. However, real-world experimentation with higher fidelity models such as physical prototypes is often limited by time and budget constraints as the following quote from Team New Zealand lead designer Peterson illustrates very well:

> The tank and tunnel method is a design process where experimentation occurs in bursts. Every couple of months, you get back the results of your experiments. As a result, there is a limit to the number of design iterations you can perform. A typical project can rarely afford more than 20 prototypes, due to time and money constraints. In each design cycle, you have to rely on big gains in performance. (Iansiti and MacCormack 1997, p. 3)

The attractiveness of using computer simulation to Team New Zealand and many firms developing new products can be found in the higher speed and efficiency of carrying out experimental cycles. Within their time and budget constraints, the additional use of simulation thus offers the potential to learn at a higher rate within these cycles. That, in turn, provides great innovation potential because these cycles can run thousands of times for even a single project.

## 5 How Learning by Experimentation Works

The objective of any experiment is to learn from the experiment. The rate at which companies can learn by experimentation will depend on many factors that require strategic and managerial commitment, and organizational flexibility. While learning from particular experiments can be affected by multiple firm-specific conditions, there are several factors that are common to learning across all experimentation (Table 3).[4]

These seven factors dictate, in general, how learning by experimentation occurs (or does not occur). New technologies for experimentation have a very significant impact on all of these factors: fidelity, cost, feedback time, capacity, sequential and parallel strategies, signal-to-noise and type of experiment all influence learning and, ultimately, innovation processes.

---

[4] Please note that these factors are not intended to be mutually exclusive and collectively exhaustive. Instead, the purpose is to describe a set of interdependent factors that affect how companies, groups and individuals learn from experiments and thus need to be managed.

**Table 3** Factors that affect learning by experimentation

| Factor | Definition |
|---|---|
| Fidelity of experiments | The degree to which a model and its testing conditions represent a final product, process or service under actual use conditions |
| Cost of experiments | The total cost of designing, building, running and analyzing an experiment, including expenses for prototypes, laboratory use, etc. |
| Iteration time (all four steps) | The time from planning experiments to when the analyzed results are available and used for planning another iteration |
| Capacity | The number of same fidelity experiments that can be carried out per unit time |
| Strategy | The extent to which experiments are run in parallel or series |
| Signal-to-noise ratio | The extent to which the variable of interest is obscured by experimental noise |
| Type of experiment | The degree of variable manipulation (incremental versus radical changes); no manipulation results in observations only |

## 5.1 The Fidelity of Experimentation Models Affects Learning

Experimentation is often carried out using simplified versions (models) of the eventually intended test object and/or test environment. For example, aircraft designers usually conduct experiments on possible aircraft designs by testing a scale model of that design in a wind tunnel – an apparatus that creates high wind velocities that partially simulate the aircraft's intended operating environment. The value of using models in experimentation is twofold: to reduce investment in aspects of the real that are irrelevant for the experiment, and to 'control out' some aspects of the real that would affect the experiment in order to simplify analysis of the results. Thus, models of aircraft being subjected to wind tunnel experiments generally include no internal design details such as the layout of the cabins – these are both costly to model and typically irrelevant to the outcome of wind tunnel tests, which are focused on the interaction between rapidly moving air and the model's exterior surface.

Models used in experimentation can be physical in nature, as in the example just given, or they can be represented in other forms, e.g., by computer simulation. Sometimes designers will test a real experimental object in a real experimental context only after experimenting with several generations of models that isolate different aspects of 'reality' and/or that gradually encompass increasing amounts of model complexity.

In Team New Zealand's case, the design team – a multi-disciplinary group of naval architects, designers, engineering researchers, analysts, and sailors – relied on complementing 'tank and tunnel' tests with computer models and simulation. Structural characteristics were analyzed using Finite Element Analysis (FEA), the flow of water over the yacht's critical surfaces were optimized using Computational Fluid Dynamics (CFD) and the velocity of the boat design under particular wind and sea conditions was predicted by Velocity Prediction Programs (VPP). Originally developed for the nuclear and aerospace industries, these tools allowed for cheaper and faster experimentation cycles than partial or full-scale prototype boats. Equipped with these tools, the team realized that the experimentation bottleneck had shifted from step 2 (build) of a cycle to step 4 (analysis) where most of the learning happens. They also realized that the tools fundamentally changed how they learned; alternative design choices could be compared by looking at color pictures of pressure distribution and flows around a hull and keel which, in turn, could be linked to the drag of a design alternative. In contrast, tank and tunnel tests would give information about a boat's speed under specific conditions but not at the level of detail and ease provided by simulation tools. Moreover, results from scale models introduced bias when applied to full-size boats because of the chaotic nature of fluid flow, which was very sensitive to the size and shape of a surface. Simulation did not suffer from such a bias.

Of course, while models and prototypes are necessary to run experiments they do not represent reality completely (if they did, they would be the reality they are meant to represent!). *Fidelity* is the term used to signify the extent to which a model does represent a product, process, or service in experimentation. Perfect models and prototypes, those with 100 % fidelity, cannot usually be constructed because an experimenter does not know or cannot economically capture all the attributes of the real situation, and so could not transfer them into a model even if doing so was desired. Lower fidelity models can be useful if they are inexpensive and can be produced rapidly for 'quick and dirty' feedback, which is often good enough in the early concept phase of product development, when experimentation itself is in 'early development'.

As the experimentation process itself unfolds, however, higher fidelity models become increasingly important, first because the learning from experiments is increasingly vital to understanding how close to a solution the effort is, and second, because modeling errors can get 'carried along'. Not surprisingly, Team New Zealand would still rely on some tank and tunnel tests because, according to chief designer Peterson, "Even with all the simulation in the world, no one is going to commit $3 million to a yacht without towing it down a tank first." (Iansiti and MacCormack 1997, p. 4). The problem is that while simulation has proven to be quite effective at optimizing design, the team's computers were not fast enough to simulate complex architectural changes affecting the hull of a boat. Instead, the team found the simulation especially effective at incrementally optimizing the hull's and keel's shape. For example, CFD was particularly effective in improving the performance of a yacht through the design of aerodynamic wings attached to the bottom of a keel. Refining such appendages had a very significant impact on overall boat speed.

Table 4 lists the two classes of unexpected errors that can result from incomplete models. While type I errors can lead to wasted resources by overdesigning a product (i.e., designing for failure modes that will not occur), it is the errors of type II that can have dramatic consequences and are therefore of compelling interest to experimenters. The failure to detect the relationship between primary and secondary O-ring blow-by *and* low temperatures, in spite of extensive and documented testing, had catastrophic consequences for the Challenger Space Shuttle and the

**Table 4** Possible outcomes from the use of incomplete models

| Error Classes | Description | Example | Result |
|---|---|---|---|
| False negative (type I) | Experiment detects false problem | Crash test barrier is more rigid than actual obstacle | Over-design |
| False positive (type II) | Experiment fails to detect true problem | Crash does not test toxicity of airbag gas | Design failure |

Note: Error classes that can result from incomplete (or inaccurate) models of the object and/or environment

U.S. space program. One of the most dramatic – and highly publicized – Type II errors, this is a reminder that common to all good experimentation is the development of increasingly accurate models as the process proceeds.

## 5.2 Less Expensive Experiments Mean More Iterations and Learning

Conducting an experimental cycle typically involves the cost and time of using equipment, material, facilities, and engineering resources. These costs can be as high as millions of dollars, in the case of a prototype of a new car used in destructive crash testing. They can be as low as a few dollars for a chemical compound used in pharmaceutical drug development and made with the aid of combinatorial chemistry. In general, firms facing high experimentation cost will be more reluctant to try radically new ideas or to depart significantly from existing know-how. They will also try to economize; many design changes will be combined in a single experiment, which will make learning more difficult. There will be fewer errors vis-à-vis the number of trials to learn from.

Consider the four-step experimental cycle defined above. The cost of building (step 2) an experimentation model depends critically on the available technology, the maturity of knowledge about the phenomena,[5] and the degree of accuracy the underlying model is intended to have. For example, modern computer-aided design (CAD) tools sometimes have an interface to computer software that converts a design directly into a simulation model. In such cases, building a model is relatively inexpensive; the cost represents primarily the investment in conversion tools, which is fixed, and the time required to operate them, a variable cost. Furthermore, experimentation models can have varying degrees of fidelity with respect to reality. As noted, the rationale for using 'incomplete' models in experimentation is to

---

[5] Jaikumar and Bohn (1986) noted that [production] knowledge can be classified into eight stages, ranging from merely being able to distinguish good from bad processes (but only an expert knows why) to complete procedural knowledge where all contingencies can be anticipated and controlled and production can be automated. Building models for experimentation will in itself force developers to articulate and advance their knowledge about systems and how they work, thus elevating knowledge to higher stages.

reduce investments in 'real' aspects that are irrelevant to the experiment, and to simplify the analysis of the test results (step 4). Sometimes a model is incomplete because one cannot economically incorporate all relevant aspects of the 'real' or does not know them. The incompleteness of a model, however, can result in design errors when it is replaced by higher fidelity product or process models in the actual use environment for the first time.

The cost of analyzing (step 4) results from step 3 (run) depends to a significant degree on access to test-related information and the availability of tools that aid in the problem-solving process. Consider the discovery of an error during prototype testing and the series of subsequent diagnostic steps to identify the error cause(s). Sometimes a designer has a thorough understanding of a tested prototype and finds the cause of the error quickly. Very often, though, subtle errors make the analysis difficult, especially in cases of great complexity and poor knowledge of causal relationships between system inputs and outputs. As a result, designers have to rely on diagnostic tools and problem-solving methods to aid in their analysis of error symptoms. A very effective analysis tool is the use of computer simulation since it gives a designer quick access to virtually any information within the realm of the underlying simulation model. By contrast, an analysis of data from prototype testing is more difficult since access to error-related information is typically limited. A real car crash happens quickly – so quickly that it is difficult to observe details even with high-speed cameras and well-instrumented cars and crash dummies. By contrast, a computer can be instructed to enact a virtual car crash as slowly as one likes and can zoom in on any structural element of the car to observe the forces acting on it and its response to them during a crash (Thomke et al. 1999).

## 5.3 Rapid Feedback Is Critical to Effective Learning

People learn most efficiently when their action is followed by immediate feedback. Imagine that you were learning how to play the piano, but the sound of your keystrokes took a day to be heard! How would you ever learn how to practice, much less learn how to produce anything that could be performed? Yet, far too many experimenters must wait days, weeks or months before their ideas can be turned into testable prototypes. Time passes, attention shifts to other problems, and when feedback finally arrives, momentum is lost and the link between cause and effect is severed. Moreover, time-to-market pressures do not allow people to wait around until results from an experiment become available. They usually continue with their work and more often than not, the delayed feedback is no longer relevant or used primarily for verification rather than learning.

This is precisely what still happens in some automotive development projects where prototype build times can be several months while overall lead times are being reduced, forcing managers to make project decisions faster than ever before. From the time that design data is made available for building physical prototypes until feedback is received, the project progresses and decisions (such as design freeze) have to be made. In some cases, the data even comes too late to contribute to planning the next round of tests. The result? Feedback contributes little to learning and improvement and is more or less used for verification that certain standards are met. Only when test results point towards major problems (such as not meeting minimal government safety standards in the case of crashworthiness) do they have a major impact.

When Edison planned his new West Orange (New Jersey) laboratory in 1887, he designed supply and apparatus rooms and the machine shop to be very close to the experiment rooms. The laboratory provided a larger space in which a system of experimentation could be put to work, where libraries and storehouses of common and not so common materials could be established. This workplace design in turn helped transform Edison's approach to invention. The result was the 'invention factory' – a physical arrangement that supported a more systematic and efficient definition, refinement, and exploitation of his ideas. In fact, Edison firmly believed that all material, equipment and information necessary to carry out experiments needed to be readily available since delays would slow down his people's work and creativity. When he or his people had an idea, it had to be immediately turned into a working model or prototype before the inspiration wore off (Millard 1990, pp. 9–10).

Similarly, Team New Zealand emphasized rapid feedback from experiments integral to its boat development process. After the hull design had been made robust and performance improvements had begun to diminish, the team's focus shifted towards optimizing the keel appendages for minimal drag. Through design enhancements and the placements of wings, they were hoping to increase boat speed much further. For all of these experiments, they would operate on a 24 h iteration cycle that guaranteed rapid feedback. The entire team would generate hundreds of improvement suggestions for the keel appendages which were analyzed by the simulation team. The most promising one or two design alternatives that emerged from simulation were prototyped overnight and tested the next day on a full-size boat by the crew. Only they could determine if in fact the boat 'felt' faster and real performance improvements were evident. Their feedback also drove the generation of new improvement ideas. David Egan, one of the team's simulation experts, recalled the importance of rapid feedback:

> Instead of relying on a few big leaps, we had the ability to continually design, test, and refine our ideas. The team would often hold informal discussions on design issues, sketch some schematics on the back of a beer mat, and ask me to run the numbers. Using traditional design methods would have meant waiting months for results, and by that time, our thinking would have evolved so much that the reason for the experiment would long since have been forgotten. (Iansiti and MacCormack 1997, p. 6)

## 5.4 More Capacity Avoids Learning Bottlenecks

The ability to provide rapid feedback to a developer is in part affected by an organization's capacity for experimentation. Not surprisingly, when the number of experiments to be carried out exceeds capacity, the waiting time will grow very rapidly and the link between action and feedback is severed. What often surprises people, however, is that the waiting time in many real-world queues increases substantially even when not using the total capacity. In fact, the relationship between waiting time and utilization is not linear – queuing theory has shown that the waiting time typically increases gradually as utilization goes up, and then the length of the delays suddenly surges (Thomke and Reinertsen 2012).

Moreover, when people expect long delays, they tend to overload queues, slowing down the system even further. More experiments are submitted in the hopes that one makes it through quickly but without any sense of how it may affect the overall innovation process. Or simply, firms often lack the right incentives and organization to remove queues and speed up experimentation. Building sufficient experimentation capacity is therefore not only important but essential for effective learning. With new technologies bringing down the cost of experimentation dramatically, the opportunities to bring capacity in line with an organization's need to experiment rapidly now exist, but they need to be taken advantage of.

## 5.5 Sequential or Parallel Experimentation Strategies Affect Learning

Most large-scale experimentation involves more than one experiment, and, as we have seen, usually requires multiple iterations within that effort. When the identification of a solution involves more than a single experiment, the information gained from previous trials may serve as an important input to the design of the next one. When learning from one cycle in a set of experiments is incorporated into the next cycle, experimentation has been conducted sequentially. By contrast, when there is an established plan of experimental cycles that is *not* modified by the findings from previous experiments, the experiments have been performed in parallel. For example, you might first carry out a pre-planned array of design experiments and analyze the results of the entire array. You might then run one or more additional verification experiments, as is the case in the field of formal design of experiments (DOE) methods. The experimentation cycles in the initial array are viewed as being carried out in parallel, while those in the second round have been carried out in series with respect to that initial array.

Between November 1993 and May 1994, Team New Zealand built physical prototypes for tank and tunnel testing three times, resulting in 14 scaled-down models. There simply was not enough time to build and test all prototypes sequentially and feed the learning from each round into the next. The advantage of building multiple prototypes per round enabled them to test different alternatives more quickly, drop the least promising directions and experiment further on the best alternative. Similar approaches can also be found in early car design. *BMW,* for example, usually considers a large variety of styling concepts in parallel, ranging from evolutionary to revolutionary directions, and are whittled down in a sequential process where fewer and fewer parallel alternatives are built during each round until one is chosen for engineering and production.[6]

Parallel experimentation clearly can proceed more rapidly, but it does not take advantage of the potential for learning between and among trials. As a result, when parallel experimentation is used, the number of trials needed is usually much greater – but it is usually possible to get 'there' faster. In comparison, getting 'there' takes longer with a sequential approach; the number of trials conducted depends very much on how much a firm expects to learn between each round. For example, trying 100 keys in a lock can be done one key at a time, or all keys at once, as long as enough identical locks are available. Since little can be learned between experiments, a sequential strategy would, on average, require 50 trials and thus cost only half as much – but also take 50 times longer.[7]

Typically, parallel and sequential approaches are combined, depending on the experimentation strategy chosen. In turn, that strategy depends on many factors: cost of experiments, opportunity cost of time, the expected learning between experiments and how firms envision the 'value landscape' they plan to explore when seeking a solution for their problem. Not surprisingly, a dramatic decrease in the cost of experimentation – the kinds of changes that new technologies provide – will make parallel strategies much more attractive to managers. The result will be a shift in many industries toward innovation processes that emphasize parallelism to explore greater experimental space and bring product and services to market more quickly.

## 5.6 More Radical Experiments Invite Different Learning Opportunities

Not all experiments, structured or trial-and-error, are alike. Tweaking independent variables usually results in smaller changes in output – the kinds of changes that are desired in the incremental improvement of product and processes.[8] Alternatively, large variable manipulations or introducing new variables can foster a much wider search, thus increasing the probability of discovering more radical improvements and, at the same time, inviting more failures. More radical experiments can point us in new directions and take us into unknown territories that may or may not result in

---

[6] 'Set-based' design approaches advocate a similar approach where parallel alternatives are pursued simultaneously (Sobek et al. 1999).

[7] Loch et al. (2001) formally model the trade-off between sequential and parallel experimentation strategies and derive optimal policies for decision-makers. Thomke et al. (1998) show the essence of this trade-off with a thought experiment.

[8] An exception is highly non-linear systems where small changes in independent variables can result in large changes in dependent variables. Optimizing such systems can be challenging but experience has shown that increasing robustness, rather than a single point performance optimization, via Monte Carlo-type methods appears to be promising (e.g., in improving automotive crash safety). However, in many areas of engineering design, this will require much more experimentation capacity than is available to development teams today.

more radical innovations – one has no way of knowing in advance. As a result, real-world innovation needs to strike a healthy balance between incremental and radical experimentation.

Again, Team New Zealand's yacht development process illustrates this balance very well. The team knew that experiments with its hull design could result in the most important improvements in performance but at the risk of breaking apart under real sea conditions. After spending several months of experimenting in parallel with different hulls and testing them using scale-models in tank and tunnel tests, the improvements from each iteration started to diminish significantly. With the beginning of the race only 8 months away, their strategy then shifted to sequential experimentation during which rapid iterations accumulated into significant changes, one small step at a time. Experimenting with different hull designs so close to the race would have been too risky since 'home runs' came at the cost of 'strike-outs' which cost its competitors from Australia a full-size racing yacht when their boat broke apart and sank. Therefore, Team New Zealand shifted its development strategy from more radical, parallel experiments affecting the boat's hull structure to more incremental, sequential experiments that optimized the boat's keel.

## 5.7 Noise Impedes Learning

A final factor, one often overlooked, is how ambiguous or excessive feedback 'noise' can block learning. In a study of learning in semiconductor manufacturing, research found that production plants with low noise levels could potentially learn much more effectively from their experiments than high noise plants (Bohn 1995). Using data collected at five plants, Bohn estimated that the probability of overlooking a 3 % yield improvement – a large number as first year improvements are usually between 0.5 % and 3 % – was about 20 %. The study concluded that brute-force statistical methods are ineffective or too expensive to deal with these high noise levels.

This noise occurs either when certain variables cannot be controlled, or when too many variables are being manipulated – because the design of the experiment itself is poor or because the aim is to reduce the number of experiments overall (and too many variables are 'stuffed' into one or few attempts). In either case, it is not possible to discern what is actually happening to the experiment. What is interacting with what? The sad result is that rather than being cost-cuttingmaneuvers, experiments loaded with too many variables often need to be redesigned and rerun, making the whole endeavor more expensive than it would have been in a better designed state. Alternatively, 'noise' can be a problem if the independent variable itself has too high a variability when observed. In this case, the experiment has limited value since the connection between cause (a variable change, procedure or policy) cannot be linked to the observed effect (change in performance). Under such circumstances, effective learning cannot take place.

The presence of noise was a big problem for Team New Zealand when it was testing changes to its racing boat under real conditions. While tank and tunnel laboratory tests and computer simulations allowed the team to control external conditions such as wind and sea movements, putting a full-sized yacht with a real crew and constantly changing wind and weather into the sea made learning from experiments very difficult. Racing one yacht with the design change and then racing it again without it would only be possible if they could control all the other conditions that would affect performance – a nearly impossible task since they had to detect changes in the order of 2–3 seconds over the entire course. The impact of a minor change in wind speed between the two trials could easily swamp the effect of the design change and thus make the experiment worthless. The crew would have to sail multiple times to average out the effect of noisy wind, sea, and crew conditions on performance which would have slowed down the team's experimentation cycles significantly.

To maximize learning from keel experiments and speed up iterations, Team New Zealand decided to build two yachts that could be used in combination to test iterations on the keel wings. Unique among the few teams that opted to invest precious resources into two racing boats, Team New Zealand chose to construct two very similar boats that allowed them to test design changes side-by-side. With one boat as an experimental control, they could put two keels with different wing designs on each boat, race them, and then see how much difference there was. To minimize the effect of the crew, they could swap the keels and test if the difference still held up. The advantage of using an experimental control was that the effect of noise was now minimized since the two boats were operating under the same noisy conditions. Following this experimentation strategy was more costly for the team but ended up maximizing learning and performance improvement in the 6 months before the first race.

### Conclusion

On the surface, the experiments run on world-class yachts, in car companies and in the entire integrated circuit industry could hardly look more dissimilar. Yet, they share a basic iterative process of four-step experimentation cycles and can be organized to maximize learning. How learning-through-experimentation occurs (or does not occur) is affected by seven factors: fidelity, cost, feedback time, capacity, sequential and parallel strategies, signal-to-noise and type all enhance the power of experimentation. New technologies for experimentation amplify the importance of managing these factors, thus creating the potential for higher R&D performance, innovation and ultimately new ways of creating value for customers.

# Dancing with Ambiguity: Causality Behavior, Design Thinking, and Triple-Loop-Learning

Larry J. Leifer and Martin Steinert

## 1 Design Thinking at the Front End of Innovation

This chapter is based on results from a workshop sponsored by the US Air Force, where we were tasked to help *Understanding and Influencing the Causality of Change in Complex Socio-Technical Systems*. We are coming from a product development and design background where we build smart products, systems and services and research the dynamics of the involved development teams and the underlying *engineering design* paradigms. We believe that our insights, commonly referred to as *Design Thinking*, can help the fuzzy front end to innovate faster (rapid prototyping and iterations), for a better market fit (human centric design principles) and generally create more radical innovations. Ultimately, rather than seeing uncertainty as a threat that needs to be pseudo quantified or abstracted away we invite you to embrace ambiguity and to leverage it in order to create better innovations faster. The key lies in *letting* your innovation teams be truly creative and in focusing on iterative learning and redesign rather than on optimizing on concrete, but potentially ill fitting requirements.

Over the past 30 years, a powerful methodology for innovation has emerged from engineering and design thinkers in Silicon Valley. It integrates human, business and technical factors in problem forming, solving and design: 'Design Thinking.' This human-centric methodology integrates expertise from design,

---

This chapter is an adapted reprint from Leifer, L.J., & Steinert, M.: *Dancing with ambiguity: Causality behavior, design thinking, and triple-loop-learning*. In: Information, Knowledge, Systems Management, 10(1), 151–173, Copyright (2011), with permission from IOS Press.

L.J. Leifer (✉) • M. Steinert
Center for Design Research at Stanford, Mechanical Engineering Design Group, Building 550, Room 125, Stanford, CA 94305-4021, USA
e-mail: Leifer@cdr.stanford.edu; steinert@stanford.edu

social sciences, business and engineering. It is best implemented by high performance project teams applying diverse points-of-view simultaneously. It creates a vibrant interaction environment that promotes iterative learning cycles driven by rapid conceptual prototyping. The methodology has proven successful in the creation of innovative products, systems, and services.

By courting ambiguity, we can let invention happen even if we cannot make it happen. We can nurture a *corpus of behaviors* that increase the probability of finding a path to innovation in the face of uncertainty. Emphasis is placed on the balance of the questions we ask, and the decisions we made. A suite of application examples and research finding will be used to illustrate the concepts in principal and in action.

## 2 Why Designing Products and Systems Translate into Changing Human Behavior

While creating and testing technical prototypes is traditionally highly analytical in nature and driven by system decomposition with a focus on sub-problem solutions, the complexity of system integration is often underestimated.

Over the past two decades, the isolated optimization of sub-systems has, however, given rise to more holistic system approaches. Consulting companies like *IDEO* and *Frog Design* have achieved notable success in a wide variety of industries through the use of adaptive design thinking and semi-formal use of a 'coaching' model that has some members of each development team explicitly focused on the team's behavior pattern with an eye to focusing activity on the critical tasks from a system integration point of view. A parallel movement in the software industry operates under the tag line, 'agile systems development', and our favorite protocol 'scrum development' (MacCormack 2001).

While still grounded solidly in engineering principles and construction, design thinking understands the meta level issue of customer adoption as the defining parameter for measuring the success or failure of a new product or system. Changing the actual behavior of the user in positive ways determines success

In other words, the successful introduction of any new product, service, system, organization or process requires the solution to overcome the inherent behavioral barriers to change. Barriers might arise on behalf of sometimes unexpected parties, stakeholders or system advocates that have not been identified a priori as critical to the particular new design. The cost of behavior change can define market success (Aquino et al. 2011).

### 2.1 Translating the Change Problem

A good starting point for any new design or product development cycle is a problem formulation or, as we call it, design challenge, e.g., 'Redesign (read re-invent) the driver-car interaction.' For radical new solutions, it is important to tackle customer pull challenges of a complex nature that are positioned in the future (3–5, sometimes up to 20 years). This allows one to truly understand and (re-define) the problem, a prerequisite to ideation and a stimulus to interesting new solution concepts.

Technology drives scenarios; ready solutions such as new materials with hitherto unseen attributes that are still looking for an appropriate problem or application area are a much tougher nut to crack. For example, what application domain will provide graphene[1] with its first commercial application?

In order to address socio-technical, complex systems and the *causality* of change from an engineering designer's perspective, the following basic principles apply:

### 2.1.1 Technical

The origins of product development and design lie in classical engineering, both mechanical and electronic, all solidly grounded in classical Newtonian physics. Thus, by definition, all products that engineers develop are fundamentally of a technical nature.

### 2.1.2 Systems

No physical product can be designed without taking its context into account. The user has to act, and even more so today interact, within a certain contextual environment. The trend is beyond embedded electronics and mechatronics, towards connected smart products. As most design challenges require more than a stand-alone mechanical solution, most of today's products should be more accurately described as systems. Due to the inherent system integration problem, it is necessary to establish diverse, pan-disciplinary product development teams.

### 2.1.3 Social

All products or systems interact with or influence directly or indirectly certain user and stakeholder groups. Thus, essentially, all products and systems aim to facilitate a certain user adoption and behavioral changes. The social acceptance of new solutions is also essential. One must design for the social context, and the product itself must behave in socially expected (acceptable) ways. Engineering design has therefore integrated user centric design and need finding. It even borrows from cultural and physical anthropology in its quest to understand the usage context.

---

[1] The Nobel Prize in Physics for 2010 was awarded to Andre Geim and Konstantin Novoselov for groundbreaking experiments regarding the two-dimensional material grapheme. Graphene is a flat monolayer of carbon atoms tightly packed into a two-dimensional (2D) honeycomb lattice, and is a basic building block for graphitic materials of all other dimensionalities. It can be wrapped up into 0D fullerenes, rolled into 1D nanotubes or stacked into 3D graphite (Geim and Novoselov 2007).

### 2.1.4 Complex

The combination of the technical, social and systems dimensions provides for a vast and ambiguous solution space. Radical new designs may completely solve the existing challenge by redefining existing solutions. Of course they may also create an entirely new set of problems and challenges (to be addressed with the next generation). When we are trying to understand the initial assumptions and boundary conditions as well as the interface, integration and adoption issues, not to mention the implications of radical new solutions to existing challenges, we realize that we are almost always dealing with complex systems.

### 2.1.5 Change

Success ultimately depends on the willing adoption of the new system by the user. This translates into the solution's capability to overcome resistance to change. Though, as a discipline, engineering design may not contribute to the fundamental understanding of the causality of change, we believe, that our grand task of creating better designers and a better design process produces better products and services, and increases the chances of success for change. Change does not come naturally to the majority of people, engineers and users alike. That is why we attempt to understand the underlying principles and why we are developing techniques to overcome barriers to change.

In essence, a product developer and designer is constantly struggling to improve the outcome of a complex socio-technical system interaction, as well as the project organization and the work process of a development team.

## 2.2  A Product Development Knowledge Model

In order to structure possible insights that we have gained in change, let us introduce a product development knowledge model. Initially based on empirical data from a large US auto producer (Eris and Leifer 2003) and later tried and tested in our ME310 teaching environment, we differentiate between three simultaneously occurring loops of knowledge acquisition or learning. ME310 is a three quarter project based mechanical engineering graduate course at Stanford University that teaches ten global teams based on a real industry challenge (Fig. 36).

Learning or knowledge acquisition is a prerequisite for change, whether on a design or organizational level. Based on this product development knowledge model, we have identified three learning loops.

*Learning Loop One* is based on explicit knowledge. It stretches beyond the informal product development team into the formal organizational structures. Due to its explicit nature, it can be collected, managed and synthesized into formal processes. It is mostly comprised of quantitative technical data such as business processes (BP), computer aided designs (CAD) files and workflows, data warehouses, algorithms, repositories, etc. Learning in loop one is mostly aimed at retaining project knowledge (facts, syntax, what, and how). It may comprise analytical activities and tools such as databases, and may involve simulations.

# Dancing with Ambiguity: Causality Behavior, Design Thinking, and... 145

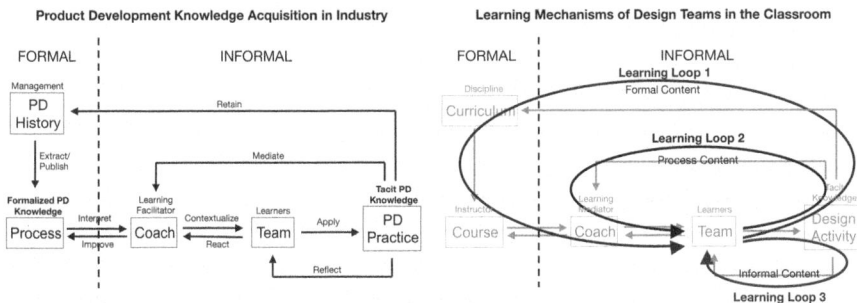

**Fig. 36** A product development knowledge model based (Eris and Leifer 2003)

Learning loop one usually allows for optimization and incremental change. However, due to its institutional character the change process is relatively slow. The vast majority of academic and business literature lies in the conduct and architecture of Loop one. (Argyris and Schon 1978).

*Learning in Loop Two* takes place within the informal space of the product development team and comprises the informal process content. Learning occurs during the exchange between the product development team members and the team's coach(s). The coaching role is often tacit itself. Coaches act as facilitators between the formal organization, its formal protocols, and actual team processes (typically unspoken, undocumented, and profoundly tacit). Loop two is based on concepts, semantics, and architecture and asks questions like when and why. It is the locus of application for many of the design-practices that allow faster learning and better output, to be explained in Sect. 4.

The least studied *Learning Loop Three* concerns tacit knowledge embedded in the teams themselves and the established practices. Team members learn from each other and prior team's experiences by applying, reflect upon and improving informal practices. Changes stimulated through learning in this third loop are fast paced. It forms the rational that explains the success of activities such as fast iterative cycles and rapid prototyping. In Sect. 5, we will present some of the research findings that shed light onto the underlying principles of design-thinking based activities.

We do not explicitly aim to understand the causality of change in design research. However, the process, practices and environmental setting of creating substantial change by means of new products or services, ultimately translates into changing behavior. Internally, we explicitly change the behavior of designers with the intent to augment their ability to change user behavior. How to achieve these changes is at the core of our research and design activities. We believe that our insights into how to initiate, support, and facilitate the creation of radically innovative new products and system designs may help to understand how to trigger and sustain change in complex socio-technical systems at large.

## 3  Formal Institutional and Procedural Learning (Loop 1)

The institutional challenge begins with capturing what is known. Procedural implementations are best suited to capturing the facts. They are least successful at capturing the behavior, how knowledge was used or ignored. Institutional efforts to extend knowledge capture to the tacit level of design-development team activity are largely unknown. One exception to this rule lies in the design thinking research program at Stanford University wherein student design development teams are accessible and open to observation within the limits of human-subject study protocol. Many of the research finding we are about to share come from this organizational behavior 'flight simulator.' There is a substantial body of literature that address issues ranging from project team size and setup (self-organizing open source projects vs. centralized R&D team) to supportive tools (knowledge management systems, wikis, computer supported cooperative work (CSCW) tools, etc.) and supposedly innovation enhancing processes (stage gate model, lean approaches, design for X).

Based on our particular experience, any one specific design-development organizational recipe can work for incremental innovation but most are not appropriate for enhancing disruptive product and system development. For this overview on Learning Loop One, we would like to focus on the factors that seem to influence the ideation and creative energy and output the most: physical space, the absence of fixed processes, and an overarching institutional practice of letting change happen. This last point stands in sharp contrast to trying to make change happen.

### 3.1  Space

Space has emerged as a key factor in facilitating change. Through adapting the physical environment, organizations are able to lower hierarchical boundaries, enhance ideation and creativity, foster and accelerate prototyping and generally increase the rate of learning and change. The key concept for the spatial setup is flexibility (adaptive/agile work places). Space ought to allow for and support any kind of ideation and prototyping activities. As will be described in Sect. 5, going through a number of rapid iterations, testing ideas and the boundaries of the solution space via prototypes, allows the project teams to significantly increase their rate of learning. Founded in 2004 by David Kelley, Professor of Mechanical Engineering and co-founder of *IDEO Product Development*, the Hasso Plattner Institute of Design at Stanford, more commonly referred to and globally famous as the d.school, is continuing to shape and incorporate lessons learned. Having already moved to its fifth building in 6 years, continual changes and experimentation with space and furniture alike has become the normal procedure. The key concepts include:

- Use flexible room separators instead of fixed walls. Move them daily as needed, or just maybe to stir things up

- All furniture is easily movable and modular to serve multiple, often previously unexpected purposes
- All furniture has evolved with a focus on enhancing creativity and lower barriers to ideation

The following key concepts further support the notion of flexibility and continual change:

- Avoid expensive solutions that bind infrastructure investment; instead use minimum commitment prototypes to facilitate rapid redesign/learning
- Building furniture and support infrastructure in-house as needed by the project teams.

We believe that we cannot correctly foretell the usage scenarios and therefore we do not want to preplan the space requirements in detail. Instead we focus on maximizing flexibility while minimizing financial requirements. We use standard modules repurposed from other products and good, but not high-end materials in order to minimize costs. The idea is to have only very limited equity tied into infrastructure that only depreciates over years. Instead we know that we are going to redesign the space, furniture and equipment constantly, we know that we learn and change constantly. This allows the project teams to redesign their space and infrastructure as needed. It allows teams to veer onto unexplored paths in pursuit of generating great new products, system ideas and concepts.

## 3.2 Absence of Fixed Processes

Throughout our joint research on radical engineering and system innovations with the US Defense Advanced Research Projects Agency (DARPA), the work was guided by the notion of projects and people being vision driven, the focus on workshops and prototyping, the absence of a formal process documentations or project management rules, and last but not least, the leadership driven decision model that does not rely on peer review or communal decision building mechanisms.

The selection of DARPA programs depends on the creation of a vision. It is the starting point for any program, and the project champion, the program manager, embodies it. The ideation and iteration of this vision serves as the central focal point for the usually dispersed sub-projects, teams and stakeholders. Envisioning a certain technological future does not define or limit the future projects; rather it serves as an indicator of the current direction of the organization's efforts.

Interestingly, the main instruments for generating, iterating and re-formulating such a vision are workshops and the creation of proof of concepts or prototypes at various stages. The first allows the socialization and evolution of the visionary ideas amongst all participating stakeholders, while the latter allows tangibly communicating and even testing the vision at various critical junctures.

The program and project managers also enjoy a remarkable freedom from established processes and rules. No established system or documentation requirement is forced upon his or her activities. Prototyping is the norm and the specific activities follow the actual demand of the specific task at hand. No institutional models force people and their behavior into fixed corsets for the sake of generating economies of scale. Innovation and change is the generation of the new – the primary goal is the best outcome at certain budget constraints, not its process efficiency in terms of minimal resource allocation.

Another point to consider is the ways that go and no-go decisions are prepared and executed. Instead of relying on a peer review processes, or committees or other group-based decision tools (not to mention pseudo quantitative stage gate filter variables) decisions are taken by the leaders who ultimately bear the responsibility. Failure is accepted and preplanned. The underlying rationale is that peer review and committees are in fact a hedging mechanism for taking tough decisions at the extreme end of the possible solution spectrum. They will inherently favor outcomes close to the sample's median opinion. Hence, traditional decision tools would prevent DARPA from actually attempting to deliver radical innovations. All of these activities, and most importantly the absence of fixed processes, serve to generate change that complies with the idea of DARPA, as described above.

## 3.3 Ability to Let Change Occur Rather than Manage It

As the prior discussion of space and flexibility as well as of the absence of fixed processes indicated, the first, critical step in supporting change and learning, and the generation of radical new product and system solutions, is allowing change to happen. We do not assume to have control over the existing solution space, so we cannot preplan, but we can facilitate it. Indeed, the concepts that challenge the established dogmas have a higher chance to deliver radically improved value. Any systematic and fixed support system, inhibiting the creative use of space and the employing and combining of new processes seems to counter the notion of change. Therefore we attempt to provide the physical, organizational, procedural and mental environment that allows the project teams to experiment and prototype. This becomes especially difficult when proposed solutions counter the experiences and knowledge models of the professors and coaches. Instead of prematurely ending the iteration processes at this point, we support the testing of these ideas and concepts. Very often, a failed prototype test, the hitting of the boundary of the possible solution space, generates the winning insights for either an extreme solution along that line or, even better, a new way that allows circumventing the existing limitations. To generate this kind of change, we attempt to minimize institutional, organizational and procedural boundaries. We emphasize and support flexibility, and we force ourselves to let change happen. Hence we do not prescribe procedural recipes. We teach skills and moves, and demand tangible prototypes rather than requirements or specification lists. Allowing and even fostering for this kind of ambiguity is difficult and demanding for the coaches and requires a conscious effort especially on behalf of individuals who have to unlearn their organization skills to a certain extent.

## 4  Design Process Learnings (Loop 2)

Almost all of our activities aim at accelerating ideation and creation; in other words, we aim to accelerate learning. One of our dogmas or design rules for our design processes is to initiate very rapid instances of change. Change, embodied in tangible prototypes, can be tested on or against user behavior. To foster the cross-team ideation activity, we deploy collaborative tools and create a physical space that encourages an even faster rate of learning or change.

### 4.1  Design Process

Contrary to the classical and rather analytical design process applied for the development of incremental changes, the design process aiming at radical changes can be seen as an iteration of divergent and convergent activities (Alexander 1964). Banathy describes the divergent activity as "... consider[ing] a number of inquiry boundaries, a number of major design options, and sets of core values and core ideas. Then we converge as we make choices and create an image of the future system" (Banathy 1996, p. 74). This divergent-convergent process may be depicted as slowly closing funnel, linear over time (Cross 2000; Ulrich and Eppinger 2008), or as repeating design cycles, spiral like, that iterate through the generic prototyping phases of design, build and test (Thomke and Fujimoto 2000) (Fig. 37).

The classical convergent phase is about optimizing the answer; it is deductive and inductive in nature and may comprise simple tools or complex model simulations and optimizations.

The design thinking approach that we are favoring not only emphasizes the circular or spiral nature of the process (feedback loops were common but limited in the classical process models), but it clearly identifies the need of divergent search activities. Developers are constantly and rapidly going through design-build-test cycles. In each cycle, during the divergent phases, we are focusing on the problem rather than on the solution, trying to understand who really is the user, which elements are truly involved, how many other ways are there to solve the problem, can we rephrase the challenge, can we circumvent the problem? We are generating concepts. These divergent activities usually result in a number of ideas or concepts that are, in the next step, built and then down-selected by testing. The underlying principles of Generative Design Questions (GDQs) and Deep Reasoning Questions (DRQs) will be explained in Sect. 5.

### 4.2  Rapid and Tangible Prototyping

In our design process we concentrate on creating prototypes as fast as possible in order to be able to test particular ideas, the design hypothesis behind the prototype. Speed, the acceleration of learning is key. As a result, our prototypes tend to be of low resolution, and physical or tangible rather than virtual. Depending on the design

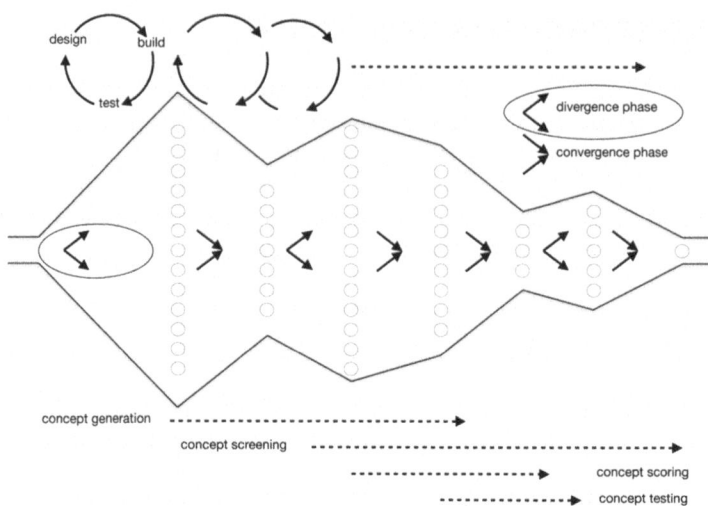

**Fig. 37** Design process as iteration of divergence and convergence steps or prototype cycles, adapted from Leifer and Steinert (2011)

stage, prototypes might be created on whiteboards, out of simple cardboard and duct tape, made from wood or clay, etc.

Each prototype is built to test a specific idea and/or a system interaction. They range from simplistic rough artifacts that merely resemble an idea (communication prototype), to lookalike prototypes (conveying certain external property ideas) to critical functional and functional prototypes (technical proof of concepts), to alpha and beta prototypes. It must be noted that later stage prototypes cost an order of magnitude more in resources, both in time and money than early prototypes. It is therefore essential to concentrate on the early stage or fuzzy front end of the new product design. The choice of the prototype material or environment, directly influences the amount and degree of the generated alternatives (Edelman et al. 2009) (Fig. 38).

The breadth and depth of the solution space explored seems to relate to the sophistication or resolution of the prototyping materials employed. A sophisticated CAD prototype is least likely to be considerably changed in following iteration cycles. The product architecture is implicitly fixed and the software and its capability limits possible ideation changes.

Tangible 3D prototypes allow the creation of more alternatives with relative ease. These types of lookalike prototypes are especially good at conveying ideas and form factors to non-specialist users. However, as can be intuitively seen by the foam model example depicted in Fig. 38, once this level of resolution has been reached, changes tend to be rather incremental.

If we contrast this to using very basic prototype material, simple cardboard or even just a sketch (see Fig. 39) the possibility for more radical and faster iterations and thus learning is obvious. As a rule of thumb, the early stage product

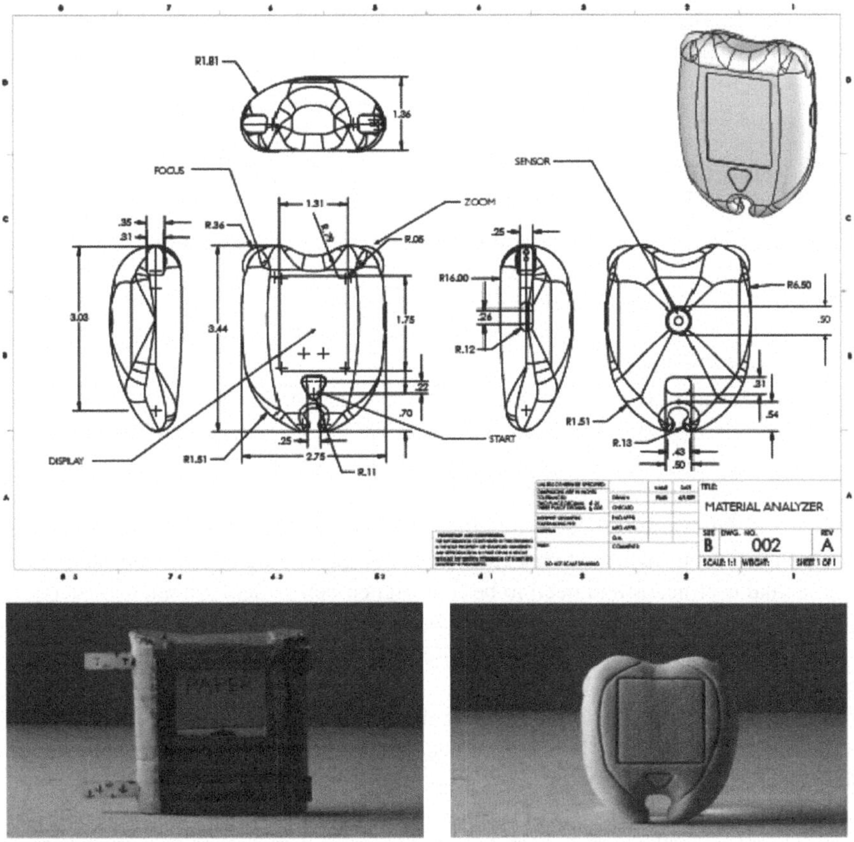

# Dancing with Ambiguity: Causality Behavior, Design Thinking, and... 151

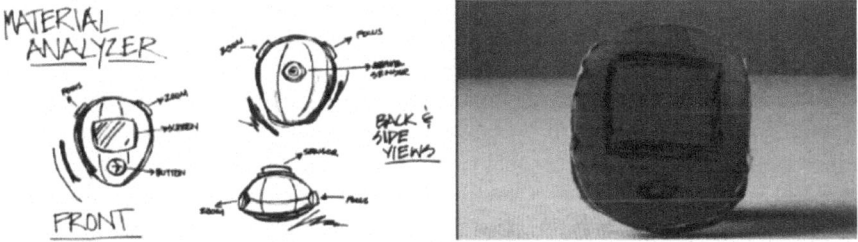

**Fig. 38** Details specifications and computer-modeled prototypes inhibit ideation and rough, low resolution, prototypes facilitate ideation (Edelman and Currano 2011)

**Fig. 39** Tangible 3D prototypes, right, facilitate associative memory, analog thinking, and exploration better than 2D sketches (Edelman and Currano 2011)

development determines the level of radicalness of the final solutions. We therefore advise product development teams to stay in this early phase for more than a third of the entire available project time. We have to force ourselves to abstain from entering solution optimization in order to gain intimate awareness of the problem space. This increases the chances for us to generate the real breakthrough idea we are looking for.

## 4.3 Need Finding, Bodystorming, User Experience Enactment

Whereas the late stage development is focused on the optimization of performance, cost, and reliability, the early phases of the product design process require a different set to tools and behaviors.

### 4.3.1 Need Finding

Central to the early stage of the new product or system concept design is an intimate understanding of firstly, who is the user actually, and secondly, what the real user's needs that we aim to satisfy with the solution? Often, projects start with a fixed set of specifications and requirements. This approach, very suitable for incremental change and innovation, focuses the attention and resources onto the optimization and execution of the selected concept. Time and again though, final solutions do not meet actual user needs and need to be re-designed before deployment can succeed. This costs significantly more money and prestige than conducting more exploration early on. Therefore we are concentrating on the first phase of the design process in particular.

The first challenge lies in identifying the actual user being target in the design process. For example, in medical device development, it is not the patient, the obvious user, and his needs that are central for the success of a new product. Though any new solution must at least be equal in terms of patient value added, the real litmus test lies in the value gained by the hospital and insurance companies, in relation to the change required by the practicing medical doctors. Who is the user for whom we have to design for in this case (Aquino et al. 2011)?

Once we identified a single target user or a user system, we attempt to gather information on the underlying needs that ought to be satisfied by the new solution. While surveying and interviewing users does give valuable information, very often, users are themselves not capable of expressing their needs. Indeed when confronted with something absolutely new, for example a device based on a new technology or material, users can only draw from analogies and not answer from experience. Even if they can, very often their personal perspective is too limited to truly understand the problem. We find that observations, especially when analyzed systematically using video interaction analysis (we code videos frame by frame and quantitatively analyze the resulting process data (Tang and Leifer 1991)) result in a better understanding of the process and behavior we attempt to improve. To achieve the same, our community has been borrowing heavily from anthropologist. We attempt to immerse ourselves into the problem, trying it out ourselves. As the literature of knowledge management shows us, this direct tactile involvement with the problem is often the only way to transfer implicit procedural knowledge. As Nonaka and Takeuchi (1995) famously describe, to build a home bread-baking automate that also kneads the dough, it was necessary to practice kneading with a baker. The development team would not have been able to uncover the complexity of the compress, pull, and twist action necessary to create dough that rises just right.

### 4.3.2 Bodystorming, User Testing and User Experience Enactment

Bodystorming is the second important tool employed. It is a design-inspired technique that challenges the designer or user to imagine what it would be like if the product existed, and act as though it exists, ideally in the place where it would be used. Rather than creating post-its notes and bullet point lists, we aim to engage an idea with improvised artifacts and physical activities to envision a possible solution. In fact the aim is to employ bodystorming instead of using classical brainstorming. The underlying idea is to tangibly create ideas and translate them into super rough communication prototypes. Very often the act of creating such a prototype storm generates artifacts with which we can better experience, test and improve the proposed solution. Also, bodystorming helps to convey and iterate the concept idea amongst the team; it allows the team to use the artifact and to enact a usage scenario. Finally, it allows it to actually go to users and to get their direct feedback on possible solutions. Last but not least, the tradition of fast prototyping, immediately during or after user testing, enables us to redesign the prototype on the fly. User testing, learning and iteration are thus combined seamlessly into fast cycles of change.

## 5 Underlying Design Principles, Lessons from Learning (Loop 3)

In this section on learning loop three, the focus on the informal creation and transmission of explicit and implicit knowledge. Combined with insights and information from the other learning loops, this area forms the Center for Design Research (CDR) at Stanford University's core research agenda. We target designers and the design process. The primary aim is to understand how designers and developers work, why some are more creative and some more analytical, how team composition and interaction can be improved, and how we can quantify and structure, or better not structure the design process. We believe our biggest opportunity for change lies in better education and in improving the support of designers and their design process including the contextual environment and support tools. On the specific topic of change, some key insights are presented that might contribute to the understanding, measuring and maybe even fostering of change, namely importance of wording, questions, gestures and emotions.

### 5.1 Noun Phrases as Change Indicators?

The first key insight is based on extensive research conducted by Ade Mabogunje from 1993 to 2007 at CDR (Mabogunje 1997). He analyzed design teams with a special focus on the usage of language amongst design teams in various settings and projects. The first series of experiments analyzed team interactions in a simulated teaching environment. The second series included the design problem context and introduced time development minimization as a dependent variable. The subsequent experiments included a focus on the generation of alternative prototypes and also on a parallel design process. A common result amongst all experiments was the emergence of the creation of noun-phrases as an interesting surrogate variable for describing the design process. In fact, based on design documentation analysis and in a setting that favors radical new solutions, the number of distinct noun phrases created correlates positively with performance. Other meta-document data, such as total length or readability, do not add insights. Mabogunje's research suggests that the creation of new words, noun phrases, is an indicator or maybe even a driver of creative activities inside the design team. On the one hand this relation might allow creating 'speedometers' of change by counting distinct noun phrases in real time; on the other hand, this might provide yet another lever to support ideation amongst the teams. By actively promoting and encouraging team members to strive for new wordings, we might be able to enhance the chances to break out of the established solution space that only allows for incremental changes. Of course this relationship is sure to be context-, language-, and culture-dependent. Verb phrases also correlated positively with innovation outcome. Interestingly, the new phrase amounts to a 're-representation' that, in agent based software systems, is often taken as a measure of learning. Please note that we have only established the existence of a correlation, not a causal relation; existences and direction of the latter remains to be explored.

In sum, it may be noted that the creation of new language correlates with the degree of change achieved. Any change beyond the existing frameworks may thus have to be accompanied by new language.

## 5.2 Change, a Question-Driven Process?

Besides noun phrases, questioning has a special place in understanding the fundamental design process. Based on Eris's research at CDR (Eris 2003, 2004), we have identified that design is in fact a question-driven process. Eris identified and developed a taxonomy of questions asked while teams of three to four designers were engaged in designing a Lego-prototype that differentiates between *Deep Reasoning Questions (DRQs)* and *Generative Design Questions (GDQs)*. The first reflect convergent, the latter divergent thinking. Based on this metric, a real time analysis of the design process is made possible.

Much like noun-phrases, on a combined level, questions may act as a design performance metric. There is a general positive correlation between the numbers of questions asked during design activities and the project team performance.

Looking at the two established subcategories, GDQs are prevalent in the conceptualization phase whereas DRQs were mainly used to reduce the number of

**Fig. 40** The role of generative design questions (GDQs) and deep reasoning questions (DRQs) in the design process

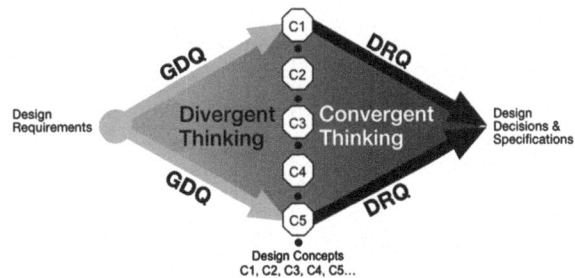

generated alternatives. Figure 40 depicts this central paradigm for radical design processes.

As depicted in Fig. 40 all design is question based, but there are two types of questions, one divergent (GDQ) and one convergent (DRQ). One might define a cycle of design thinking as a period of generative divergence. How many ways can we do this? Followed by an analytic convergent, given what we know, what is the next step? Real life is accordingly made up of thousands of loosely associated, one might even say fractal, elements that in aggregate become, for example, spacecraft, groceries, or banking enterprise software (Eris 2003).

In sum, GDQs are essential for preserving ambiguity. They generate alternatives, reframe needs and drive the creative negotiations amongst the design team. DRQs are more prevalent during concept assessment and implementation, and generally aim to reduce the number of alternatives.

DRQs are based on tools and analytical skills that are rather classical and taught and implemented abundantly. GDQs however very often contradict an organization's natural tendency to control and manage. They are central for the creation of change as they allow the opening up of the solution space in which radical new ideas can emerge. Basing change mainly on DRQs will lead to incremental learning only, as no substantially new concepts are created. Combined, GDQs and DRQs constant iterate and establish causality between possible pathways and tangible prototypes.

## 5.3 Knowledge Transmission, the Power of Additional Channels

Product development and design projects are team based. In addition there is a close and continuous interaction with other stakeholders, such as users, suppliers, etc. The success of these interactions, in our specific case the rate of learning which we want to increase, depends on how well we are able to communicate, to transmit knowledge. Besides the already mentioned impact of language (noun phrases and questions), we have (since the early 1990s) and still are studying the impact and facilitation of collaborative interactions (Tang 1989; Tang and Leifer 1991; Ju 2008). He focused on analyzing the designer interaction through video analysis for shared workspace process activities: listening, drawing and gesturing. Additionally,

**Fig. 41** Analysis of designers' interaction for workspace process activities and purposes (Tang 1989)

| Function | Text Activity | Draw Activity | Gesture Activity | |
|---|---|---|---|---|
| Store Knowledge | 40 | 19 | 1 | 27% |
| Express Ideas | 2 | 63 | 33 | 43% |
| Mediate Interaction | 0 | 21 | 46 | 30% |
| | 19% | 46% | 35% | |

Tang separated the following activity purposes: storing information, expressing ideas, and mediating interactions.

As a result, we can show that purposes and process activities intermix fluently (Fig. 41).

When collaborating, various activities call for different media, one has to be conscious about the limitations and direction induction effect of the media choice. Tang (1989) analyzed the medium (text, drawing, gesture) used to mediate the function (store knowledge, express ideas, mediate interaction between designers) during a software development project dealing with human-computer interaction design. Rows are functions. Columns are media. The diagonal features the strongest function-to-medium associations. The most notable finding was that gesture plays a very important role in mediation and is one of the most difficult media to capture, store, index or re-use design knowledge. Numbers on the perimeter reflect the net percentage of dialog transactions that took place in the medium or function.

For storing explicit knowledge, text is the medium of choice, whereas drawing is especially important when expressing ideas (a picture in fact says more than a thousand words?). Additionally, the eminent status of gesturing becomes obvious for conveying and supporting communication. Any workspace that is to foster change must support all three activities, allow common access for all participants and convey a sense of close proximity. This holds true not only for our tangible workspaces such as shop equipment, but also for our virtual project platforms and computer supported cooperative work tools. It has been shown again and again that barriers created by awkward user interfaces inhibit the exchange of information. We need to choose, not the most sophisticated platform, but the one actually accepted, adopted, and used the most. Especially in the age of rapidly evolving cloud services and apps, the tools most in demand may be outdated quite fast. Therefore, the call for flexibility and low investments made during the discussion on physical space earlier applies.

Based on this research we are currently exploring the possibility of enhancing computer-supported communication by introducing additional information transmission, for example gesturing. David Sirkin has shown that the imitation of body language on behalf of the computer terminal in a human computer interaction scenario significantly improves collaboration (Sirkin et al. 2009).

In such a way, virtual collaboration, though distant and computer-mediated, becomes more natural and more productive. In 2009 a team composed of Stanford and Swiss students were able to also increase the level of subjective proximity by introducing 3D audio. Sound seems to allow us to sense the spatial presence of our faraway counterparts more naturally, and hence lets us focus more easily at the task at hand. Creating proximity through all channels of knowledge transmission (audio, visual, gesture) can thus facilitate change and its prerequisite, collaboration.

## 5.4 Team Interaction and Especially Intra Team Conflict Is Emotional

The last lesson learned that we would like to introduce at this point concerns emotion. It is a dimension that ought to be included in collaboration and change projects. Malte Jung has focused onto the topic of team conflict for the last 3 years at CDR (Jung et al. 2010). The initial inspiration stems from Gottmann and Levenson's work that predicted long-term outcome of marriage. Their predictions are based on the affective interaction quality during a 15 min video sample of a couple engaging in a problem discussion with an impressive accuracy of 93 % (Gottman and Levenson 2000). A similar problem presents itself when looking at the functionality or dis-functionality of a development team. A positive self-sustaining cycle of iterative prototype based learning assumes a working team interaction.

Using video observation and the Specific Affect Coding System (SPAFF) (Coan and Gottman 2007) coding scheme, in this special case of pair programmers confronted with a programming design challenge, we were able to quantitatively code the affective interaction dynamics. Based on analyzing positive versus negative facial expressions, body gestures, semantic context, and tone of voice, and based on events over time and putting them into relation with team satisfaction and performance indicators, it becomes obvious that negative events do, in some teams, overwhelm positive events. This may escalate up to a pivoting point where the entire team interaction becomes irreversibly negatively loaded. At this stage it is hardly possible to innovate or collaborate creatively. In fact this pivoting creates internal friction and inertia that inhibit any kind of positive learning loop. Similar to Gottmann, based on Jung's work, we were able to predict such outcomes through the video coding and analysis of 15 min of team interaction between the pair programmers (Fig. 42).

If team dynamics develop in this negative direction and pivot, it is necessary to intervene or to change the organizational setup. Positive change cannot be ordered from above. Emotions between team members must be taken into considerations.

**Fig. 42** *Top*: VCode interface showing a coded 16-s section. The squares in the *upper rows* indicate speaker turns. The *squares* in the lower part mark occurrences of negative and positive behaviors respectively. *Bottom*: Example point graphs of a regulated and a non-regulated programming pair. The graphs always show the emotion trajectories for each programmer separately. The *left graph* is drawn from a pair that scored amongst the lowest in the sample and the *right graph* is drawn from a pair that scored amongst the highest of the pairs studied

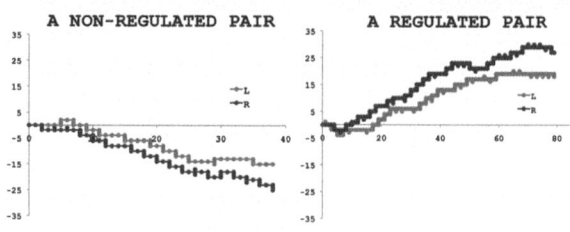

## 6   Dancing with Ambiguity: Summary and Discussion

With this chapter we have attempted to summarize and translate the lessons we have learned from studying designers and their activity into the broader context of complex social technical systems in general. Of course, as our research was never intended specifically for probing into this question, external validity remains to be discussed and seen. However, the aim was to show analogies and to open up both opportunities for learning and another perspective on managing the fuzzy front end.

After translating the problem into design specific language, we have introduced learning loops that are instrumental for setting the stage for radical innovation at the front end of innovation. Staying within our dogma of constant redesign, and knowing that we are attempting to create the unknown, we do not claim that our findings are laws in the scientific method sense, but we experience their success daily. We are working on improving the techniques employed, and strive to better understand the underlying principles.

We have primarily learned that in order to facilitate change, we have to let change happen. We have to remove institutional and procedural barriers, create a maximum of flexibility and support divergent activities. We have had to learn not only to live with change, but also to promote it. In this sense we would like to encourage the reader and invite him or her to join our dancing with ambiguity.

# Leveraging Creativity

Sascha Friesike and Oliver Gassmann

## 1 Introduction

Creativity is the basis of innovation. Being able to rethink existing solutions, to combine existing ones with solutions used in other fields, or to imagine a new way of doing things is creativity and as such the necessary foundation of innovation. The imaginative ability to come up with something new is unique to humans and it is the driver behind our technological advancements and our competence to develop tools to help us. Yet, often creativity is understood as an ability only a few among us share. They are labeled *the creatives* and they are clearly distinctive from the rest of us – they are the John Cleeses, the Andy Warhols, the Bob Dylans, or the Stanley Kubricks, and they clearly are not us. Firms have a way of promoting this kind of thinking, they divide the workforce into the creatives and *the suits*. The creatives wear T-shirts and jeans, come late and can regularly be seen walking around deep in thought through the company's courtyard. The suits, on the other hand, come early, sit in a cubicle all day and – well, as the name already suggests – wear a suit. This division of labor has successfully crafted the notion that creativity is a gift only a few share and all others are well advised to shy away from. This is not only a notion that is demotivating and downright catastrophic to any innovative firm, it is also not at all true. In his famous speech about creativity for Video Arts, John Cleese pointed out that creativity is not an ability, but rather a state of mind, one that many employees are simply hesitant to enter.

S. Friesike (✉)
Alexander von Humboldt, Institute for Internet and Society (HIIG), Bebelplatz 2, 10099 Berlin, Germany
e-mail: friesike@hiig.de

O. Gassmann (✉)
University of St. Gallen, Institute of Technology Management, Dufourstr. 40a, 9000, St. Gallen, Switzerland
e-mail: oliver.gassmann@unisg.ch

The present chapter is not a compilation on how creativity works, which brain functions are involved, and what cognitive behavior can teach us about it. It is a chapter that simply presents the most common and popular *creativity techniques*. These techniques represent ways of encouraging creative activity, they help to break the magic seal that locks the idea of creativity away from many. Creativity techniques can be used in all kinds of scenarios ranging from problem solving, to developing a corporate vision, to generating new product ideas. They are helpful tools that makes it easier for anyone to enter – as John Cleese would put it – a creative state of mind.

Given a specific problem, some techniques are more suitable than others. Sometimes it is important to find one single working solution, while at other times it is desirable to create a wide range of possible solutions to understand the possible solution space.

## 2 Synectics

*"Get familiar with the strange and estrange the familiar." After a thorough analysis of problem analogies, a new solution will be sought. The return to the original problem can lead to new and surprising solutions. Synectics was developed by William J. J. Gordon in 1944.*

| | |
|---|---|
| **Aim** | Creative process, reorganization of different knowledge into new patterns |
| **Participants** | 8–12 persons from different disciplines |
| **Time** | Approximately 4 h |
| **Advantage** | Particularly innovative and creative solutions |
| **Disadvantage** | Highly demanding on the moderator, the many steps are time-consuming and take some getting used to |

**Procedure** Synectics is divided into the following ten steps:
1. Analysis of the problem and problem definition.
2. Developing first spontaneous solutions.
3. Reformulation of the original problem.
4. Forming direct analogies and choosing the best one: For technical problems analogies from nature (biomimetics) or social sectors are helpful.
5. Forming personal analogies and choosing the best one: Participants are supposed to put themselves into their personal analogy and describe how they feel.
6. Forming symbolic analogies and choosing the best one: The analogy is to be described as concisely and clearly as possible.
7. Forming direct analogies and choosing the best one: Finding examples from nature or technology that match the statements in item 6; with this step the alienation from the original problem reaches its peak.
8. Describing the chosen analogies in a manner as detailed and accurate as possible.

9. Reconnecting to the original problem (*force-fit*): Is it possible to deduce any solutions from the chosen analogies?
10. Saving the developed solutions and evaluating them.

**Tools**
- Workshop material
- Blackboard, flip chart, projector

## 3 TILMAG Method

*TILMAG represents the transformation of ideal solution elements through matrices of association and similarities* (Fig. 43). *The method is a modification of synectics and was developed by the Battelle Institute in Frankfurt, Germany.*

| Aim | Identifying new possibilities of solution |
|---|---|
| Participants | 2–25 persons |
| Time | Approximately 2 h |
| Advantage | Target-orientated approach to finding an ideal solution |
| Disadvantage | Ideal solution must already be distinguishable |

**Procedure** The TILMAG method is divided into the following eight steps:
1. Analysis and definition of the problem statement.
2. Identification of the 'ideal' elements of a potential solution: These elements can either be concrete structural parts of a solution itself, or be deduced from the important basic conditions of the problem as well as from general requirements. They can be acquired by other creative ideation techniques, such as brainstorming.
3. Definition of the 'ideal' elements, preferably briefly and concisely.
4. Formation of associations using a paired combination (*association matrix*). The analogies that arise get listed in an association matrix.
5. First step of ideation through a transfer to the actual problem.
6. Paired confrontation of the associations (*similarity matrix*); searching for similarities between associations and possible solutions. Only positive similarities are meant to be captured, i.e. structural elements which actually display both of the associated terms!
7. Connecting similarities to find solutions.
8. Renewed ideation for finding the final solution.

**Tools**
- Flip chart, blackboard or presentation board
- Markers

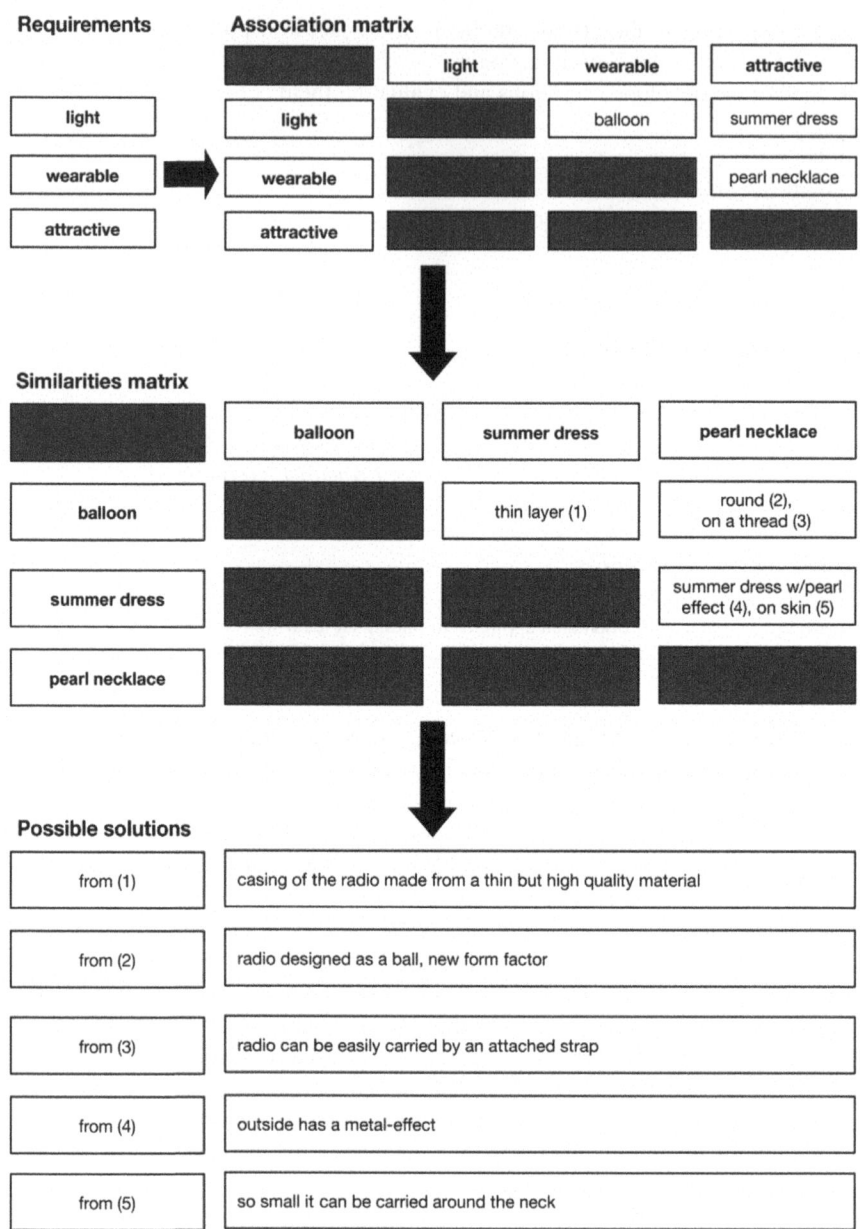

**Fig. 43** Procedure of the TILMAG method using the example of searching for new ideas for a portable, lightweight and attractive radio

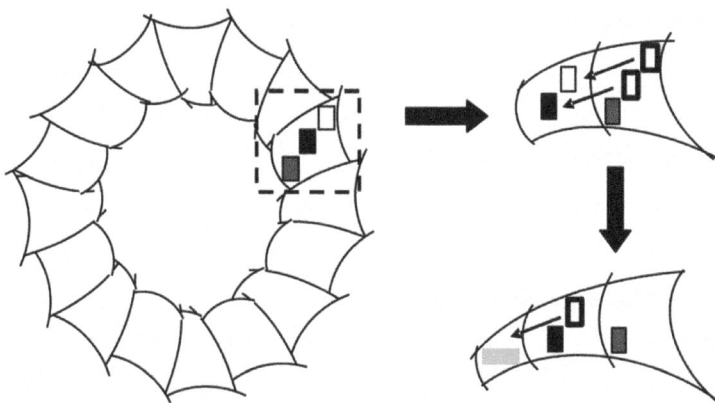

**Fig. 44** Spider meeting

## 4 Spider Meeting

*Interactive ideation through a network – from the outside (low degree of detail) to the inside (high degree of detail) – with favored solutions being chosen simultaneously. This method was developed by Barbara Widmer at the Swiss engineering firm Zühlke.*

| | |
|---|---|
| **Aim** | Ideation and evaluation within one process |
| **Participants** | Six persons and one presenter |
| **Time** | Approximately 2 h |
| **Advantage** | Developing and choosing the favored solution happens within one single meeting |
| **Disadvantage** | Not suitable for complex tasks, requires a concise definition of the problem |

**Procedure** The spider meeting is conducted using the following steps:
- Preparation: Each participant is assigned one color and receives two pens of this color. Participants sit in a circle around the spiderweb (see Fig. 44).
- First row: A total of 36 ideas are 'spun' and listed on post-it notes using keywords (no sketches). Spreading these ideas in the first row of boxes in the web: Up to three differently colored post-it notes per box.
- Selection 1: By placing his/her pens on the boxes, each participant marks two boxes which *do not* contain any of his/her own ideas.
- Second row: Of these selected ideas the labeling participant chooses two and moves them into the second row.
- Selection 2: By placing his/her pens on the boxes each participant marks two new boxes, which (if possible) *do not* contain any of his/her own ideas.
- Third row: Of the two ideas the labeling participant chooses one and outlines an appropriate solution.
- Discussion: The six solutions outlined will be discussed and, where needed, additional ideas will be described.

**Tools**
- Spiderweb (12 segments, four rows)
- Twelve pens (six different colors times two pens)
- Thirty six post-it notes
- Twelve sketch models (approximately 21 × 21 cm)
- Glue

## 5 Six Thinking Hats

*A creativity method that uses roleplaying. Participants assume six predetermined roles and consider a problem from role-specific points of view. The role allows participants to express themselves more freely than they would if they spoke as themselves.*

| | |
|---|---|
| **Aim** | Creative solution finding through discussion based on predetermined views (roles) |
| **Participants** | Six persons |
| **Time** | Approximately 2 h |
| **Advantage** | Thinking in terms of roles lowers the inhibition threshold for voicing honest criticism, because the role, not the person, is criticized |
| **Disadvantage** | Not everyone is comfortable with playing a role |

**Procedure** Each team member puts on a colored hat and plays a specific role to tackle the problem from a different point of view. The roles can be alternated, or the entire team discusses the problem using the same role. The role playing allows for a certain degree of anonymity and lowers the threshold for voicing constructive criticism, as this relates to the role and not the person behind the role. The colored hats represent the following roles:
- White hat: The analytic, who is the objective and neutral type, his/her opinion is based on facts and figures.
- Red hat: The emotional, who is the subjective and personal type, shows emotions and acts on hunches.
- Black hat: The pessimist, who is the objective and negative type, acts as *the devil's advocate*, an alarmist.
- Yellow hat: The optimist, who is positive and sees possible opportunities and benefits.
- Green hat: The creative, who is provocative, thinks laterally and has 'crackpot ideas'.
- Blue hat: The presenter, who is the realistic, structured type and is able to see the big picture.
  Thoughts are written down in the form of keywords to keep track.

**Options** The *Disney method* is a similar method which uses only three roles: dreamer, realist and critic.

## Tools
- Six colored hats
- Form set (e.g. available as download at www.zeitzuleben.de)

## 6  Bisociation Method

*Breaking up mindsets and collecting ideas through associations by using pictures which have nothing to do with the original problem.*

| | |
|---|---|
| Aim | Creativity method, determination of new ideas for solution |
| Participants | 10–25 persons and one to two presenters |
| Time | Approximately 45 min |
| Advantage | Suitable for problems that require unusual ideas and solutions |
| Disadvantage | Not suitable for finding technological solutions |

**Procedure**  The bisociation method is conducted using the following steps:
- The group agrees on one problem, which is precisely defined and written down.
- Now three to five pictures or photographs which have nothing to do with the topic/problem, i.e. are as remote from the context as possible, are handed out.
- All participants agree on one picture with which they want to deal and on which they would like to comment.
- The selected picture is hung up so it is visible to all. Now all participants are invited to freely utter associations they can come up with in regard to the picture. The presenter records the thoughts on cards. The cards are pinned on a board.
- As soon as the association round is finished, the initial question is presented again. By trying to connect the initial problem and the associations, participants are asked to name suggestions referring to a solution of the initial problem. Usually very creative and unconventional proposals arise, which are also listed on cards.
- Now the proposals suggested are hung up as well. And the participants discuss their feasibility.

**Options**  Conducting the procedure in sub-groups and then comparing the proposed solutions.

### Procedural Notes
- Pictures that are interesting and trigger associations should be selected.
- Additionally, the pictures chosen should be of varied content and completely unrelated to the real problem.
- Participants should be encouraged to express extraordinary statements; often the best solutions develop from unusual and unconventional proposals.

**Tools**
- Three to five pictures
- Pin boards/metal pins
- Cards
- Pens for the presenters

## 7 Mind Map

*Extremely multifunctional and graphically supported method that stimulates both hemispheres of the brain thanks to the visual representation and thus activates the entire creative potential.*

| | |
|---|---|
| **Aim** | Collecting and organizing thoughts, inspiration through graphics |
| **Participants** | One, also suitable for teams |
| **Time** | 1–2 h, depending on the topic |
| **Advantage** | Complex information can be structured playfully so that new ideas are generated |
| **Disadvantage** | Loss of intelligibility where structure is too detailed and terms are vague |

**Procedure**
- The structure of a mind map is similar to that of a tree.
- The central topic is noted at the center of the map. Surrounding this core, branches lead to sub-topics. This continues several times, depending on the complexity of the topic.
- By means of this presentation technique, even unsorted thoughts can be structured and the user is not forced to think in a strictly systematic way, he/she can let his/her intuition run wild while looking at the diagram and just add a new idea to the appropriate branch.
- To keep a mind map understandable, no more than seven subcategories should emerge from each node. It is also essential that the user keeps the entire picture in mind at all times. This is especially important when software tools are used to create mind maps.
- The mind map can be supplemented with images, dependent relationships can be highlighted with lines. The use of colors improves the overall quality of a mind map and makes it easier to understand.
- The method is useful in many settings, such as the recording of brainstorming sessions or meetings, the preparation of speeches and reports, collecting requirements for product designs, studying, and much more.

**Tools**
- Paper and pencil, colored pens
- A multitude of software tools is available for developing mind maps

# 8  TRIZ

*The 40 innovation principles contain recommendations for the change of technical systems. They support the process of gaining useful properties and removing unwanted properties.*

| | |
|---|---|
| **Aim** | Support finding technical solutions by applying basic principles |
| **Participants** | No restrictions |
| **Time** | 1–2 h |
| **Advantage** | Helps dissolving mental blocks and finding new solutions |
| **Disadvantage** | Unknown |

**Procedure**  TRIZ is the internationally accepted Russian acronym for the theory of solving inventive tasks (Russian: Teorija Rešenija Isobretatelskih Zada). It was developed by the Russian scientist Genrich Altshuller and his colleagues from the 1960s to the 1980s.

The main feature of problem solving with TRIZ is the identification and elimination of technical and physical contradictions in technical systems. An analysis of approximately 40,000 patents revealed that – regardless of industry – the inventing step can be found in a limited number of basic principles (procedures). The result was one of the best-known and easy-to-use tools for anyone searching for a technical solution: the 40 innovation principles.

**40 Innovation Principles by TRIZ**
 1. Decompose or segment ♣♦♥
 2. Separate from harmful ♣♦♥
 3. Adjust quality locally ♦♥
 4. Use asymmetry ♦
 5. Unify similar elements, coupling ♦
 6. Increase universality ♦♥
 7. Interleave (matrjoschka, telescope) ♦
 8. Use the counterweight or lift ♦
 9. Score the counteraction previously
 10. Score the effect before ♣♥
 11. Put a pillow under before
 12. Keep the same energy potential
 13. Reverse the functions ♣♦
 14. Use the ball similarity
 15. Make it more dynamic, more agile ♣♦
 16. Score a little more or a little less ♥
 17. Use higher dimensions (1-D, 2-D, 3-D) ♣♦
 18. Use mechanical vibrations
 19. Conduct actions periodically ♣
 20. Use continuous actions ♥

21. Rush through processes and situations
22. Convert harmful into useful
23. Implement feedback
24. Use an intermediary ♦
25. Implement self-service ♥
26. Use copies or images ♥
27. Use cheap, non-durable, replaceable materials ♥
28. Replace the mechanical system ♣
29. Use liquids or air
30. Use flexible shells and thin films ♦
31. Use porous materials
32. Change the color or transparency ♣
33. Make something similar or homogeneous
34. Remove or regenerate parts
35. Change the physical or chemical properties ♣
36. Use phase transitions (solid, liquid, gas)
37. Use thermal expansion
38. Make use of responsive means
39. Make use of inert, insulating media
40. Make use of composite materials

**Procedure**
- To work with the innovative principles preferably the following groups are selected:
- ♣ 10 best principles for brainstorming
- ♦ 13 best principles for design
- ♥ 10 best principles for creative cost reduction
- If no satisfactory solution is found, all 40 innovation principles are used.

## 9  Imaginary Brainstorming

*In this form of brainstorming, the conditions are changed in order to free participants from entrenched ideas and predefined ways of thinking.*

| | |
|---|---|
| **Aim** | Creativity method, widening the solution space |
| **Participants** | 4–15 persons and one presenter |
| **Time** | 45–90 min |
| **Advantage** | Suitable if ordinary ways of thinking are to be abandoned, participants look at problems from different perspectives |
| **Disadvantage** | Not suitable for finding specific solutions |

**Procedure**
- The entire method works like the classic brainstorming procedure.
- The presenter announces the rules for brainstorming (best visualized on a poster).
- The presenter announces the problem in a changed/altered version (as an imaginary problem to which new conditions apply).
- The whole group or sub-groups try to develop as many creative solutions as possible within 5–10 min. They note them on cards.
- The solutions are collected, pinned to a board and explained to the entire group.
- Now the 'real' problem is introduced, visualized and set in relation to the solutions to the imaginary problem.
- The group examines the solutions under the new conditions and develops them further.

**Tools**
- Poster with brainstorming rules
- Pinboard, cards and pens

## 10  Semantic Intuition

*In semantic intuition terms are combined to form new meanings. The newly created words bring ideas to light that can help find innovative solutions to a given problem.*

| Aim | Innovation method, new solution to a given problem |
|---|---|
| Participants | One to seven persons |
| Time | 45–90 min |
| Advantage | Suitable for finding new and innovative product ideas |
| Disadvantage | Not suitable for finding a specific solution |

**Procedure** Semantic intuition aims to reverse the usual process from *invention* to *naming*. Hence, it proceeds from naming to invention. Through free association, new terms are formed. For this purpose, words are collected from the problem environment. These words then get combined randomly. They are paired and it is then discussed which product or solution might be hiding behind a given combination of words.

For example, words like microwave and refrigerator can be combined while searching for a new kitchen appliance and a device that cools food quickly could be deduced.

**Tools**
- Notepad and pens
- Cards
- Pinboard

| Parameter | Characteristic | | | | |
|---|---|---|---|---|---|
| Form | 1.1 Sedan | 1.2 Station Wagon | 1.3 SUV | 1.4 Convertible | 1.5 Mini-Van |
| Engine | 2.1 Combustion Engine | 2.2 Electric Motor | 2.3 Hybrid | 2.4 Muscle | 2.5 |
| Fuel | 3.1 Gasoline | 3.2 Diesel | 3.3 Natural Gas | 3.4 Alcohol | 3.5 Electricity |
| Wheels | 4.1 Two | 4.2 Three | 4.3 Four | 4.4 Six | 4.5 |
| Pwertraion | 5.1 Rear Wheel | 5.2 Front Wheel | 5.3 4x4 | 5.4 | 5.5 |

**Fig. 45** Morphological box

## 11 Morphological Box

*Systematic and structured analysis to find solutions for a given and complex problem. Solution parameters are displayed in a matrix.*

| | |
|---|---|
| Aim | Finding solutions to complex problems in product development |
| Participants | One to six persons |
| Time | Approximately 2 h |
| Advantage | Through systematic combination of individual features, a large number of possible solutions arise. |
| Disadvantage | None yet found |

**Procedure** The procedure is divided into four steps:
1. Precise description, respectively definition, and generalization of the given problem.
2. Definition of the main characteristics (parameters). At this point it must be considered that these are independent of each other and relevant to the problem. Their number should not be higher than seven to keep the analysis manageable. Each parameter is entered into the first column of a table (see Fig. 45).
3. For each parameter, possible characteristics are searched for and entered into the corresponding row of the table. Again, it is important to limit the number of entries. It is often helpful to divide the problem into sub-problems and to create several tables.
4. Synthesis of solutions through combination of different characteristics of each parameter. At this step, the fields are connected with each other by a zigzag line. Assessment of the combinations found to identify the optimal solution.

## 12 Method 6-3-5

*The method 6-3-5 is probably the best-known brainwriting technique. Participants write down their ideas and take note of their neighbors' thoughts.*

| | |
|---|---|
| **Aim** | Creativity technique, brainwriting for problem with low to medium complexity |
| **Participants** | Six persons |
| **Time** | 30 min for 108 ideas |
| **Advantage** | Many ideas generated within a short period of time, ideas are not 'talked to death', easy to use |
| **Disadvantage** | No direct feedback, the rigid procedure might disrupt creativity |

**Procedure** The numbers 6-3-5 indicate that this method calls for six persons to note or sketch three ideas in 5 min. The ideas are entered into form sheets, each of which passes around the group in a round robin procedure. The 6-3-5 method is divided into the following steps:
- The presenter introduces the problem and discusses it with the team. The result is a precise definition of the problem.
- Each of the six participants enters three ideas in the top row of his/her form sheet. They have 5 min for this task.
- Each form sheet is passed on to another person.
- Now everyone looks at the first three ideas his/her predecessor has written down and then adds three new ideas to the form sheet (another 5 min). These additional ideas can:
  - Represent an addition to a previous idea,
  - Be variations of a previous idea,
  - Represent completely new ideas.
- The forms are forwarded to the next participant. The procedure is repeated until a complete round has taken place.

**Options** Sketching in addition to writing ideas, in which case the timeline should be extended.

## 13 Gallery Method

*Creativity technique that combines individual and group work.*

| | |
|---|---|
| **Aim** | Associations through visual representation |
| **Participants** | Five to ten persons, one supervisor |
| **Time** | 2–4 h |
| **Advantage** | In design questions, effective communication with the aid of sketches, easily evaluable documents |
| **Disadvantage** | None yet found |

**Procedure** The gallery method has its name from hanging ideas on a wall just like pictures in a gallery. The method solves a problem by presenting ideas, eliminating unfeasible ideas, creating new ideas based on the ones presented previously, and finally choosing the most practical solution.

The entire method can be broken down into five distinct steps:
1. Initiation: A supervisor presents the problem and provides instructions.
2. Idea phase I: Every participant creates a single solution on his/her own.
3. Association: All ideas are hung to the gallery walls, jointly presented and discussed. At this point, some ideas are tossed out or combined with others.
4. Idea phase II: Inspired by the other solutions, the participants refine or change their solutions.
5. Selection: In the final phase the group jointly selects the one idea that seems most feasible and solves the problem best.

This method is especially useful for solving problems that can be presented visually. Such solutions go well with the idea of a gallery. On the other hand, this method often fails when the solutions cannot be presented visually, but rather are presented in text form. In such a case, all participants have to read through all other solutions, which can be demotivating in some group constellations.

**Tools**
- Movable walls, posters, pens, pins

## 14  Collective Notebook Method

*A brainwriting technique that facilitates collecting ideas over a longer period of time (2–4 weeks).*

| Aim | Problem solving/ideation |
|---|---|
| Participants | Two or more persons |
| Time | 1–2 h preparation (leader) 2–4 weeks collecting ideas 1–3 h evaluation (team) |
| Advantage | Collecting ideas from each team member, possible anytime and regardless of location |
| Disadvantage | Motivation and discipline, long duration |

**Procedure** The collective notebook method (CNB) functions as a written brainstorming, i.e. a brainwriting. The method is very suitable for large groups. It is carried out over a longer period of time and is therefore right for long-term or strategic problems. Every team member can add his/her ideas regardless of place and time.

The CNB works as follows:
- Preparation (team leader): The problem is formulated precisely and stated in writing at the beginning of the notebook. Selection and instruction of team

members, definition of the time horizon. Each team member receives his/her own notebook.
- Implementation phase: During this phase participants keep their notebooks and write down ideas or sketch solutions. The notebook is supposed to accompany the team member at any time and at any place. At the end of the implementation phase, every member creates a summary of his/her ideas.
- Evaluation (entire or reduced team): The summaries are compared and the notes are studied. The team develops proposals to solve the problem and designs a concept. The rules at the evaluation stage are analogous to those used in a brainstorming session.

**Options** Instead of personal notebooks for each team member, a single notebook can be kept at a central place.

**Tools**
- Notebooks
- Description of the problem in each notebook

## 15 CATWOE

*CATWOE is a checklist used to define problems which was developed by Peter Checkland and Jim Scholes.*

| Aim | Definition of problems |
|---|---|
| Participants | One, also usable for teams |
| Time | 1–2 h |
| Advantage | Structured approach |
| Disadvantage | Unknown |

**Procedure** CATWOE is an acronym used for problem definitions. The main idea behind CATWOE is to focus on the problem's context and not on the problem itself. The context of a given problem is named 'system'.

The following steps are carried out:
- C = Customer (At this point the role of the customer in the system is discussed. Who is the customer? What is his need? How is he using the current product? ...)
- A = Actors (Here all actors of the system are looked at. Who is involved in the system? What are their interests?)
- T = Transformation process (What does the system do? What is its input, What is its output?)
- W = World View (Here the system is looked at in the context of ideology.)
- O = Owners (This aspect discusses who the key stakeholder in the system are and how these are motivated.)

- E = Environmental constraints (This aspect looks at the constraints the system is facing: legal, personal, economic, ethical, etc.)

**Tools**
- Notepad, pens

## 16 Provocation Technique

*This method developed by Edward de Bono serves ideation. Provocations challenge existing assumptions.*

| Aim | Ideation |
|---|---|
| Participants | 2–25 |
| Time | 1–2 h |
| Advantage | Highly innovative ideas |
| Disadvantage | Fails if provocations deviate too far from reality |

**Procedure** This technique challenges existing assumptions through provocation. Consequently, new ideas arise which would normally not have been considered. For this method, different approaches exist which can be chosen depending on the problem and the setting:
- Abolishing an existing assumption
- Describing an ideal/desirable situation
- Facts or relationships are reversed
- A given quantitative attribution is exaggerated
- A randomly selected term is used
- A given qualitative attribution is distorted

Following the provocation technique, ideas for new solutions are developed and recorded.

## 17 Quick-and-Dirty Prototyping

*Spare parts and unused materials are used to illustrate forms or interaction.*

| Aim | Concept development |
|---|---|
| Participants | Small groups up to eight participants |
| Time | 1–2 h |
| Advantage | Easy way to communicate a concept |
| Disadvantage | Concepts are not necessarily feasible |

**Procedure** Quick-and-dirty prototyping can be applied for the purpose of developing a concept. Available materials are creatively used to discuss forms, relations, and ideas. It is a simple and rapid method by which to communicate a concept and develop it further within a team. Even though the method originated in the context of design, it is transferable to other fields of application.

**Tools**
- Diverse materials like paper, clay, glue, scissors

## 18  Five 'Whys'?

*'Why?' is asked and answered five times in a row. In this way, root causes are determined which may have been unknown previously.*

| | |
|---|---|
| Aim | Uncovering root causes |
| Participants | At least one asking and one answering person |
| Time | Less than 30 min |
| Advantage | Causes that were previously unknown are discovered |
| Disadvantage | Method does not always lead to useful answers |

**Procedure** Questions are used to reveal a root cause. One person questions a circumstance, and once an answer is presented, the fundamentals of this answer are questioned, too. This is done until 'why' has been asked up to five times. By means of this method the underlying causes of a problem are uncovered. If these underlying causes are found, solutions can be developed which have not been considered before the application of the method.

**Options** An extension of this method is the Ishikawa or cause and effect diagram. With this method, the 'why'-questions are asked in the six areas (6 Ms) of machine, method, material, man power, measurement, and mother nature.

## 19  Extreme User Interviews

*Creative input through atypical interviews with people who are very, or not at all, familiar with the core problem.*

| | |
|---|---|
| Aim | Finding core problems and appropriate solutions |
| Participants | Several users and at least one presenter |
| Time | Approximately 2–4 h |
| Advantage | Other perspectives can be incorporated into the creative process |
| Disadvantage | Time-consuming |

**Procedure** Extreme user interviews are interviews with persons that either have appropriate technical expertise in the field or are completely alien to it. As a result, new approaches are developed. This method can also be used to test new products or services.

The two specific groups of people are often able to identify problem areas that were not considered beforehand. Furthermore, they can suggest previously unknown solution statements that would not have been considered without extreme users.

## 20 Long-Term Prognosis

*The long-term prognosis is a creativity method in which future scenarios are developed in order to obtain new solution statements.*

| Aim | Developing new ideas from solutions |
|---|---|
| Participants | Two or more persons |
| Time | 1–2 h |
| Advantage | Current trends lead to unusual solutions |
| Disadvantage | Unknown |

**Procedure** Within the team, future visions are developed that are based on today's social and technological trends. The impact these trends may have on human behavior as well as on dealings with products, services, and environments is considered.

The technique of long-term prognosis helps to understand user behavior. As such, this method can help to find new solutions that differ from those used today.

## 21 World Café

*World café is a dialogue and workshop method that is suitable for large groups and generates collective knowledge in a relaxed atmosphere. The method was invented by Juanita Brown and David Isaacs.*

| Aim | Creating collective knowledge |
|---|---|
| Participants | 12–2,000 persons |
| Time | 2–3 h |
| Advantage | Participation of many people |
| Disadvantage | Unknown |

**Procedure** Participants sit at tables in groups of four to five. The tables are covered with paper. A question is asked and the groups collect corresponding ideas for 15–30 min. Afterwards, groups are reshuffled, hosts who always stay at their tables instruct their new groups and provide a seamless transition. Motivating hosts and insightful questions are crucial for the success of the method. It is also important that the questions are phrased simply and tailored to the backgrounds of the participants.

World café is most suitable for an introduction to an important topic, for the creative search for options, and for use in large heterogeneous groups.

**Options** After two to three rounds the presenter can also pose further questions or provide clarification.

**Tools**
- Tables, paper, pens

## 22   Remarks

Creativity techniques are a way to help people use their imagination and break free from the constraints of their usual work environments. Many of these techniques use a supervisor who explains the setting and helps the participants through the process. The supervisor is essential to the success of the technique. It is his/her responsibility to engage the participants in the process and to encourage them, while not coming across as overly didactic. This is a fine line, which explains why creativity consultants are in such high demand.

As the techniques show, creativity is not a rare gift only a few of us possess, but the outcome of a certain situation we are in. It is a rather childlike playing with thoughts, an open engagement in the exchange of ideas that stirs our imagination. Creativity techniques are a way of creating a playing field for imagination and yet there is something oddly contradictory about the idea of creating rules on how to be creative. As such, the presented techniques should be considered as guidelines rather than rules. Following the rules of a creativity method precisely might lead to the exact opposite of what it is supposed to achieve.

# A Design Perspective on Sustainable Innovation

Markus Kretschmer

## 1 The Long Way from Artistic Creativity to Sustainable Design

Design, as a profession, has come a long way from artistic creativity to strategic problem-solving. Based on a self-conception that is strongly influenced by arts and craftsmanship, the history of design is inextricably linked to the industrial production processes, entrepreneurship, and individual consumption. The power of design is steadily growing due to its lasting influence on consumer preferences and thus has a tremendous impact on key interdependencies of modern economies. Despite the many positive aspects of this influence of design, phenomena such as the climate change also indicate that our industrial product culture with all its designed artifacts has very decidedly evolved into a massive global problem with far-reaching negative consequences for all of us.

Issues such as pollution and scarcity of resources affect design at its core as some of the visionary designers started recognizing by the end of the 1960s. Back then, the first conceptual problem solving processes were developed as an answer to the ongoing waste of resources by products. Viktor Papanek's groundbreaking book, Design for the Real World, published in 1971, portrays design as 'one of the most harmful professions'. For a long time, little attention was paid to many of the progressive approaches that today are incorporated in the concept of *sustainable design*. Even though the environmental consequences of mass consumption could already be foreseen back in the early 1970s, a repositioning of the design profession did not take place at that time – "One ignored or denied that design ever has any responsibility, least of all a social one" (Rams 1994a). Later, and closely linked to the reflection of the entire life cycle of a product (life cycle thinking), the concept of

M. Kretschmer (✉)
University of Applied Sciences Upper Austria, Innovation and Product Management,
Stelzhamerstr. 23, 4600 Wels, Austria
e-mail: markus.kretschmer@fh-wels.at

ecodesign was developed. Ecodesign describes a systematic approach which aims to incorporate environmental considerations in the process of product planning and design development as early as possible (Tischner et al. 2000). Consistently applied ecodesign ultimately reduces the negative environmental impacts of products. Since the 1990s, various methods and tools emerging from the concept of ecodesign have found their way into the early stages of product development and product design to minimize the consumption of resources and the environmental impact caused by products. Thus the interest in resource and environmental protection has left traces in design – although often too superficially, and sometimes even too radically as demonstrated by the cradle-to-cradle concept, which is based on a complete circuitry of all materials used (Braungart and McDonough 2008).

In the four-stage model by Charter and Chick, basic strategic approaches to achieving progress regarding resource and environmental protection are described: Re-pair, Re-fine, Re-design, and Re-think (Charter and Chick 1997). *Re-pair* describes the use of end-of-pipe solutions, mainly for technically driven measures of environmental protection at the end of an existing process chain. *Re-fine* aims for improving eco-efficiency of existing processes, for example, through a production process that is less harmful to the environment. *Re-design* strives for the design of products that are environmentally friendly (mainly by methods of ecodesign) and thus resource-efficient, recyclable, durable and free of pollutants. The highest level of the model that Charter and Chick describe is *Re-think*. In addition to merely improving the product, Re-think searches for new strategies for the development of systemic infrastructure to improve the flow of energy and material resources in production systems. On the one hand, this model indicates a development from end-of-pipe solutions towards holistic-systemic innovations. On the other hand, it shows that in order to minimize the environmental impact of products, not only the product itself must be rethought, but also the higher-level systems and frameworks in which the product is embedded. The highest level within the model signifies that the three dimensions of social, economic and environmental sustainability must ultimately be transferred to the entire designed product environment. It is obvious that this is no easy task and will require an immense effort on the part of those in charge of designing a truly sustainable product culture, because many contradictory factors must be considered and balanced: First, at the very least, social sustainability is to ensure equivalent living conditions for future generations compared with today, and must generally minimize existing inequalities. Second, environmental sustainability is to practice environmental protection by reducing emissions in production, by not wasting precious resources, etc. Last, economic sustainability is to establish economic strength (such as that of a company or a region) that can be maintained permanently.

In the last few years, social issues have increasingly gained importance: Fairtrade and ethical issues have already become factors of success in the food sector. But similar developments can also be observed in the field of products. Thus, low social standards and relatively poor working conditions at *Foxconn*, the world's largest contract manufacturer of consumer electronics, are an issue with large repercussions in the public. It is not mainly the negative impact on Foxconn itself, but especially the impact on companies that have their products manufactured there. For example, one of these companies is *Apple*, whose brand image is beginning to suffer because of such issues, among others. Another issue that is becoming more and more relevant for design and for *sustainable innovation* is the empowerment of communities or individuals by collaborative design methods and emerging technologies. Signs of this evolution can already be seen in replicating machines, such as the self-copying 3D-printer RepRap, or in designs that are available for download for free. In line with the 'democratization' of certain steps of the innovation process, our standards for assessing social and economic sustainability also begin to adjust to these changes. Because, isn't a product that individuals can design and produce on their own perhaps the most socially sustainable one we can imagine?

Unlike the solely ecological discourses that took place from the late 1960s through to the end of the twentieth century, the contemporary discourses about environment, climate and sustainability (and also the discourses on design) focus less on the pure product level and more on fundamental systemic changes (i.e., from the use of products to the experience of product-service systems). All of these discourses are ultimately driven by the search for sustainable mitigation and adaptation strategies related to the global climate change (Huber 2001; Minx and Kollosche 2009). In addition to the emergence of the predominantly product-related ecodesign, more holistic approaches are coming to the forefront, focusing much more on social, ethical and economic factors of product life cycles – the global vision of sustainable development is translated into the product level.

In the context of sustainable development, it is not surprising that over the last decades design has moved away from end-of-pipe-solutions ('beautification' and improvement of products) towards more holistic and systemic approaches (*strategic design*) (see Fig. 46). Many examples of such a systemic understanding of design can be found in the history of design. A particularly prominent example is *Apple*'s strategic approach to design – although Apple certainly has many deficits as regards genuine sustainability. The success of the Apple design is based on different levels of design. The first level of design is the only level which concerns the product itself. At this level, Hartmut Esslinger and later Jonathan Ive combined some of Dieter Rams' 'functional' abstractions of the Braun 1950s and 1960s designs as well as Mario Bellini's and Ettore Sottsass' avant-garde Olivetti designs with their own interpretation of high-tech design. The second level of design is about the system. Apple has long followed the end-to-end system idea as reflected in the concept of the Apple iPod, which is successful not simply because all individual elements of design and use work together in a comparatively optimal way, but because all of the elements of the overall systems were designed for music listeners. The product for playing back music (the iPod) and its interplay with other products (Mac and PC platforms) as well as the concepts of buying music (iTunes store) and using music (iTunes) were all optimized and carefully designed. As the ongoing success of the iPod shows, this interaction not only creates a holistic user experience, but also high added value from the viewpoint of the user. The third level of systemic design is about the perception of the whole system. Apple's

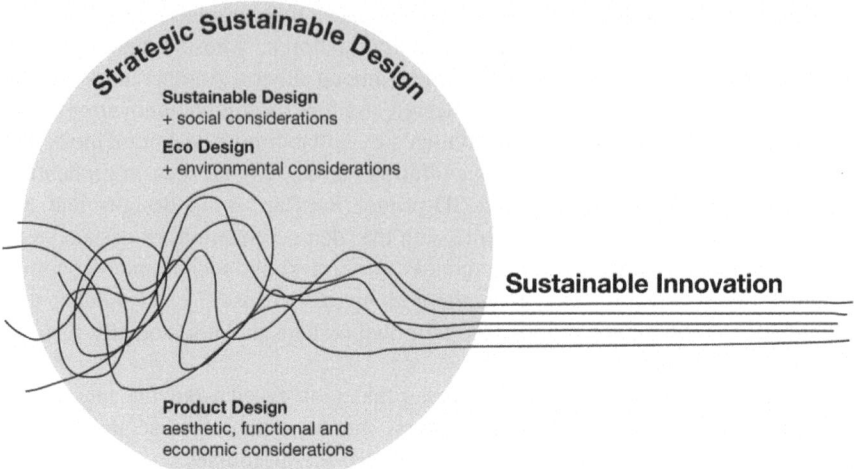

**Fig. 46** The different layers of strategic sustainable design in the fuzzy front end of sustainable innovations

consistently communicated concept of the digital lifestyle has become the DNA of the company; and therefore, the success of the Apple design cannot be limited solely to the esthetic, semantic aspects of conventional product design, but must be seen as the optimized interaction of the three levels of design along with the consciously designed user experience. Lucius Burckhardt once called this a design of tomorrow that is able to consciously consider the invisible complete systems (Burckhardt 1995).

In order to successfully help shape an era of sustainability, an expanded systematic understanding of innovation is essential in design, and not only in design. As in the past, the term innovation is still strongly marked by the notion that progress is primarily something technical and particularly always associated with new products. A much more holistic understanding of innovation is urgently needed: one that should not remain limited only to the product level, but that must also include the cultural level (since, at the very least, sustainable development concerns cultural transformation processes). Such an understanding of sustainable innovation is not just limited to products and their design, but extends to human needs, to ways of satisfying them, and, if appropriate, to transforming them into artifacts. As described above, all three levels of design must be consciously shaped. Thus, sustainable innovation involves societal system innovations, usage innovations, organizational innovations, etc. Even product innovations may – but need not necessarily – represent the best solution to a problem. This knowledge should guide the beginning of any design process: it should be anchored in the front end of innovation. After all, this is the only realistic chance of establishing reduced resource consumption by fewer products. Or, as Dieter Rams puts it with reference to a sprawling and increasingly haphazard environment of goods: "Less, but better!" (Rams 1994b).

Some fundamental dilemmas arise for design if it is to shape an era of sustainability: Especially in saturated markets, design has been explicitly assigned the task of keeping consumption levels high or even increasing them by creating more new products. It is obvious that with regard to sustainable development this role is extremely questionable. This is the first dilemma of design in the context of sustainability. The second is that design is indeed fundamentally integrated into the structures of complex globalized product origination and even dependent on them, but due to its traditional role as 'beautifier and provider of ideas' has little to no influence on these structures, which actually are rarely sustainable. The third dilemma lies in the fact that entrepreneurs and designers still do not adequately recognize the potential that a strong strategic alliance between entrepreneurs and strategic design has for shaping a sustainable future. The approach toward a sustainable future still primarily centers on questions of purely technical efficiency and effectiveness, but fails to make sustainability as a concept actually come to life.

With its basically esthetic, artisan nature and its strong dependence on industrial production and technically driven development processes, the common conception of industrial design seems to have reached its limits and is consistently losing influence in the face of global problems that arise from a non-sustainable product culture. The commonly accepted role of design as a cheap source of ideas, as a profession of beautification, and as a powerful marketing 'tool' must therefore be questioned seriously and urgently. Too often, and especially for reasons of differentiation in largely saturated markets, design is still used as a mere 'tool' – as Raymond Loewy described already in the 1950s: "Between two products equal in price, function and quality, the one with the most attractive exterior will win" (Loewy and Weseloh 1953). However, esthetics as a selling point was used strategically much earlier. In 1927, *General Motors* introduced the annual model change – and the new concept of brand hierarchy – to boost vehicle sales. The key aspect of this strategy known as 'artificial aging' or 'planned obsolescence' clearly was a noticeable visual change between the individual production years ('style obsolescence') – despite only minor technical changes. The secret of success is that repeatedly renewed products make the predecessor appear outdated and obsolete from the consumers' viewpoint, without this actually being so in objective terms. The purely esthetic and semantic aspects of this strategy ultimately reduce the impact of design on the product to fashionable and stylistic statements. Combined with appropriate marketing activities, this strategy provides a simple, yet powerful means of promotion. Over the last six decades, in a constant flood of newly created product features and product types, design thus has mutated into probably the biggest and most powerful differentiator in the market.

The reduction of design to purely esthetic aspects that is associated with 'style obsolescence' may still be economically beneficial for some companies today. If, however, strategic design is recognized as an important catalyst for a sustainable future – which is urgently required – one can no longer proceed with such a waste of creative resources. Instead, creative talent must be used selectively to build a more sustainable product culture. Viewed realistically, we are still far away from such a culture, but more and more companies realize that a consistent focus on the principles of sustainability also makes economic sense. Across all industries, sustainability is becoming a success factor. But such 'neo-green' activities performed by companies have little to do with the 'deliberate restriction' of the 1980s and 1990s (Nachtwey and Mair 2008). One reason for this is that *corporate social responsibility* (CSR) as a holistic business concept is now in the focus of many companies' corporate decisions. Another reason is that consumers are now much better informed, not only about the products themselves, but also about systemic relations (Carbonaro 2008).

For large companies such as *BMW*, the continuing commitment already pays off. Relatively early BMW began to develop and communicate a comprehensive sustainability strategy, to embed strategic design in the organization, and to invest in environmentally friendly production and energy-saving technologies. Consequently, BMW established a sub-brand for electric mobility early on. As a result of all these efforts, BMW has consistently been ranked as the most sustainable automobile manufacturer since 2005 and has the highest brand value in the industry. Another example is the US-based furniture manufacturer *Herman Miller*. Herman Miller not only has a long tradition of cooperation with pioneering designers (including Charles and Ray Eames and George Nelson), but since the early 1990s has also been one of the most innovative pioneers in the industry regarding activities of environmental and social sustainability (Braungart and McDonough 2008). In products such as the Mirra office chair, the consistent implementation of environmental criteria in product design becomes apparent: The Mirra chair was consistently developed according to standards of ecodesign. According to Herman Miller, it is made of 42 % recycled materials, and 96 % of the chair can be recycled.

Just as BMW has become the most sustainable automotive brand, Herman Miller understands sustainability as part of its long-term corporate strategy. *Puma* has the top position as the most sustainable brand in sport and lifestyle. One element contributing to this rank is a packaging system for footwear, marketed as the 'Clever Little Bag'. It is made out of a flat piece of cardboard and a bag made of recycled PET. According to Puma, approximately 8,500 t of paper are saved per year, which represents 65 % of the paper used for shoeboxes annually. Furthermore, 20 million mega joules of electricity, one million liters of water, and one million liters of fuel are saved per year. Compared to traditional shoeboxes, Puma promises a decrease in the overall $CO_2$ emissions of the company by 10,000 t annually. The 'Clever Little Bag' is one of many contributions to a fundamental change taking place in industry and the economy (Wirtschaftswoche Green Economy 11/2010), according to Yves Béhar, whose *Fuseproject* design agency not only designed the packaging system, but actually re-designed the entire transport logistics and distribution life cycle to make Puma a more sustainable company.

It needs to be noted that examples such as the Puma shoebox do not mean a change of paradigm towards a radically sustainable product ecosystem. But such examples show that leading companies are at least aware of the economic benefits of a sustainability strategy which is implemented consistently. They have recognized that successful corporate sustainability strategies require strategic design decisions, which pick up on latent user needs. This is first and foremost a matter of corporate culture, because strategic design must always be a serious top management priority. However, above all, these examples show that within a relatively short time leading design agencies have evolved from mere contractors of the R&D departments to accepted strategic partners that define, develop and implement major parts of the companies' business models. A development which, incidentally, also reflects on the strategic direction of leading design agencies like *IDEO* or *frog* – and on the whole field of design. Design truly has come a long way from artistic creativity to strategic influence on sustainable innovations.

## 2  The Contribution of Design to Sustainable Innovation

Sustainable innovation is something other than just green products that sell well. For a start, sustainability – and *sustainable innovation* – is an abstract concept which is difficult to understand due to a complex interaction of conflicting aspects. It has to be transferred to coherent artifacts and routines to become internalized by people.

In the history of mankind, nonverbal communication of information (for instance communication intended to emphasize status) has been an important function of all man-made objects. At least in the context of industrial mass production and consumption, the communication function of products is designed deliberately as it is often the most important, if not the only, differentiator in the battle for market share. Moreover, the character of products always corresponds with fundamental notions of the culture in which they have originated: While, for example, in pre-industrial times most products had to be extremely durable, this is not essential in a product culture which is essentially characterized by mass production. Products are always media for communicating cultural information. They indicate fundamental values and norms of a culture. Also man-made institutions, like products, are always media of cultural information. The sociologist Lucius Burckhardt describes a hospital as an institution that is first and foremost a system of relations between people (Burckhardt 1995), and as such communicates cultural norms of human coexistence.

Design has a severe impact on shaping cultural values and norms, especially via its influence on artifacts (i.e. design at the level of the product), and on institutions (i.e. design at the level of the company). In the context of sustainable innovation, the greatest potential of design lies in influencing the perception of sustainability, thus making sustainability as a concept actually come to life. Finally, it has always been the mission of design to give products a character, a cultural relevance and importance. Over a period of more than 100 years, design gave the era of industrial modernism a consistent face. And there is no reason to assume that it would be unable to give the era of sustainability that has just begun an even more convincing and charming face. In fact, there can hardly be a more exciting task for designers.

Products obtain cultural relevance through the history of their origins, through the use of cultural symbols, and thus, through their innate 'invisible' qualities. The entire history of the origins and use of a product is decisive for a large fraction of its cultural value, its importance in a culture, and also its ability to influence a culture. Just think about the sentimental value of heirlooms. As many products are becoming more and more equal, both technically and esthetically, this 'hidden history' will become increasingly decisive for competition, and today it is necessary to rise above the competition by the manner and style in which the business is operated, or in other words by the how (Friedman 2008). Consequently, the customer's trust in the system, beyond that of the product itself, is gaining ever more weight in the competitive arena. This development is also related to the fact that the users of a product usually have no say whatsoever in defining the product and the method of its production. However, in order to change to a sustainable product culture, joint determination by the people for whom products are created is an unconditional prerequisite, since: "No artifact can survive within a culture without being meaningful to those that can move it through its defining process" (Krippendorff 2011, p. 413). Therefore, the contribution of design to sustainable innovation cannot be restricted to the product level, but must consequently include the redesign of the involvement of the people in the product development process. Design must be a catalyst of technology, economy, ecology, and social sustainability to establish processes of product development, production and use which are culturally sustainable in a consistent way. For this purpose, design depends on close cooperation with the users already very early in the fuzzy front end of the innovation process. During this phase, new ways of networking with customers and users are needed to enable them to contribute to sustainable innovations. A promising approach consists in *crowdfunding*, in which many money lenders support a particular project that they want to see implemented, usually with relatively small sums and via social media platforms. Crowdfunding, which at first appears to be no more than a form of financing, opens up completely new opportunities for the cooperation of designers and users, for the identification of the stakeholders' needs, and for the establishment of shorter and more regional value chains. Nevertheless, internet portals that merely offer customizable hand-knitted woolen hats can be considered just the tip of the iceberg. 3D-printing will be the 'next big thing' that will radically change the way design is implemented in the early stages of innovation processes. In this respect, 3D-printing will open up completely new possibilities for sustainable innovations, because it will make the stakeholders more independent of the structures of industrialized mass production, which only rarely are truly sustainable. Perhaps this will then constitute what the author Ivan Illich called conviviality: He once chose this term 'to designate the opposite of industrial productivity' and considered it to be 'individual freedom realized in personal independence' (Illich 2012).

Many aspects of the traditional model of production and consumption have to be rethought in order to establish a sustainable product culture. This is an essential contribution of design to sustainable innovation. It is not enough that a few specialists have proposed solutions for environmentally friendly products. Instead, strong partnerships need to be forged with designers, managers, executives, users, and even with politicians in order to design a more sustainable product culture.

Sustainable products and business models are already absolutely crucial to success in many cases. Beyond the transition from a production-oriented to a service-oriented economy, companies are also facing the challenge of having to rethink their business strategies fundamentally and develop innovative service-based business models (Hamel 2002). A few years ago, *Daimler* launched the car-sharing project Car2Go as a user-friendly mobility service ('share vehicles instead of owning them'). In addition to significantly changing attitudes of many young people regarding ownership of a vehicle, the problem of private transport in the public space also contributes to the acceptance of this service. But the most important strategic reason for the development of Car2Go is the desire to reduce the dependence of the car manufacturer Daimler on its own traditional business model – and the fear of disruptive innovations from inconspicuous start-ups. Even though many traditional business models still work fairly well, the development of innovative, service-based business models that rely on fewer resources will become a key factor for future business success (Hamel 2002). Strategic design plays a crucial role in this respect, as it can create not only products, but also a holistic user experience – if design is strategically implemented in the fuzzy front end of innovation processes, and if it is not considered a disturbance to the engineering department.

Companies and their business models play a central role in the design of sustainable innovation. The traditionally close relationship between entrepreneurship and design – just think of all the successful 'entrepreneur-designer alliances' in Italian design since the 1950s – is both an opportunity and an obligation for the design of sustainable innovations. However, the full potential of this alliance for the design of sustainable innovation is often utilized only in a very rudimentary way. In the face of massive global challenges, the willingness of companies to anchor strategic design in the innovation process early and comprehensively often appears poorly developed.

Obviously, many companies have to develop creativity as another core competency in addition to their well-developed analytical skills. All too often, managers fail terribly when it comes to appreciating the intangible cultural potential of brands and even of companies. In this way, plenty of the potential for sustainable innovation is wasted, because a company's success can only be secured permanently through design concepts that fit into the corporate culture and business strategy. However, unfortunately there are too many cases that illustrate that this fact was neglected. The cultural decline of the Swedish car brand *Saab* is a striking example. Even the iconic *Braun* design has not been immune to such mismanagement, which demonstrates that a deep understanding of cultural contexts and of the intangible strengths of a product, a brand or a company are absolutely necessary for successful and sustainable innovation. In this case, efficiency is only one half of the story. The other half, or rather more than half, of the story is empathy, creativity and – of course – strategically applied design. This demonstrates that, in the case of brands, a sustainable path can only be followed successfully in harmony with the

**Fig. 47** The three decisive skills of designers

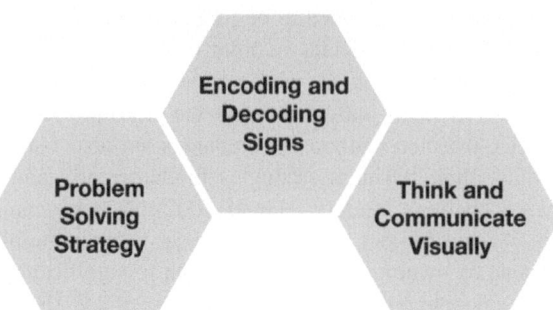

historic core of a brand, its intangible values, and their incarnation in the brand's design and processes. The complex challenge is that far more creative talents have to be deployed in order to create a truly sustainable product culture, and not just to ensure the evolution of a brand.

Owing to their education and predisposition, designers usually have the cultural-creative potential and the capacity for network thinking and for developing holistic solutions which go far beyond esthetic, artistic questions. Designers are also much better qualified than most other professionals to solve complex problems (Cross 2007) and deliver creative approaches 'off the cuff'. A holistic mindset and the creative artistic skills that designers master are fundamentally necessary to designing truly sustainable innovations. Design education imparts three decisive skills (see Fig. 47) that are essential to the development of this mindset and abilities: The first decisive skill of designers is the mastery of a particular kind of *problem-solving strategy*. The kinds of problems that designers have to solve are often referred to as 'wicked problems'. These design issues are essentially characterized by the fact that neither the causes of a problem, nor the way in which the problem is solved can unambiguously be identified as 'true'. The rigorous change of consumption patterns and consequently that of cultural values necessary to achieve sustainable development is such a 'wicked problem'. For solving 'wicked problems' efficiently – one might even say for solving complex problems with an extremely large number of variables – during their training designers need to learn appropriate problem-solving strategies that are fundamentally different from those used to solve scientific problems. Characteristically, these strategies are not primarily based on verbal, numerical and writing skills, but mainly on visual skills of thinking and communication. This clearly distinguishes the strategies relevant in design from those helpful in the sciences and humanities. Hence, the design process usually appears unstructured, arbitrary and difficult to understand to persons with a scientific or rational predisposition.

The second decisive skill of designers is the mastery of *encoding and decoding the signs of a product culture*. In simple terms, one can say that designers must be able to read and write. But of course these capabilities go far beyond the writing and understanding of texts. Rather, 'reading' and 'writing' refer to the competence to decode and encode the complex sign systems that surround us. This competence

# A Design Perspective on Sustainable Innovation

entails far more than the purely visual sign systems that have traditionally been at the center of the education in esthetic-artistic design – culture as a whole must be understood as a complex system of signs (Eco 2002). Design must see its role in helping to actively shape such a system of signs in order to become a driver for 'setting new realities' (Bonsiepe 1992). In the course of their training, designers learn a kind of language which enables them to decode the signs of culture, thereby detect human needs, and translate them into solutions in the form of artifacts (Cross 2007). The third decisive skill of designers is the mastery of *thinking and communicating visually*. Similarly to the world of numbers in math, which is necessary to process and solve mathematical problems, the world of three-dimensional prototypes and products is necessary for communication and for solving complex problems in design (Cross 2007). Thus design not only helps to solve such complex problems using visual representations, but the problem-solving strategy also helps to communicate the sustainable solutions very clearly to others. This is of immense importance to fostering sustainable innovation, and, therefore, these skills are to be welcomed so that methods and processes that are characteristic to design can be – and in fact are already being – applied to other fields of expertise: 'Rational' managers and executives experiment with design thinking as a process of inspiration. Universities such as Stanford, St. Gallen, or Potsdam have already started programs on design thinking. However, the results are still limited because of a lack of cross-disciplinary integration – e.g., with creative talents – and a lack of perspicacity regarding sustainable innovation. Nevertheless, the holistic approach of strategic design and its decisive methods have apparently already begun to change other fields of expertise.

## 3 The Perspectives for Sustainable Design in the Fuzzy Front End

By common definition, an ideal innovation process ends with successful market penetration and market success (Herstatt and Verworn 2007). However, for the definition of successful sustainable innovation this description is not enough, because the success of sustainable innovations cannot be measured solely by market share, but must also be measured by their impact on the three dimensions of sustainability and on sustainable development in general. These impacts of an innovation, categorized as social, environmental and economic, are usually determined in the defining early stages of the innovation process. As design usually occurs in these phases, decisions in design have a significant effect on sustainable innovations.

The fuzzy front end of innovation is characterized by high uncertainty regarding the outcome of the project, by its non-linearity, and the difficulty of structuring tasks (see Fig. 46). Probably the biggest challenge for design in the fuzzy front end lies in envisioning and designing a desired future on the basis of little – or even no – information about the driving forces behind the actual future context. Creative talents, such as designers, who are involved in the early stages are able to deal with these challenges. Moreover, they are used to them and are thrilled when asked to act with such creative freedom, their major motivations for action simply being the creative challenge, the opportunity to create something new, and the opportunity to contribute to the improvement of general or specific circumstances (Esslinger 2012). Designers flourish under such conditions. Amazingly, uninspired individuals often cannot stand this creative uncertainty of the fuzzy front end at all.

Perhaps this is a major reason why design activities in the early stages of innovation processes often receive little appreciation: Many managers and executives still do not understand why designers should be properly paid for their 'creative romp'. This perception is even more annoying if one realizes that designers carry responsibility for millions of products and billions of Euros. Rather than being a creative service of no merit, design is a strategic discipline with far-reaching consequences for the entire life cycle of a product – and beyond that for our product culture in general. Designers – as creative generalists or as specialists – can make crucial contributions to the success of an innovation process. The great abundance of creativity provided by creative talents such as designers is especially required for successful sustainable innovations, because in many cases uncharted territory must be entered after conventional solutions have obviously failed. As one of the leading researchers on creativity, Mihaly Csikszentmihalyi, points out, creativity is the ability to create something truly new that is considered so valuable that it is added to the culture (Csikszentmihalyi 2007). The value of creativity and design should be measured accordingly. The driving force for achieving sustainable development is therefore creativity – and not efficiency.

Despite its enormous influence on the success of an innovation process, in most innovation process models design is treated as rather subordinate. By contrast, the process model of integrated innovation management and product management described by Gaubinger (2009) recognizes the significance of design management as a third core process of innovation management, in addition to R&D-/technology management and marketing management. This model also emphasizes that all activities in a company that are related to design play a (co-)critical role for the success of innovation management. In this respect, design has become an integral factor in the innovation process and is critical to the company's success.

Presently, design must be seen in a larger context of innovation, just as sustainable innovation cannot be limited to product innovations, but must clearly affect all types of innovation. In such a larger context of innovation, design is a core component of sustainable, enterprise-driven innovation processes. It significantly contributes to the success of the early phases of innovation – and thus to innovation success in general. As can be seen in many examples, ranging from collaborative design methods to fab-labs (Gershenfeld 2005), a new culture of design is already emerging. This shift may strongly affect the role of designers as specialists who merely determine the appearance of the products. Because it is evident that in its traditional role design can only be incremental and not have much of an impact. The emerging exertion of influence on more than just products is exactly what the designer Tim Brown means when he urges designers to 'think big' again and to tackle systems solutions instead of proceeding with mere product design.

In this position, strategic design can play an important role in establishing a common understanding of sustainable innovation and in creating a path towards sustainability using visual and tangible representations of a desired future. Design helps to communicate a common vision very clearly to others. Strategic, sustainable design enables transferring the CSR strategy into the DNA of the products and the processes of the company or organization. If the field of design clearly positions itself as a communicator of this vision, design can take on the role of a change agent for sustainable innovations. Only then will the function of strategic design go far beyond pure design management (the management of all design-related activities during the innovation process), and design will become a catalyst of technology, economy, ecology, and social sustainability.

# Part II
# Practical Cases

# 3M: Beyond the 15 % Rule

Stephan Rahn

## 1   3M's Logic of Innovation: Knowledge, Ability and Desire

The invention and sales of sandpaper marked the beginnings of *3M*, which was founded in Minnesota, USA, in 1904. Today, 3M is still considered one of the world's most innovative companies. Each year, 3M brings to market more than 1,300 new products, the company holds 45 technology platforms and nearly 27,000 patents and generates nearly $ 30 billion in annual sales.

At 3M, all employees can spend 15 % of their work hours on *dream time* – time to play with their own ideas and projects that have no relation to their current projects and tasks. This elementary rule is only the starting point for the logic that builds the very distinct and specific *innovation culture* at 3M. The decisive parameters for 3M's innovativeness are the factors knowledge, ability, and desire. 3M understands *knowledge* as the ability to combine technological competencies with specific market opportunities. Most of all, it is a matter of identifying *Customer Pain Points*, which are then used as the starting points for developing innovative new products. Customer pain points are perceived, articulated or non-articulated, weaknesses of either product performance or manufacturing performance. Possible solutions are created by granting the company's 7,000 researchers and development engineers unrestricted access to all of the company's 45 technology platforms, which are not considered to be the 'sole property' of any particular division. In this way, the company's greatly diversified technological capabilities can repeatedly be recombined in new ways in order to develop an innovative product specifically tailored to the customer's particular needs. Of special importance in this respect are the company's 'platform technologies', such as 'microreplication' – which refers to the creation of tiny, precisely shaped, three-dimensional structures arrayed to alter the physical properties of a surface. These

S. Rahn (✉)
3M Deutschland GmbH, Carl-Schurz-Str. 1, 41453 Neuss, Germany
e-mail: srahn@mmm.com

structures make it possible, for example, to enhance the light-guiding effect of films, but can also be used in abrasives to greatly improve the cutting performance and service life of the products. Microreplication technology is found in a wide variety of 3M products intended for the most diverse markets and applications. This example illustrates particularly well that the utilization of existing technological competencies across divisional borders is one of the decisive aspects of 3M's logic of innovation. 3M devotes a great deal of energy toward determining the customer's pain points. The company has more than 40 *Customer Technical Centers* worldwide, which are highly application-focused development labs. Here, customers can experience cutting-edge products hands-on in the context of real-life, 'demo lab' applications. It thus comes as no surprise that the Neuss-based Customer Technology Center of the German 3M company alone welcomes 7,000 visitors every year. Above all, this dialog-oriented manner of maintaining customer contact serves to discover requirements and needs which the customer himself is often not able to clearly articulate. In order to identify future areas for innovation, 3M has even established its own separate organization named *3M New Ventures*. With a reporting line directly to the CEO, one of the primary tasks of this organization is to find promising young start-ups, in which 3M then invests.

The factor ability describes 3M's process competency – particularly with respect to its innovation process, which employs a stage-gate procedure that structures the path from the initial idea all the way to final marketing of the new product. The goal of this process management is, above all, to reduce the time to market, to ensure market success and also transparency of the entire pipeline. But even more importantly, all project ideas must be subjected to a constructive 'stress test'. The criteria for meeting the requirements in the stage-gate reviews are to be so stringent that, in accordance with the 'kill early – kill cheap' principle, the greater number of projects do not enter the cost-intensive development phase.

3M understands *desire* to mean the direct effect the company's *corporate culture* has on the innovation-relevant attitudes and conduct of its employees. The goal in this context is to create a climate that promotes a continual desire for new developments and a creative restlessness.

## 2  Innovation-Promoting Corporate Culture

A corporate culture can either foster or stifle innovation. This is exactly why the term 'culture of innovation' is misleading, because it suggests that, in addition to the general corporate culture, there is also a kind of 'subculture' aimed specifically at increasing the employee's innovation performance. This perception alone clearly illustrates how complex the network of cultural characteristics relevant to innovation is. Employees perceive the atmosphere within a company in its entirety. And these very different impressions and experiences must be consistent in their overall array. Therefore, it does not make sense to promote risk-taking on the one hand, when management, on the other hand, is strictly focused on avoiding mistakes at all costs. By the same token, creative freedom for new ideas would be fruitless, if the resulting ideas were stamped as 'nonsense' already during the early phases of the innovation process. It is ultimately a question of the overall perception of how a company works, beyond any individual, isolated 'cultural artifacts'. The crucial factors are the liveliness and unrestricted freedom of the corporate culture. This must be of universal relevance to every employee, and people must be able to feel it in their daily work. They must be able to rely on this culture, and it must be highly stable over time. Only then does it have a chance of being considered by all employees as the 'natural' way of working and living within the company.

## 2.1 'Courageous Decisions'

One example is 'courageous decisions' – which is one of the indispensable prerequisites for a dynamic generation of ideas. This aspect is even included in 3M's employee performance reviews. The performance of all 85,000 employees is assessed yearly based on six so-called 'leadership attributes'. One of these performance attributes is 'makes courageous decisions'. De facto, this means that every employee is judged based on the ability he/she shows for taking such decisions. This, of course, is always put into perspective against the background of the particular person's job profile. However, the concept encourages employees to pursue even those ideas which appear to be very risky at first glance.

## 2.2 The 'It's Up to You': Principle

"3M does not have just one innovation manager – it has 85,000!" Though somewhat exaggerated, this claim reflects that 3M has an entirely unique understanding of the responsibility for innovation. One could describe it as the 'it's up to you' principle. Anyone who has a new idea for an innovative product should also be able to champion this idea personally. By promoting this principle, 3M eliminates the often paralyzing delegation of responsibility for innovation. No innovation manager, no innovation department, no suggestion box – every employee is to feel him/herself personally responsible for new ideas, innovation and growth. Functional silos play virtually no role in daily work life. Anyone who contributes new ideas, regardless of whether they are related to his/her regular work area, is greeted with esteem, respect, and support. And 3M does not suddenly relieve the originator from his/her responsibility and delegate the matter to someone else. He/she remains the champion of his/her idea – which also applies to the meetings and reviews held throughout the entire innovation process.

## 2.3 'Strength-Focused Leadership'

3M has an employee leadership approach that focuses exclusively on the strengths and talents of its employees. The logic is persuasive – it makes little sense to make enormous investments in training in order to turn an employee's supposed weaknesses into strengths. This can only make weaknesses maybe a little less weak. True strengths, however, can never be developed through training courses. It makes much more sense to identify existing strengths and talents, and subsequently to utilize these to the best interest of the company. This is exactly one of the tasks for which managers are responsible. Together with employees, they attempt to work out where the employee's particular strengths and talents lie. In case the current position does not optimally fit the employee's strengths, it is usually easy to find other job profiles within the company that are much better suited to the employee's capabilities. This form of leadership supports the feeling of honest respect and appreciation, while also opening up new possibilities, such as giving especially creative employees the opportunity to use their strengths directly for innovative ideas.

## 2.4 Communication and Networking

Cross-functional networking is a catalyst that is especially important for new product development. There is hardly ever an important meeting at which employees from the most diverse functions are not discussing a new development project. Only by taking advantage of the various perspectives, experience and expertise brought to the meeting by the different employees do the opportunities for true innovation come into view. Readiness for networking and the ability to communicate are, therefore, an integral part of the job profiles for 3M employees. And 3M always finds new ways to put this willingness for networking into practice. For one thing, several 'Open Lab Days' are organized every year, at which a particular lab is given the opportunity to present its latest development projects within the scope of an in-house trade fair. These events often form the basis for cooperation on new product ideas – initially on an informal basis, and often spanning across lab borders. It is exactly this kind of networking and exchange of knowledge and experience that makes up the fertile soil for new ideas.

## 2.5 Passion

'Passion fuels innovation'. It is in this vein that 3M promotes the passion with which '3Mers' devote themselves to new product ideas. Already when recruiting new employees, 3M signals this stance through its employer brand claim of 'diversity and passion'. In their interviews, recruiters pay very close attention to whether the applicants show heart and commitment. How else will they later be able to stir excitement for their innovative ideas and to win others as fellow combatants? Incidentally, this desire for passion is also reflected in the leadership attributes, with 'leading with energy and passion' being one of the criteria of the worldwide employee performance review.

## 3  Balance of Fostering and Demanding

The characteristics of the company's corporate culture delineate 3M's vision for its employees. Personal development and esteem, freedom and building on strengths are the vital basis for striving for the spirit of innovation. However, these 'fostering' elements of the corporate culture must also be accompanied by a goal-oriented organization that provides rules. Clear rules of business conduct, measurable innovation performance, and growth goals are demanded of the employees. Personal freedom thus unequivocally reaches its boundaries where the high ethical standards of the company could be impaired. Innovative performance is consistently measured against the 'new product vitality index'. The activities in the innovation process are transparent at all times in order to enable corrective intervention. Clearly defined sales and margin expectations prescribe goals that are binding for everyone. Only by achieving a good balance between the freedom provided by the corporate culture, on the one hand, and setting performance and conduct goals, on the other hand, is it ultimately possible to ensure the lasting and healthy innovative performance of the company.

# ABB: Integrating the Customer

Patricia Sandmeier Kahmen and Petr Korba

## 1 A Market for Energy Storage Systems

Governments worldwide are focusing on reducing carbon dioxide ($CO_2$) emissions and on increasing the utilization of renewable energy sources – e.g. wind and solar energy – to meet stringent global environmental targets. According to the *International Energy Agency* (IEA), the portion of renewable, $CO_2$-free energy will increase to as much as 15 % of global power generation by 2035 – in 2009 its share amounted to 3 %. As a result, power generation is becoming more and more weather dependent and intermittent, and correlates less and less with the actual power consumption needs. This inherently leads to an increasing demand for storage systems as a possible solution to the evolving energy landscape. One such solution is based on the integration of battery energy storage systems into the grid. As no viable business case has been established for battery energy storage so far, *ABB* set up a project to develop a prototype and new applications with the intention of analyzing and testing the new product in cooperation with a pilot customer in a real-world environment (Coetzee et al. 2012; Korba et al. 2012). This customer integration project served to explore the potential of the energy storage market.

The integration of the customer in each of the process phases was vital because the customer possessed broad know-how as a grid operator. This knowledge was essential to guaranteeing that the system development is in line with the market needs in each phase.

---

P. Sandmeier Kahmen (✉)
ABB Switzerland Ltd., Bruggerstrasse 72, 5400 Baden, Switzerland
e-mail: patricia.sandmeier@ch.abb.com

P. Korba
Zurich University of Applied Sciences, School of Engineering, Technikumstr. 9, Postfach, 8401 Winterthur, Switzerland
e-mail: petr.korba@zhaw.ch

## 2 The Customer as Initiator

Based on a simple declaration of intent, the project team – composed of participants from both the customer and ABB – faced the challenge of handling a project with a high degree of freedom. The target of the project was to develop a system that optimizes the usage of batteries within electric power systems. The customer suggested the rough framework, namely ideas about the dimensions and power performance of the system, and that a photovoltaic plant as well as an electric car charger should be integrated. Besides this system setting, a battery storage system offers several functionalities that could potentially lead to a business case in their appropriate grid environment. Since there is such a wide variety of these functionalities, the team started with open idea generation for system applications. Approximately 15 useable ideas were generated. Two examples of applications in this context are (1) load leveling by storing excess energy produced when demand and the time-variable energy prices are low and then making it available when demand (energy price) is high, or (2) peak power shaping to help customers avoid high rate charges associated with exceeding contractual supply limits (grid connection fee) (Coetzee et al. 2012).

For each idea, cost-benefit calculations were carried out first. For example, for peak power shaping, the cost of investing in a battery station system and the benefit of retrieving cheaper energy from these sources were compared with the cost of buying expensive peak power energy from external sources. This process revealed that selecting one single application would hardly lead to a commercial business case at this point in time when the battery technology for large commercial applications – e.g. lead-acid, lithium-ion or sodium sulphate – is not mature yet, but will be subject to significant improvement and price reduction due to economies of scale. This led to the decision that the pilot project needed to be dimensioned in such a way that several applications could be combined in a system. As a consequence, the number of ideas was not restricted at an early stage, but options were kept open for a screening and selection process to be performed at a later point in time. In this later selection stage (idea selection in Table 5), four ideas were chosen for realization.

## 3 Tackling Internal Challenges

Internally, the early ABB project team members – until then composed of business developers and corporate researchers – found that they needed additional resources and competencies for the task of framing the general concept in the development departments of the company. It meant that a means had to be found to, as a first challenge, identify the internal experts who were able to design a platform although the final applications had not yet been defined, and as a second challenge, get their commitment to contribute. In ABB, which is organized along business units with specific technological competencies, this is a challenge since business units have a profit and loss responsibility and therefore cannot invest unlimited resources in

**Table 5** Front-end process stages with customer and ABB involvement

| Front-end process stage | Customer | ABB | Outcome of process stage |
|---|---|---|---|
| Project initiation | Based on analysis of near-future needs and requirements and the identification of an ideal partner company, customer contacted ABB with specific intention for prototype project | Positive ABB response. Base: existing internal previous work due to own interest in the topic in the research center | Project officially started identification of system boundary conditions<br><br>Project manager defined on customer side<br><br>Patent search carried out in ABB Research Center to clear starting position |
| Idea generation for target setting | Brainstorming meeting with ABB's early project team members | Brainstorming with customer on new intelligent applications potentially required for optimal battery integration into the power grid | Ideas for possible system applications, customer and ABB fully involved and strongly committed to the project |
| Idea pre-screening | Internal analyses, discussions with ABB | Internal analyses, discussions with customers and potential battery vendors | Insight that final decisions for applications cannot be taken at this stage due to technology uncertainties |
| Idea substantiation | Learning about ABB analyses results, customer's own business case considerations | Identification of additional ABB internal experts/contributors, further business case analyses and technology clarifications. Identification of responsible business unit | Definition of applications to be considered further for implementation by customer and ABB. Internal responsibilities defined at ABB, project management defined |
| Idea selection | Based on substantiated ideas: idea selection meetings with ABB | Based on substantiated ideas: idea selection meetings with the customer | Applications selected for realization – basic system specification defined |
| Start of base design | Workshops with ABB every 2 weeks | ABB in the lead, start of actual development, workshops with customer every 2 weeks | Detailed system specification, adjustment and tuning with customer throughout base design process. Customer strongly involved in decision making process and committed to the resulting product |

developing a prototype for which success and final business unit allocation are uncertain.

Facing the first challenge, namely internally identifying the right people to contribute, a method similar to the process of lead user identification was applied. The early ABB project team members brainstormed which engineers known to them

- Had the expertise in the technological areas that were needed,
- Had the ability to think of a design in an entirely new context where the final functionality of the system was not clear,
- And also had the skill to discuss with the customer team although they were not used to working at the interface to the customer.

Having identified two developers – one in the area of converter technology for battery connection and one control system expert – the original team discussed the general concept to be developed with the newly found experts, but also asked them for further referrals to colleagues they knew who worked in an area where more of the required competencies would be available. In this way, additional contributors could be identified and involved, which brought the team further and closer to the core of the expertise required for the actual start of the development process. It should be noted that at this stage, the individuals contributed through their personal motivation and curiosity – approved by their superiors – rather than by fulfilling a task allocated to them in a top-down decision.

This leads over to the second challenge: finding the resource commitment from a business unit. At this project stage, the top management committed themselves to co-financing the prototype with the customer. It was internally agreed upon that a certain financial amount was to be provided out of a central growth fund to cover internal expenses for hardware, software, the project manager eventually to be identified in the business unit willing to take responsibility, as well as corporate research expenses. There were three possibilities regarding a 'home' for the new system. What helped making the decision with the relevant managers was that the contributors identified were concentrated in one business unit. In view of this knowledge base, the manager responsible for the local unit was convinced of a real chance for success. He agreed – in spite of capacity issues – to assign one of his project managers, a hardware developer, as well as a software manager to the development of the battery energy storage system. A crucial role was also played by the research center. With the assignment of an additional project manager on the research side, the project manager from the business unit had a counterpart and sparring partner in the ABB corporate research center. This project manager on the research side was the one with the crucial power and control systems know-how required to solve the optimal battery integration task by keeping options for the system applications open, and was therefore the person integrating the know-how from the project team.

## 4 The Customer Interaction Process for Energy Storage Systems

During the entire front-end process of this project, the customer was continuously involved. The customer team met with the ABB representatives on a very regular basis. Whereas on the customer side the project manager was defined from the stage of project initiation, his counterpart on the ABB side followed at a later stage once the internal responsibilities had been clarified. The actual specification of the system, which is the first of its kind in this market, was elaborated on with the customer in the base design phase. In half-day workshops which took place every two weeks, the specific details were discussed with the customer, and commitments and decisions were taken. The workshops were initiated by the ABB project manager and the project manager from the research center. In spite of a lack of international standards and proven technologies, the challenging base design phase resulted in a convincing specification.

A summary of the phases during the front end of this innovation project which also illustrates how ABB and the customer interacted is shown in Table 5.

Close cooperation between the customer and ABB and joint efforts of all sub-suppliers helped to accomplish delivery of the system and completion of commissioned work. After the system had been realized according to the customer's boundary conditions and the jointly defined applications, the collaboration continued and an additional advanced battery control system, developed by the ABB Research Center, was tested and implemented in line with the customer's needs. The algorithm allowed the forecasting of local power consumption and generation for the next few days on the basis of historical data, actual measurements and the weather forecast. With this built-in intelligence, the system could be operated at maximum efficiency in terms of commercial and technical aspects.

Besides using the installation for system tests and development after completion, ABB was entitled to visit the plant with other potential customers or other interested third parties. Given the know-how and the competencies built up during development of this prototype system with the customer, ABB has now established the global center of competence for battery energy storage systems in the local business unit that took the lead in this project.

## 5 Lessons Learned and Checklist

For exploring a business case based on prototype projects together with customers, the following aspects should receive special attention:
- Target setting: Given a situation of open targets at the project's start the commitment on both the customer's and the supplier's side has to be emphasized and continually ensured for all decisions to be taken throughout the entire front-end process.
- Defining responsibility: Spending some early resources on clarifying 'how to share the cake before it is baked' is vital.
  - For a supplier like ABB, which is organized in business units that each have a profit and loss responsibility, the question of where a new business case project should be embedded is challenging, but needs to be addressed as early as possible. Only the early commitment of the right business unit ensures that know-how built up in corporate research is integrated in the business in an optimal way and that resource requirements for the prototype project will be covered until project completion.
  - For the customer and the supplier, intellectual property aspects need to be defined at the very beginning of the project.

- Composition of the team: Consistency and continuity in team composition on the customer's and on the supplier's side are crucial in order to solve complex tasks and take decisions efficiently. In the case described, also geographical closeness helped to ensure progress at continuously high speed. It enabled frequent and partly spontaneous meetings when input was needed or site visits provided clarity.
- Communication and progress sharing: Continuous communication about the progress of the project to all experts and specialists involved at the various stages during the front-end phase of the project proved to be essential. Having contributed out of personal motivation and curiosity, everyone finds it very rewarding to be kept in the loop. Such communication fosters an innovative spirit in the company and encourages individuals to share know-how beyond their current responsibilities. In the case described, it was the project manager's task and merit to ensure an excellent information flow with weekly project updates to all people involved on the customer's and on the supplier's side at each stage of the project.

# Bayer: Strategic Management of the Early Innovation Phase

Wolfgang Plischke, Jürgen Heubach, and Stephan Michael Maier

## 1 Introduction

Innovation is a key driver of future growth at *Bayer*, playing a vital role in overcoming global challenges. Mindful of its corporate growth objectives, Bayer is constantly working to rejuvenate and expand its product portfolio. Bayer's R&D portfolio is closely aligned with future market needs and requirements and therefore subject to a continuous process of review and adjustment.

As a core part of its innovation strategy, Bayer heavily invests in R&D. As a group, Bayer spent about €3 billion per year on R&D over the past 3 years, *Bayer HealthCare* accounting for about two thirds of the overall expenditures, *Bayer CropScience* for about one quarter and *Bayer MaterialScience* for about 8 %.

## 2 The Strategic Management Process at Bayer

Bayer is a strategic management holding, where the group board of management is responsible for the group portfolio strategy and the operative business responsibilities are assigned to the three subgroups. As a logical consequence of this segmentation of Bayer's overall portfolio into subgroups, the strategic management of the individual portfolio segments is carried out within the respective subgroups. This particularly applies to the management of *early innovation* projects.

Hence, decisions regarding the R&D portfolio are made on a subgroup level. The subgroup-specific R&D strategies – once they have been agreed with the group

---

W. Plischke (✉) • J. Heubach • S.M. Maier
Bayer AG, Building W11, 51368 Leverkusen, Germany
e-mail: wolfgang.plischke@bayer.com; juergen.heubach@bayer.com; stephanmichael.meier@bayer.com

**Fig. 48** The strategic cycle at the Bayer Group and its interaction with Bayer's subgroups

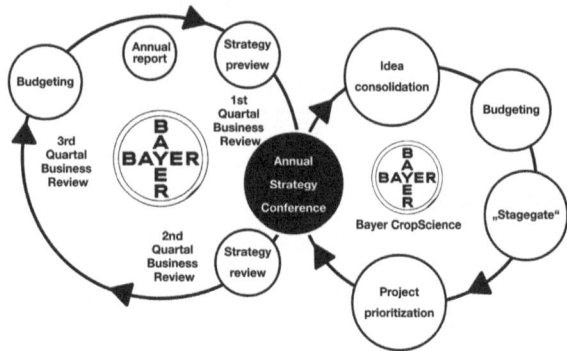

board of management – remain fairly stable on a mid-term to long-term basis so as to allow the necessary specific expertise and capabilities to be built up and maintained. The R&D project portfolios, however, require ongoing management decisions regarding new product developments and, where appropriate, transformation or termination of projects.

In order to align the group strategic goals with the subgroup goals, an annual strategy and resource allocation cycle has been introduced. This cycle is driven by the annual fiscal reporting cycle and drives the annual strategy development and resource allocation cycle at the subgroups. The ways these cycles interact and an exemplary scheme for Bayer's CropScience subgroup are shown in Fig. 48.

By virtue of the interface between the Bayer group cycle and the subgroups' cycles, as effected by the annual strategy conference that takes place after publication of the business report of the first quarter, effective strategic alignment can be ensured. Projects in the early innovation phase are embedded in these strategies and usually address strategic objectives that are distinct and specific with regard to Bayer's subgroup businesses. That is why there is no uniform strategic management of the early innovation phase at Bayer on a granular level – and none is planned. What really matters is that the principles and criteria defined are consistently applied across the subgroups.

For the purposes of this chapter, two of Bayer's subgroups share sufficient similarities with regard to management of the early innovation phase. These are our *life sciences* subgroups Bayer CropScience and Bayer HealthCare. Both demonstrate a very similar innovation model, with research spending at a double-digit percentage of sales and a high level of structural development risk, but they also possess overlapping research areas and opportunities to jointly use technology platforms. Bayer MaterialScience runs a rather different innovation model, with process and application technology innovation being at the heart of the business.

## 3   The Strategic Management Process of the Early Innovation Phase in Bayer's Life Science Subgroups

Both life science subgroups use a stage-gate process for continuously assessing the progress, value and probability of success of their early innovation projects. Referring again to the aforementioned cycle and taking the example of Bayer CropScience, an annual R&D stage-gate decision process takes place for the early innovation projects about 4 months after the strategy conference. Staying with the example of Bayer CropScience, it is important to understand that the stage-gate process is divided into four major stages. Stages 0–2 address the early innovation phase stages, while three and four address the later phases of product development and launch.

Stage 0, as the earliest innovation phase, involves discovery for the detection of new (e.g. molecular) concepts to address strategic development targets. On a molecular level, this involves screening of potential new chemical/biological entities in the order of magnitude of about $10^5$ to generate 'hits'. A 'hit' – in the context of the present chapter – is any new chemical or biological entity or a variant of an already known entity that might have a desired effect on a selected biological target. A biological target is any biological moiety (for CropScience e.g. in a plant or an insect). Usually a target is a protein, particularly commonly an enzyme.

It is the aim of stage 0 to generate 'leads' from 'hits' found during concept generation. Such leads may then be promoted into stage 1. The criteria to be applied in promoting a concept or hit to a lead candidate comprise first and foremost that the concept (gene, compound, microbial strain) has shown the potential to address the relevant research target. Furthermore, initial assessments of human and environmental safety profiles, as well as the IP situation, need to be positive.

Looking again at a molecular level, the application of the above decision criteria typically results in a reduction in molecular concepts from $10^5$ to $10^2$. Such rigorous reduction of concepts pursued or hits is also necessary, as the subsequent development costs of just one or two leads in stage 3 outweigh the costs of the earlier stages by a factor of between 2 and 4–1.

From stage 1 onwards, a research project manager is allocated to pursue the new lead candidates further. Throughout stage 1, more properties of the respective leads are determined and safety and efficacy data are collected. Furthermore, a more thorough determination of the IP situation including a freedom to operate analysis is undertaken at this very early stage, and first requests for regulatory approval are submitted. Along with the above-mentioned milestones, appropriate manufacturing technologies are sought and first market studies are made to estimate the economic potential at this stage.

To promote a lead candidate from stage 1 to stage 2, basically all of the aforementioned deliverables must have been completed successfully. The extensive catalogue of criteria results – on a molecular level – in a further condensation of lead candidates down to only two or three. Once a lead candidate is promoted to stage 2, the research project manager becomes dedicated to this one lead candidate project and is supported by a project team. Throughout stage 2 the stability of performance and possible optimization approaches are technically assessed, while the IP situation (own as well as third party) is clarified in detail and a production process/supply chain is determined. Finally, stage 2 comprises the establishment of a global regulatory strategy for stage 3.

If all of the foregoing criteria have once again been fulfilled, the lead candidate leaves the early innovation phase for stage 3, which mainly comprises obtaining product approval and executing all necessary regulatory procedures up to the market launch. Throughout stages 0 and 1, the stage-gate process is detached from the annual cycle to allow for a continuous influx of new concepts. As ideation is a continuous process not linked to the aforementioned strategic cycle, this idea-consolidation process is crucial to the overall process to ensure a constant flow of attractive new projects while staying focused at a very early stage.

Strategic management of the early innovation phase in the life sciences therefore requires two key capabilities:
- The skills and tools for sourcing additional early lead candidates;
- The organizational and technological capability to master complexity.

Considering the high probability of failure at an early stage, the first key requirement is of the utmost importance. As already outlined, good internal ideation is valuable in achieving this. However, it has been found in recent years that internal sourcing alone is no longer sufficient to fully leverage the innovation potential needed to maintain the required output of new products.

The *collaborative insourcing* of new projects and project ideas from outside the group has therefore gained considerable significance and attention. In recent years, Bayer has specifically outlined the number of collaboration projects with company participation in its annual reports (Bayer Annual Reports 2010, 2011). Therefore, the strategic management of the early innovation phase nowadays also incorporates the strategic management of beneficial collaborative approaches. The most common collaborative approaches employed in the pharmaceutical industry are shown in Table 6, along with the associated strategic tasks and objectives. As can be seen, five of these approaches directly address the early innovation phase and, particularly, ideation. Bayer participated in more than 800 collaborative projects in 2011; some 70 % of these collaborations were in the life sciences.

To fully exploit the value invested, the strategic management of these activities not only requires sufficient scientific know-how but also an organizational and technical capability to master the inherent complexity. A now well-established although intrinsically novel approach that particularly requires such organizational and technical capability is *crowdsourcing* (see also Table 6).

In May 2009, Bayer Healthcare introduced its *Grants4Targets* initiative to discover new therapeutic options by bringing together knowledge of potential novel targets in academia with the company's drug development expertise. The initiative is backed by a web-based application and submission procedure as well as by a predefined review process that ensures timely review and response. Three types of grants are provided: support, focus and collaborative. The IP remains fully with the applicants in the case of support and focus grants. After the grant period has expired, promising targets may be pursued further via collaborative agreements.

**Table 6** Overview of the most common collaborative approaches in the pharmaceutical industry

| # | Strategic objective | Type of collaboration |
|---|---|---|
| 1 | Access to global idea pools | Crowdsourcing/platforms |
| 2 | Network expansion, development of ideas | Incubator concept |
| 3 | Sharing of ideas, joint development | Strategic partnerships |
| 4 | Access to basic technology | R&D pipeline partnerships |
| 5 | General issues of drug research | Project partnerships or consortia |
| 6 | Access to particular knowledge carriers | Consulting (advisory boards etc.) |

Since the beginning of the program, 825 submissions have been received in Europe, America and Asia. 114 of these have already been accepted by our scientists and now receive a grant (status as of 11/2012).

Such initiatives and their implementation exemplify the second key ability required for successful strategic management of the early innovation phase: the organizational and technological ability to master complexity. This ability is of the utmost importance in the life science businesses in particular. Early pharmaceutical and agrochemical innovation projects are inherently more complex than those in most other branches of industry. Complexity originates e.g. from the high number and variety of chemical compounds employed in screening and from the need for multidimensional characterization of novel active ingredients in various models. But most of all, we have to deal with the inherent complexity of the biological systems targeted, which are still not fully understood.

In response to such complexity and the inherently high probability of failure of individual projects, Bayer's life science businesses review their innovation portfolios in the aforementioned stage-gate decision process very restrictively.

To facilitate sound decision-making, the review process is, again taking Bayer CropScience as an example, guided by a strategic decision analysis process (see Fig. 49). This latter process is accompanied by a continuous assessment of the technical feasibility. In the course of this procedure, most of the probable influence variables on the portfolio decision are amalgamated with the help of various tools into assessment parameter sets, including e.g. a risk profile and a probable commercial value for each project, which are mirrored with the strategic fit and respective technical feasibility of each project. The entire process is iterative and strongly relies on international and interdisciplinary teams that collectively feed the decision process with data and assessments. The process is supported by IT infrastructure measures that also help visualize the data for easier decision-making. Depending on the respective outcomes of the probable value, strategic fit and technical feasibility ratings of each project, the decision to proceed with the project and advance to the next stage in the stage-gate process is taken.

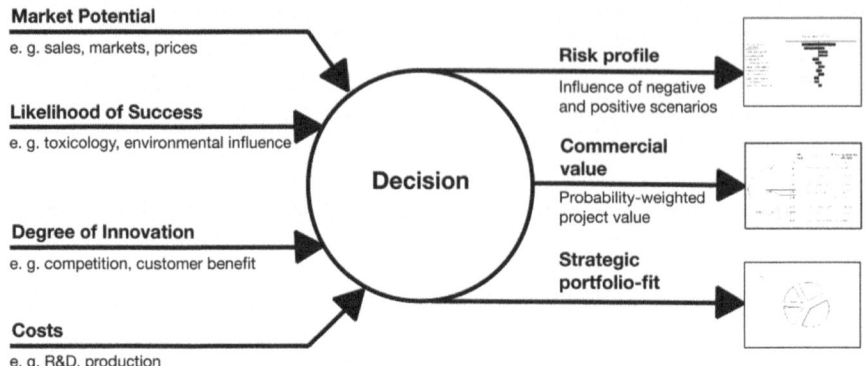

**Fig. 49** Criteria during the strategic decision analysis and possible derived outcomes

## 4    Lessons Learned

Taking the example of the strategic management of the early innovation phase in Bayer's life science areas, several overarching principles and developments can be deduced that may be applied to the management of innovation processes in other areas as well.

Generally, the strategic management of the early innovation phases requires a clear commitment to an overarching strategic goal. For Bayer this is condensed in the mission statement 'Bayer: Science for a better life', providing the overall socioeconomic framework for strategic decisions. Moreover, the number of decisions to be made is growing. In addition, they need to be made at a very early stage. The main drivers for this phenomenon are increasing regulatory requirements pertaining to environmental and human safety as well as increasingly competitive pricing. Such a growth in the number of pertinent criteria renders increasingly essential the availability and employment of suitable (e.g. software) tools and rigorous discipline in decision-making.

## 5    Checklist

- Innovation sourcing: All available innovation-sourcing tools should be assessed for their capability to achieve a given strategic goal. Each of these tools should be used to ensure an adequate supply of development leads in the long term.
- Innovation management: A comprehensive decision-making process must be implemented to handle the increasing number of decisions required within a reasonable timeframe.

# BGW: Partnering the Outside-in Process – The Expert Innovation Journey

Christoph H. Wecht

## 1 Introduction

Since the early stages of the innovation process are characterized by high uncertainty, the search for ideas should not be left to chance. A standardized approach facilitates the *idea generation process*. Regardless of company size, existing organizational set-up and the type of partners, the following procedure has been proven in more than 20 workshops for companies in Germany, Austria and Switzerland with a total of about 300 participants.

The underlying concept used to generate the expected ideas is an idea generation (creativity) workshop with internal and external participants. The overall goal of the workshop – called *expert innovation journey* – is to generate about 50 tangible ideas and five 'rough' concepts.

In this chapter, an example from a European manufacturer of engines and exhaust gas aftertreatment systems will be used to illustrate the key steps. To find fresh ideas and develop concepts for the reduction of exhaust emissions from stationary gas engines, an expert innovation journey was planned and conducted by *BGW AG* using the method described.

Involving external experts in idea generation workshops yields two main benefits: Firstly, access to knowledge which does not exist internally, and secondly,

---

This chapter is based on project work done together with Dr. Alexander Conreder. He had a major role in the development of the method and came up with the name 'expert innovation journey' during his work for BGW. In 2011, he left the company and took up his position as innovation manager for *EnBW* in Karlsruhe.

C.H. Wecht (✉)
BGW Management Advisory Group St. Gallen - Vienna, Varnbüelstr. 13, 9000 St. Gallen, Switzerland
e-mail: christoph.wecht@bgw-sg.com

the disruption of existing thought barriers and thus the revelation of new perspectives. Additionally, such workshops also serve to build a network and to get to know potential partners. We recommend conducting a 2-day workshop at a venue (e.g., a conference hotel) away from the participants' everyday environment. In our experience, the potential of the participating group can be much better utilized in such a setting. To facilitate a focused, successful workshop, the following outline is proposed.

## 2 Methodology

The detailed method – developed together with the *Institute of Technology Management* at the University of St. Gallen (ITEM-HSG) – consists of three steps, i.e., set-up and preparation phase, workshop phase, and transfer phase, which will be described in detail below. The planning, the acquisition of the external participants and the facilitation are taken over by members of BGW AG.

### 2.1 Set-up and Preparation Phase

The initialization phase marks the very beginning of an expert innovation journey. Coordination with the core team, the formulation of the right questions, the selection of the appropriate workshop participants as well as the detailed planning of the workshop with regard to process and methodology are essential activities in this phase.

A core team is formed consisting of 2–3 members from the client and 1–2 members from BGW. Through detailed discussions, the scope of the workshop is defined.

In the case of the exhaust treatment manufacturer, the core team was formed about 3 months before the date of the expert innovation journey. Specifying the topic was a complex process and involved a trend workshop that was held prior to the expert innovation journey. For example, it was necessary to focus on certain types of emission reduction (before, during, or after the main combustion) instead of looking at all of them at once.

This focusing step is about identifying and describing the real opportunities from among a very large number of possibilities. Behind this task lies a strategic approach that aims to select those areas where the chances or threats (external point of view) are highest, and fitting competencies and capabilities exist or can be built (internal point of view).

A key part of this phase is phrasing the right questions as a starting point for the idea generation itself. The right balance has to be found between topic-specific depth and overall understandability. Later, in the course of the workshop, all participants – especially the non-experts – have to be able to understand the meaning of the questions.

**Fig. 50** Activities in the set-up and preparation phase

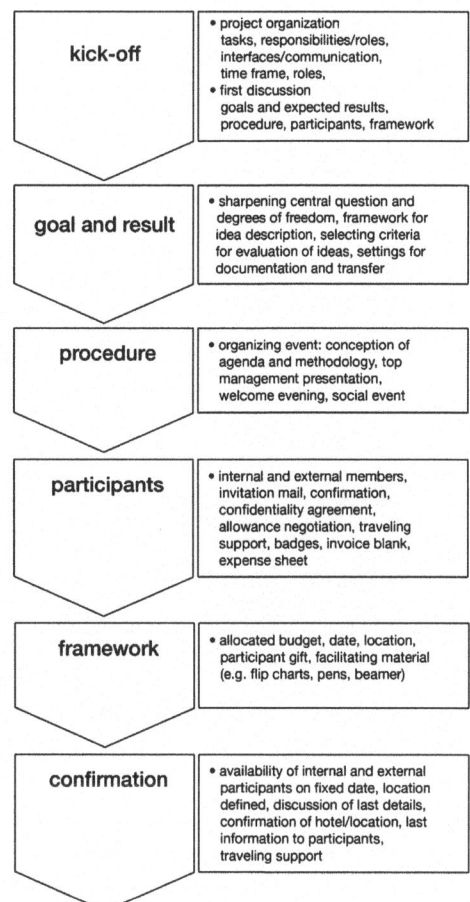

Once the overall focus has been set, the actual preparation of the workshop can start. There are two main areas of activities, i.e., partner selection and organizational planning of the workshop. Important steps and activities for this phase are listed in Fig. 50. Between the kick-off and the confirmation meeting, four relevant intermediate steps complete the outcome of the set-up and preparation phase.

In the following, the selection of the journey participants will be explained in more detail. The composition of the team follows the notion of uniting both internal and external participants. The overall credo is to achieve a high level of diversity along the following parameters: level of knowledge in the respective field of the workshop, and seniority. For choosing the external participants, the particular environment, academic or industrial sector, is important. Figure 51 shows two matrices to illustrate the selection of participants. All four fields of the matrix should be filled to ensure a broad base of knowledge and experience. It is of crucial importance that the full range of potential partners be exploited, because the

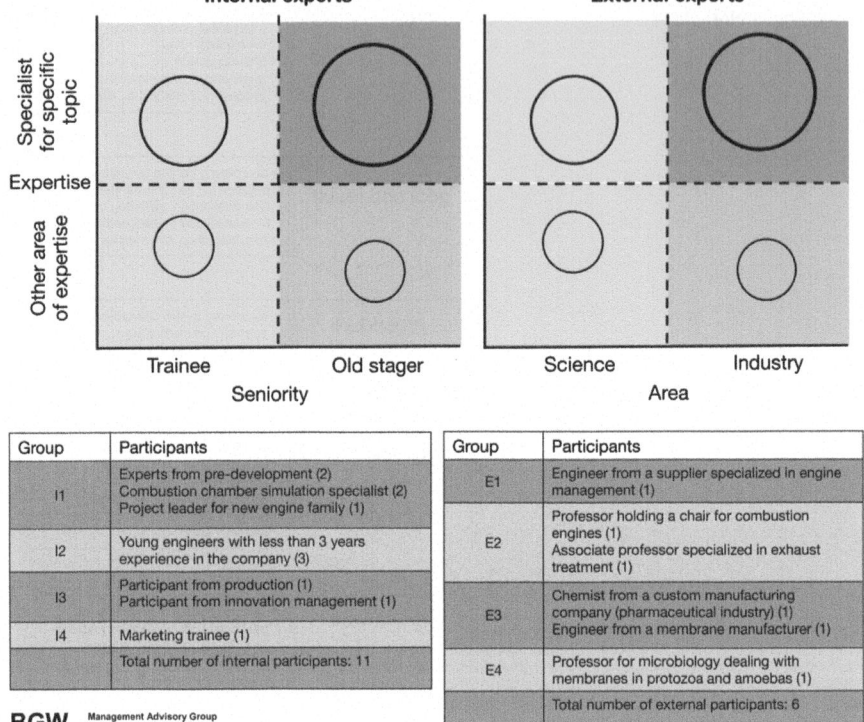

**Fig. 51** Selection parameters for internal and external participants

involvement of various partners in all phases of the innovation process leads to the greatest power to leverage the innovation output.

External participants – about half of them with industry background and the other half from the academic world – are identified drawing on the BGW network as well as on already existing customer contacts. The number of external participants should be about five to eight. The optimum total number of participants ranges from 14 to17 persons (minimum 10, maximum 20).

The overall goal is to cover all areas of the above matrices, however, with a focus on each of the two upper right panels. As prerequisite, all participants have to be creative and 'open-minded'.

## 2.2 Workshop Phase

In this phase, the workshop is facilitated by experts from BGW. After the introduction by a core team member and a senior management representative, the flow (agenda) and methodology are explained by the moderator. In the next step, it may be necessary to apply a mixture of methods to convey points relevant to clarifying the starting position. The goal is to achieve an understanding of the initial situation shared by all participants.

The central part of our expert innovation journey is ultimately the idea generation. It represents the very core and requires a selection and adaptation of methods and tools specific to the given situation. We apply a proven three-step approach to generate and develop ideas during the workshop. The opportunities are selected and sharpened during the preparation phase. This process results in questions which reflect the right level of aggregation, i.e. sufficient focus to act as starting point for idea generation, and sufficient breadth to open up the solution space. Then, first raw ideas as answers to these questions are generated – using different situation-specific creativity techniques – and captured on traditional moderation cards (first documentation stage). The traditional system of moderation cards, pin boards and pins is still the most versatile and flexible, and suits even the most forward-looking, 'wild' ideas. Ideas are then further developed using a form in A4 format (second documentation stage). After that, rough initial concepts are developed and noted on flipchart paper (third documentation stage). The concepts mark the end of the early phase and the transition into the actual development process.

A critical part of the maturation of an idea in the early phase is idea evaluation. The number of ideas has to be reduced to only the most promising and relevant ones. The selection criteria for all filtering steps should be agreed upon with the workshop clients before the start. Ideas need to be assessed in line with their level of maturity according to a multi-stage principle. At the very beginning, only qualitative criteria, such as strategic fit or basic technical feasibility, are used. During the maturation process, the criteria change in two ways. On the one hand, they become more market-oriented, and on the other hand, more quantitative. Overall, it can be stated that it is important to show the courage to leave gaps.

In the case of the exhaust treatment manufacturer, about 200 raw ideas were generated for treatment of emission reduction before and after the actual combustion. Of these, 25 were selected on the basis of the following two criteria that had been defined beforehand: strategic fit and level of innovativeness. All 25 ideas were then elaborated, presented and prioritized, with the top five of them being developed further into (raw) flip chart concepts.

## 2.3 Transfer and Follow-up Phase

In this step, the results from the expert innovation journey are digested, assessed, and hopefully commercialized. The documentation of the workshop takes place in two ways. A documentation of the process is created that highlights essential steps, intermediate results and photos to visualize the procedure for all participants. In addition, detailed minutes containing the resulting ideas and concepts are drawn up for the core team only.

The ideas and concepts identified need to be sharpened conceptually and discussed with a wider circle of persons. Thereafter, the core team has to finetune them taking into account the overall strategic guidelines (corporate strategy

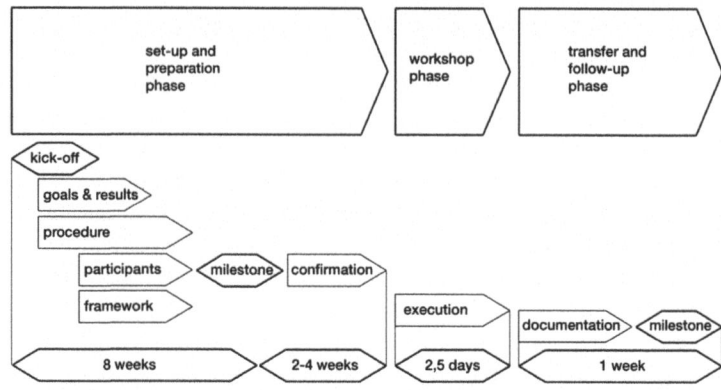

**Fig. 52** Overview expert innovation journey

and innovation field). After these steps have been completed, another prioritization is carried out. It is possible and may be necessary to involve external participants after the journey to continue working on the results more specifically. This can involve a more detailed evaluation of the final ideas as well as further development of the concepts.

The five concepts mentioned above (together with all 25 elaborated ideas) were integrated into the client's technology management process (roadmap, technology screening, idea management) later on. Two of them could be integrated into already planned or running technology projects; two new projects were started, one of them together with the two external members of the expert innovation journey (one professor and one technology supplier).

In addition to these concrete and tangible results, the methodology delivers implicit results as well. The *joint development* of concepts in a workshop setting with a mixed team fits our understanding of a 'bottom-up' approach. Participants learn about the importance of professional workshop preparation and execution. Ideally, this changes their own behavior, and they start to act as 'innovation ambassadors' and lay the seeds for efficient innovation management.

An overview of the expert innovation journey is shown in Fig. 52. The times given are rough averages, however, a key learning is the fact that careful preparation takes up to 3 months.

## 3   Lessons Learned and Success Factors

We recommend using an approach centered on a selected group of external experts. The essential core elements are:
- A diverse, multidisciplinary team consisting of external and internal participants.

- Specific questions which have to be phrased beforehand by a small core team.
- Capturing of the results in a three-step system (standard moderation cards for raw ideas, A4-size landscape paper for elaboration of ideas, and flipchart for first concept designs).

Special attention should be paid to the preparatory phase prior to the actual workshop. A clear project plan clarifying the timeline and the division of tasks between the clients and the external resources of support from BGW has to be established. The importance of the preparation before the workshop starts must not be underestimated.

Based on our extensive experience with such events, we recommend conducting a 2-day workshop around those elements using our special format dubbed 'expert innovation journey'.

The following points have emerged as key success factors:
- Create central, multidisciplinary core team as anchor point and 'engine'.
- Structure preparation phase clearly, with kick-off meeting, to-do list, and project plan.
- Select internal and external participants carefully.
- Prepare workshop thoroughly, plan enough time and choose external location.
- Ensure flexible, goal-oriented workshop moderation to establish and keep flow.
- Adapt standard *creativity methods* and tools in a way fitting to the specific situation.
- Develop clear methodological guide for the entire process.
- Schedule sufficient time for group discussions alongside the idea generation sessions.
- Communicate results in a transparent and consistent way.

# Emporia: The Merits of Online Idea Competitions

Fiona Schweitzer and Walter Buchinger

## 1  Project Setting & Focus

The Austrian mobile phone and services market for senior citizens was the setting for the *online competition* and the *focus group* workshops. While mobile phones are popular with younger consumers, 21 % of persons between 65 and 75 years of age and 43 % of seniors aged 75 and above do not use mobile phones. This is often because they consider them too small, too complicated, or too difficult to use. The competition and the focus groups therefore focused on finding new designs, functions, accessories, improvements in usability, or services that could increase the value of mobile phones for the elderly and decrease barriers to their use among this population.

The backing of gathering external inspirations for future products and services by the whole company was assured by involving top management, marketing and R&D in the process.

---

This chapter is based on the following article: Schweitzer, F.M., Buchinger, W., Gassmann, O., Obrist, M. (2012). Crowdsourcing – Leveraging Innovation through Online Idea Competitions. Research Technology Management, 55(3), 32–38, doi: 10.5437/08956308X5503055.

F. Schweitzer (✉)
University of Applied Sciences Upper Austria, Innovation and Product Management, Stelzhamerstr. 23, 4600 Wels, Austria
e-mail: fiona.schweitzer@fh-wels.at

W. Buchinger
Emporia Telecom Produktions- und Vertriebs GesmbH & CoKG, Industriezeile 36, 4020 Linz, Austria
e-mail: buchinger@emporia.at

## 2 Method

An external specialist was commissioned to design a new website for the competition and to integrate the software that would enable users to register for the competition, submit ideas, and evaluate, discuss, or complement others' ideas. Users were given the option to create personal profiles, but they could also choose to remain anonymous. Ideas could be entered either as freeform text-based posts and document uploads ('free ideas'), or via a *toolkit* that allowed the participants to easily build their own 'ideal' mobile.

The contest, which was open in Austria for 2 months in 2012, was communicated via online and offline channels, mainly newspapers and their webpages, social media (Facebook and specialized senior citizens' platforms), and the partnering telecommunication company. To stimulate participation, the best ideas were selected by a prominent expert jury (whose members also proved helpful in promoting the competition via their networks), and the winners received awards and press coverage as well as material rewards. All of the participants were included in a raffle of ten mobile phones. Eligibility requirements for the competition specified that the contributors of ideas would be mentioned if their ideas were used, but that the rights to exploiting submitted ideas remained with *Emporia*. While these requirements might be discouraging for professional designers, who usually strive for a percentage of the sales generated, this competition was more focused on involving ordinary customers (elderly end-users and children/grandchildren who buy the mobiles for their elderly relatives).

The four focus group workshops were carried out with four groups of five to six senior citizens each, who were recruited in cooperation with senior citizen associations. The recruiting process was difficult, because many individuals refused to participate, arguing that they lacked competence regarding the mobile world. Effort had to be exerted to build trust and convince potential participants that their input would be beneficial. The focus groups lasted approximately 2.5 h and included different *creativity methods*, such as the use of visual stimuli to sensitize participants for the tasks, individual and group brainstorming sessions (see Beck et al. 2008), and toolkits to enable participants to build icons and phones with different materials. The sessions also included breaks to facilitate socializing and to structure the different phases of the *idea generation* process. The workshops were enabled by the University of Salzburg and the University of Applied Sciences Upper Austria.

## 3 Results

Content analysis (Krippendorff 2004) was used to compare the ideas developed in the online competition and in the focus groups. The virtual idea competition gathered 226 ideas at an average cost per idea of €86.28. In the focus groups, 52 ideas were collected at €105.76 per idea (Table 7).

**Table 7** Online idea competition versus focus group workshops

|  | Online idea competition | Focus group workshops |
|---|---|---|
| Total number of ideas | 218 | 52 |
| Comments on ideas | 303 (average 25 words per comment) | Several per idea (full discussion of interesting concepts) |
| Active participants/ registered members | 191/4,183 (4.6 % active) | 23 (100 % active) |
| Ideas per participant | 1.14 (active participants) 0.05 (registered members) | 2.26 |
| Ideas for existing functions | 52 (24.4 %) | 20 (38.4 %) |
| Unconventional ideas (% of total ideas) | 83 (38.9 %) | 17 (32.7 %) |
| Ideas realizable within the next 24 months | 78 (36.6 %) | 15 (28.8 %) |
| Participant background | Provided by 53 % of active participants | Provided by all participants |
| Total cost | €19,500 | €5,500 |
|  | €11,500 for conceptualizing and setting up competition | All for conceptualizing, carrying out, and analyzing workshops |
|  | €8,000 for communication, expert jury, and prizes | Rooms and equipment free of charge |
| Cost per idea | €89.45 | €105.76 |

The focus groups comprised typical representatives of the senior citizen mobile phone market. Participants were willing to provide background information on their sociodemographics, mobile phone usage, and use of other technical products. This allowed for gathering *consumer insights* that increased our understanding of the target market and different segments within the market. Obtaining information on the participants in the online competition turned out to be more challenging. A high number of the participants did not disclose personal information; for example, 47 % did not provide information on their age.

All ideas from both sources were grouped together and divided into three categories: ideas for devices, ideas for functions, and ideas for services. 'Functions' was the predominant category; participants had ideas related to improving mobile phones (for instance, positioning an emergency number prominently in the menu, improving compatibility with hearing aids, providing search and find functions such as 'find my keys'), suggestions for existing functions of which they had been unaware (for instance, age-appropriate displays with respect to the size of the fonts and buttons, separate 'on' and 'off' buttons, and a function for the automatic saving of new callers' phone numbers), and ideas for developing new functions (such as capabilities for collecting biophysiological data, such as blood pressure or blood sugar, a drug-reminder function, a shopping list audio recorder, and a home automation function to perform such tasks as opening the garage door or activating a home alarm).

There was substantial overlap between the two sources of ideas; 38 % of ideas mentioned in focus groups were also submitted in the online competition. Many of the ideas collected were not new; 24.4 % of the ideas submitted in the online competition and 38.5 % of the ideas generated in the focus groups were already available in mobile phones. This indicates that participants, especially the older users in the focus groups, were often not aware of the full range of functionalities that mobile phones for the elderly generations offer. A possible explanation for the higher percentage of 'old ideas' in the focus group sessions may lie in the higher level of technological knowledge and affinity with technology in the online sample. A large proportion of the ideas were immediately usable; 36.6 % of the ideas gathered from the online competition and 28.8 % of those collected from the focus groups could theoretically be realized by Emporia within the next 24 months. Additionally, 39.0 % of ideas submitted via the online competition and 32.7 % of those collected in the focus group sample were considered unconventional. These ideas may be interpreted positively as potential, radical innovations, or negatively as not conforming to the market. For Emporia, most of these unconventional ideas were too far removed from current demand to be considered for further action. In total, 46 ideas (21.6 %) from the competition and 12 ideas (23.0 %) from the focus group workshops were included in Emporia's innovation process for further consideration, evaluation, and development. Although a greater percentage of focus group ideas were advanced, the online competition outperformed the focus group sessions with regard to the number of potential new product offerings generated.

In the online competition, ideas were presented briefly (an average of 60.7 words) and comments on or advancements of ideas from others were rare (303 in total, with an average of 25.3 words per comment). The focus groups offered more scope to elaborate on ideas very intensively and interactively; the ideas thus had the possibility to evolve during the discussions.

Moreover, the focus groups allowed for flexible adaptation and control of idea generation to prevent respondents from moving in unintended directions. For example, the first focus group's confrontation with current mobile phones for the elderly generations in the toolkit session led the participants to rebuild existing phones instead of inspiring them to build new models. The toolkit kept participants busy building 'their mobile phone', while providing poor information on their reasons for building their phones the way they did. The participants' ideas were determined by their knowledge of current models and shapes.

## 4  Lessons Learned

From the comparison of online idea competitions with focus group sessions discussed above, it can be concluded that the two techniques have different strengths and weaknesses. While focus group participants tended to interact with each other to develop ideas more fully, the online idea competition produced nearly four times as many usable product ideas as the focus groups. This suggests that the choice of tool should be guided by the following rationale: if the objective is obtaining a more elaborated set of ideas that provide insight into the needs and beliefs of a particular group of consumers, focus groups may be more effective. Online idea competitions are advantageous for generating high quantities of ideas, but less so for improving customer understanding and for evaluating the attractiveness of ideas for the target group. Generally speaking, focus groups seem to outplay online competitions when *needs-based information* is central, while online idea competitions are superior when *solution-based information* is of paramount importance. The ideas gathered were assessed and discussed inside the company by an interdisciplinary team including R&D, marketing, and top-management. Out of the 93 ideas which seemed to be realizable within 24 months, 52 were considered worth investing further time and effort in. After 24 months, the future of 18 ideas was still under discussion, while 16 ideas were in the concept phase, and 18 ideas have been realized and launched on the market.

## 5 Checklist

The following checklist may serve to guide processes of idea generation that use external sources as idea providers:

- Preparation: As in every project, the objectives have to be clarified. What result is expected, e.g., better understanding of needs (which improvements are current customers looking for) or solutions (e.g., basic ideas, detailed concepts)? How can the task and the process be defined so that valuable ideas are generated? (e.g., What should a focus group workshop look like – which phases, which tasks, which questions from the moderator should be included?) Do we need external partners to execute the idea generation process? (e.g., Which online platform will be used? Does it make sense to develop our own platform, such as the *BMW virtual innovation agency* or *Tchibo ideas*?)
- Initiation and Execution: How can participants be recruited? What motivates them to contribute ideas (e.g., reputation, financial rewards, fun)? Is the reward system transparent? How can the participants be reached effectively (sampling techniques, communication activities)?
- Assessment and Commercialization: Should internal employees be integrated in the project in order to reduce the not-invented-here syndrome? How should the diversity of incoming ideas be managed? What are the selection criteria? Are these criteria transparent? Who is involved in the selection process? How are the prizes/rewards distributed? Most important is the exploitation of the most promising ideas. How are the ideas to be implemented within the company?

# Evonik Industries: Managing Open Innovation

Georg Oenbrink

## 1 The Importance of Open Innovation

Today open innovation is standard in industrial companies, whether these companies are active in the B2C or B2B industries and markets. The reasons why enterprises prefer to open up their innovation eco-system for external partners are manifold: decreasing product and technology life cycles along with shorter innovation cycles, aggravated by a dramatic increase in the complexity of individual innovations. Besides, business model innovations are gaining importance for supporting sustainable and profitable growth of established enterprises into new markets and applications. Therefore, competencies are needed that can only be offered by external partners, who most often are active in a different position in the value chain. All these reasons have led to the importance of *open innovation* increasing dramatically. Today, innovations are generated at the interfaces of technologies and industries. A current and very good example is organic electronics: here, innovations can only be realized when chemical competence and electronic and engineering competencies are brought together in close and tight partnerships of collaborative innovation between academic and industrial partners. A prominent and well known example is the Innovation Lab Organic Electronics (www.innovationlab.de) set up by *BASF SE, Merck KGaA, Freudenberg & Co., Heidelberger Druckmaschinen AG, SAP AG*, and the *Universities of Heidelberg and Mannheim*, respectively.

But open innovation starts much earlier and does not necessarily need such complex, publicly funded and politically supported approaches as the Innovation Lab. Moreover, open innovation is not really new. Joint research collaborations with scientific institutes, universities, suppliers and customers as well as licensing

---

G. Oenbrink (✉)
Corporate Innovation Strategy & Management, Evonik Industries AG, Rellinghauser Str. 1-11, 45127 Essen, Germany
e-mail: georg.oenbrink@evonik.com

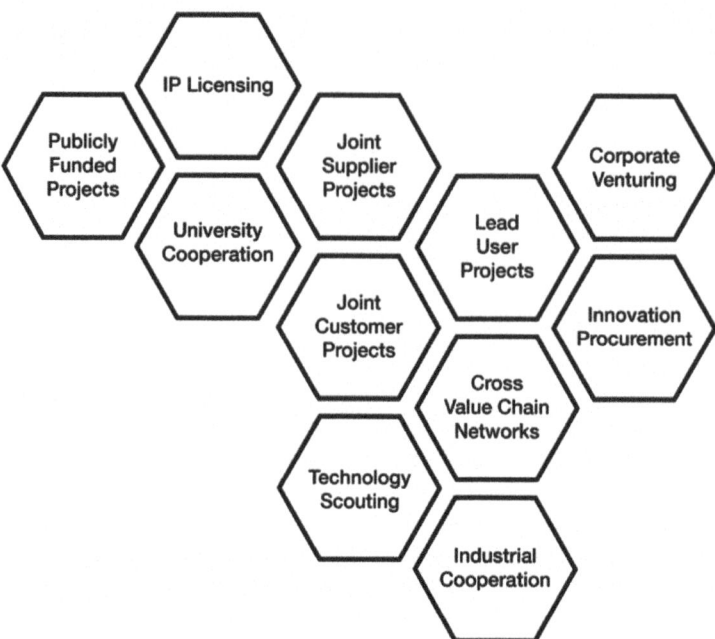

**Fig. 53** Established open innovation approaches in B2B companies

in and out of relevant intellectual property have been recognized approaches to increasing the knowledge and competence base available in research & development for decades (Fig. 53).

Many companies – especially globally active, multinational enterprises – have established efficient *technology scouting* to get early and efficient access to knowledge, expertise and competencies they do not own themselves, but that are needed for future innovations. Depending on the strategies in place, technology scouting is organized on the corporate level, searching for technologies and knowledge in strategic areas relevant to growth, or it is integrated in the new business development units of the regional organizations or operative business units. Also combinations of both approaches are well known.

## 2 Evonik Industries' Approach

*Evonik Industries* – one of the globally leading companies in specialty chemicals – has established a somewhat different approach: technology scouts have been positioned in all important growth regions globally, acting as agents for the operative business units and at the same time running a kind of blue-sky scouting for strategic topics of the company. All technology scouts have long-lasting professional experience, have been born and qualified in the region they work in, and are – of course – native speakers who easily understand the different cultural and mental influences.

Participating in publicly funded joint research projects that focus on basic research, joint research projects with universities and research institutes as well as contract research with partners in academia are common ways of getting early access to really new knowledge, technologies and competencies. In recent years, industry-on-campus approaches, driven within public-private-partnerships, have found increasing interest within industrial organizations.

*Corporate venturing* has also become a well-known and successful approach used by globally active, multinational companies to get early access to new technologies and knowledge in recent years. Investing in attractive venture capital funds that are active in strategically important areas of technology as well as direct investments in spin-offs and start-up companies that own strategically relevant expertise not only offer access to needed assets and resources, but at the same time have the potential of becoming profitable financial investments. Besides corporate venturing, direct acquisition of technology providers is becoming more and more important. Technology scouting and corporate venturing are therefore also needed to feed the M&A pipeline of industrial enterprises.

*Corporate foresight* supports all open innovation activities as it offers potentially attractive search fields for new innovation, which lead to new innovation activities within the enterprise based on developing scenarios for well-defined global megatrends.

What has changed dramatically over the past few years is the ease and speed with which potential partners for joint innovations, technology owners and providers as well as ideas and insights for new innovations can be identified and reached. Never before has it been as simple and efficient as today, in the times of *web 2.0* and *enterprise 2.0*. The number of *intermediaries*, such as *Innocentive*, *NineSigma*, *SpecialChem* and others, that specialize in supporting enterprises in efficiently identifying innovation partners, knowledge owners or technology providers needed by using their proprietary expert networks has grown tremendously over the last two decades. The real expertise of these service providers lies in their ability to file 'requests for proposals' within their online expert networks in a way that ensures that it is really understood which technology, solution or support is needed by their clients. In recent times, even completely open, non-confidential idea competitions have been facilitated on internet-based platforms like *Innovationskraftwerk* (www.innovationskraftwerk.de). Within these ideation contests, enterprises ask for new insights into how their products, solutions and technologies are used in new markets and applications, needs for new variations of their products and technologies, and so far unknown and unvoiced needs within their already existing markets. These ideas and insights will then be used internally to define potential new product or technology innovations, even business model innovations (Fig. 54).

**Fig. 54** Example of an internet-based open idea competition

## 3 Crowdsourcing and Open Ideation

All these approaches and methodologies are nowadays summarized as *crowdsourcing* or *open ideation* approaches, giving enterprises fast and easy access to previously unavailable knowledge, expertise, technologies, and product ideas.

Especially within the B2C industry, companies have begun to build up their own, proprietary and confidential crowdsourcing or open ideation platforms. One of the most prominent examples is the 'Connect & Develop' platform owned by *Procter & Gamble* (Huston and Sakkab 2006), or the 'Pearlfinder' platform operated by *Beiersdorf* (www.pearlfinder.beiersdorff.com). These approaches are actually transferred into life-science and B2B companies as well. The crowdsourcing platform 'Grants4Targets' owned by *Bayer Health Care* is just one example. Researchers can apply for a membership within this platform, introduce their planned projects in drug target research, and ask for financial support.

But with huge, multinational and globally active companies, open innovation and crowdsourcing start already within the enterprise itself. Very often, these companies are characterized by a considerable lack of transparency regarding already available expertise, competencies and knowledge within all the individual business units and organizational units. To overcome this situation, many enterprises have built their own, internal, cross-organizational innovation and expert communities. Based on enterprise 2.0 solutions, experts from different organizational units are lined up in virtual networks representing the existing technology competencies and thereby increase the knowledge and transparency about what is already available within the company. Leveraging the existing know-how of the whole company in successful innovations in individual organizational units is the great benefit of this internal approach to open innovation. Based on the impressive and successful experiences that IBM has made with their *Innovation Jams*, (Bjelland and Wood 2008) many industrial companies have started to run such internal, cross-functional and cross-organizational open ideation approaches to generate new, unconventional and appealing ideas that are usually not generated within creativity workshops or brainstorming sessions within organizational units responsible for the same business. The ideas generated in these *ideation jams* are subsequently evaluated further and transferred to the regular idea and project management processes.

## 4 Summary

To summarize, it can be stated that open innovation today goes far beyond research co-operations with scientific organizations, research institutes, joint customer or supplier projects, lead user approaches and joint innovation partnerships between industrial entities. Today, open innovation leverages the huge potentials of enterprise 2.0 solutions and the high degree of connectivity offered by the world-wide web. To survive in an increasingly global competition, it is more than essential to be able to identify and join forces with the right innovation partner, who offers exactly the missing piece of knowledge and competence that is needed for the next innovation, in order to be more efficient and faster than the competitors.

# Case: Google Ventures

Sascha Friesike

## 1 Introduction

*Google Ventures* is the *venture capital* department of *Google*. It is a hybrid, half investment firm and half corporate *incubator*. Corporate incubators are a prominent way for tech-intensive companies to fill their front end with new and innovative businesses. An incubator provides an environment that is friendly to a young and fragile new business idea in development. It is a surrounding unlike the one present in most large corporations, which is often described as process-oriented and non-entrepreneurial. Once the business idea has evolved into a promising venture, it is either reintegrated into the parent company or sold. Studies show that for start-up companies that have successfully completed an incubator program the likelihood of staying in business increases compared to other start-ups. At Google Ventures, this idea of an incubator is coupled with the methods of a classical venture capital firm. In simplified terms, a corporate incubator seeks to create innovation, while a venture capital firm seeks to create money. As such, corporate incubators select ideas that are related to the parent companies' businesses (a pharmaceutical company, for instance, might include a high-risk drug development in an incubator). And venture capital firms select their portfolio of companies based on investment opportunities. Venture capital firms are less focused on the actual field of business than on financial projections, the team, and future market opportunities.

Google Ventures tries to combine both investment strategies. Therefore, the operation funds companies which either develop a technology interesting to Google or show a promising investment opportunity. Overall, the department is designed as an investment firm and thus forced to create its own business, as it is measured by its financial returns.

S. Friesike (✉)
Alexander von Humboldt Institute for Internet and Society (HIIG), Bebelplatz 2, 10099 Berlin, Germany
e-mail: friesike@hiig.de

## 2 Broad Range of Investments

Today, Google Ventures invests in over 100 firms (103 having been the officially announced number at the time this case was written), ranging from home appliances like *nest* – a company that tries to reinvent the thermostat – to health care firms like *Foundation Medicine* – a cancer diagnostics company. Google Ventures itself explains that its investments are focused on the core areas of mobile, gaming, energy, and life sciences, whereby, in fact, the firms in the 'mobile' category outnumber the firms from the other three categories combined. This is not surprising if one considers that Google Ventures is a modern corporate incubator. Google itself is neither a gaming company nor remarkably active in the energy sector or in life sciences. Yet Google is particularly active in the market of mobile communication, increasingly so since its merger with *Motorola*. One could sum up the investments as a potpourri consisting of innovations at the core of Google's current markets and big ideas which might change certain aspects of life for a large number of people. *CoolPlanet Energy Systems*, a company that develops negative carbon fuels based on plant photosynthesis, is an example of such a company. *Transphorm* is another; the company develops ultra-efficient modules that are supposed to drastically reduce electric conversion losses.

Currently, Google Ventures invests around $100–300 million a year in young companies. The investments range from seed funding – a little less than half of all investments made – to late-stage investments in the amount of up to tens of millions of dollars.

## 3 Hands-on Environment for Start-ups

Other tech companies like *Intel* or *Microsoft* have venture arms, too. Google Ventures labels its venture operation as 'radically different' from these two competitors and bases this assertion primarily on the form and intensity of support and engagement Google Ventures offers the companies it invests in. Currently, around 60 people work at Google Ventures, fewer than ten of whom are engaged in investment decisions. The other 50 are there to work with the companies in which Google Ventures has invested. Most of the people that now support start-ups had previously worked for several years at Google. They cover all business aspects necessary to run a successful company from marketing to sales, HR and public relations. In fact, Google's former director of global communications and public affairs and number one PR-manager David Krane now works as a partner at Google Ventures. Engineers complement the business knowledge that Google Ventures offers its portfolio firms.

The support for start-ups takes place in the 'startup lab', the 'design studio', or in one-on-one sessions. The idea behind the startup lab is to offer a university for founders. The lab provides courses where professionals, experts, academics and successful founders teach start-up founders and employees and help them to implement the lessons learned in their firms. The topics taught at the startup lab are manifold and range from managing engineers to best social media practices. They are visible in the Twitter feed, which is used to promote Google Ventures and attract future founders. Most sessions at the startup lab are held for companies from the Google Ventures portfolio only. They are designed as an add-on for funded firms – and used by Google Ventures as a marketing tool. In the design studio located in the Google Ventures building on the Google campus in Mountain View, California, Google employs several designers to work with the funded companies on all aspects of their businesses which relate to design.

## 4  Keep Your Friends Close

Following one of movie history's most memorable lines, "keep your friends close, but your enemies closer", Google Ventures makes an effort to invest in companies whose aim it is to directly compete with Google's products.

One of the investments, for instance, *Airtime* (www.airtime.com), is a platform for online video conferences and hence a competitor to one of the key features of Google Plus. *Pocket* (http://getpocket.com) is another investment; the business idea consists of a platform-independent service that enables users to quickly retrieve previous search content, eliminating the need to 'google' for it again. Google Ventures openly communicates the fact that it invests in companies that might compete with Google in order to make sure that those companies are not scared away by the 'Google' in Google Ventures and still apply for funding.

Google's motivation to invest in potential competitors is twofold:
- First, Google is able to gain early insights into the technologies in development. Google Ventures makes an effort to connect founders with Google engineers for the purpose of mutual learning and for the purpose of cooperation. They encourage founders to seek possible bridges into Google, Inc.
- Second, strong ties between the founders and Google do encourage an early integration into the parent company. This happened for instance with the company *Milk* (a mobile app development firm) whose entire team joined Google within the first year of Milk's existence, with one of Milk's founders being a partner at Google Ventures today.

## 5  Exit Strategies

Given the short history of Google Ventures – having been founded in 2009 – the number of firms that have left the ventures fund is low. Yet, since Google Ventures is a hybrid of a venture capital investment firm and a corporate incubator, it offers the exit strategies of both of these two categories. Consequently, one strategy is the inclusion of the funded company into the parent company Google. As described above, this happened for instance in the case of Milk. Here, however, the integration was mainly a talent acquisition. The one product Milk had on the market prior to its inclusion in Google is no longer available. The second strategy is to sell the funded company to another investor. Google Ventures has done so, for instance, with the video game publisher *ngmoco*, which was sold to the Japanese internet service platform *DeNA*. Other examples are *Dasient*, an anti-malware technology, which was acquired by *Twitter*, and *Hipster*, which was acquired by *AOL*.

# Idea Generation in the Consumer Business at Henkel

Thomas Müller-Kirschbaum and Juan Carlos Wuhrmann

## 1 The Innovation Process at Henkel

Laundry and Home Care is one of three *Henkel*'s areas of competence. It focuses on detergents, fabric softeners, dishwashing and cleaning products for *consumer markets* throughout the world. Strong brands and innovation offering consumers added value provide the basis for Henkel's strategy of profitable growth. Through efficient management of its innovation process and based on profound insights into the purchasing habits of consumers, the company is able to quickly identify and respond to consumer trends and effectively convert these into new products. Laundry and Home Care's innovation process is steered by the *Henkel InnoGate process*. It strictly follows the stage-gate classical system of concept definition, development and validation phases (Cooper 1993) where stop-or-go decisions at the gates are taken by the executive board. Idea generation and subsequent selection mark the beginning of the whole process. Proficiency in these early stages of the new product development process is critical for an organization's overall innovation performance (Franke et al. 2006) and is therefore particularly emphasized in the Henkel innovation process.

## 2 The InnoLounge Process

The *InnoLounge tool and process* was introduced in 2010 in order to invite the whole organization worldwide to participate in ideation and at the same time guarantee a fair evaluation of submitted concepts. Within a very short period of time, Henkel's InnoLounge established itself as the company-wide hub for new

---

T. Müller-Kirschbaum (✉) • J.C. Wuhrmann
Henkel AG & Co. KGaA, Laundry and Home Care, R&D, Technology and Supply Chain, Henkelstr. 67, 40589 Duesseldorf, Germany
e-mail: thomas.mueller-kirschbaum@henkel.com; juan-carlos.wuhrmann@henkel.com

ideas. It is the first recipient for all input worldwide. More than 1,500 employees are requested to enter their ideas and solutions to their consumers' problems. Through worldwide crowdsourcing, Henkel uses the creativity of the whole organization and invites, gathers, stores, and processes ideas and insights centrally.

A basic principle of this ideas system is that contributors do not only submit an idea flash, but scrutinize the consumer problem that forms the basis of this idea. Hence, at Henkel, the innovation process does not start with an idea, but with a problem consumers have. Consumers might express their problem explicitly, or it may be deduced implicitly through consumer observation and the evaluation of data. Henkel's employees in all regions get to know their consumers through home visits, market research surveys and analysis. They visit supermarkets and shops where detergents and home cleaners are sold. They initiate and read trend reports, not only those regarding fast moving consumer products, but also reports with a broader scope. They become real specialists regarding their consumers' needs. And it is at this very early stage of the innovation process that a consumer problem is solved with a new idea for a solution. As consumers' laundry washing and home cleaning habits differ considerably throughout different regions in the world, their problems and the appropriate technical solutions vary as well. Some of these solutions might be applicable only in a specific region, others might be transferred to other regions once they have been realized successfully in one region.

InnoLounge is not only the place to submit ideas at random, but can also be used for guided creativity efforts solicited via detailed innovation requests on specific topics. Parallel to the regular InnoLounge, these contests address items like sustainability, convenience, or special issues related to Henkel's brands and their image. In order to help employees hone their creativity skills, special training workshops and materials regarding creativity are offered, such as booklets about techniques, additional sources for inspiration, trend information, or tips and tricks on how to write a good concept or success story. Each submitter has to enter some personal details first and has the possibility to name co-submitters in case the concept idea was born in a team. This is an important part of the process as very often a concept is generated, formed and formulated in a team. In this way, InnoLounge is able to register a team's output, too, and take the whole team into consideration in case the idea is voted the winner of the month. Any product-related idea for the business unit Laundry and Home Care can be entered. This includes all kinds of product ideas within the existing business portfolio, but new *business ideas* are also welcome. The product idea should address a clear consumer benefit concerning a need a consumer has or will have in the future. When submitting a concept idea, Henkel employees have to describe the idea title succinctly before describing the *consumer insight* leading to the idea. The space allocated to the description of the relevant consumer insight only comprises 190 characters. It is therefore important to describe and formulate the consumer insight in a clear and concise way. This helps evaluators read through the texts and better grasp the meaning and message of the submitted concepts. Having indicated the consumer insight, Henkel employees have to enter the consumer benefit and, in a different entry, the reason for the concept. Again, to each entry only 190 characters are allotted. After the participation rules and the terms governing data protection have been accepted, the concept idea can be submitted and is ready for review by evaluators in the InnoLounge. Submitters have access to an overview of their contributions to innovation, which covers both their submitted ideas and their drafts. In the so-called InnoWarehouse, an archive with limited access, new ideas, evaluated ideas, and the winning ideas of the month can be found.

A full-text search helps submitters and evaluators alike to navigate through the archive and – where appropriate – use older concept ideas for further development, or use consumer insights for the definition of new ideas. An additional classification by product categories helps to simplify the search for ideas or consumer insights.

## 3 Evaluation

Each so-called *idea concept* – which is basically an elaborated description of a consumer problem and a description of a possible solution to this problem – is evaluated by 30 evaluators out of a pool of about 200 employees. This measure guarantees an ever changing set of evaluators, as both the individuals differ and evaluators are pooled together in a different mix for each evaluation. This is rather unique and emphasizes the relevance Henkel attributes to this step of the process. The pool of evaluators draws on employees working in Marketing, R&D, or Market Research. When asked to submit an evaluation, evaluators receive a normal e-mail indicating an evaluation request. A practical link leads the evaluator to the evaluation page and directly to all open evaluations. He or she can read the title of the idea concept, the consumer insight, benefit and reason why, but not the name or other details about the submitter, who remains completely anonymous. This last feature had been desired already at the beginning of the design of the InnoLounge process and tool in order to prevent evaluators from being influenced by name, hierarchy, or regional provenience. This influence is well known in research literature (Reitzig and Sorenson 2010) and can lead to undesired or simply biased results. Evaluators then have to mark the concept according to five criteria: market/sales potential, novelty, relevance, feasibility, and logic of concept. Each criterion is assigned a score from 1 to 7, one being the lowest possible mark, and seven the highest. Subsequently, evaluators have to enter a comment, which is mandatory. This comment is very important for two reasons. First, it should be supportive and motivating so that submitters know how to improve their specific idea or how to enter an even better concept the next time. Second, it should help to develop the idea concept further as most evaluators have long-standing experience in the relevant market categories and in consumer research. If an evaluator does not feel comfortable with a concept idea, he can click on a 'No Evaluation' button and is thereby excused from dealing with the concept further. The rationale underlying this option can be expressed as 'better no evaluation at all than a poor one'. Typically, an evaluation takes less than 5 min, a good investment in the future which is also manageable. Evaluators are trained to make use of the full rating scale and not only the medium range. And they have exactly ten working days for evaluating an idea. When this deadline expires, the evaluation is finished, even if not all of the 30 evaluators have had the possibility to grade the concept and write their comments. All concepts rated 4.0 and above on average are reported to the executive committee of the business unit including the names of the submitters. Each month, a winning 'Concept of the Month' is determined, albeit not automatically on the basis of the highest score. All ideas with an average rating of 4.0 and above go directly to a jury board comprising top management representatives from the Marketing and R&D departments, who finally and directly vote for the winner. In this way, possible repetitions are avoided. Such repetitions could concern concepts that are already in the innovation pipeline and have proceeded some steps further down the Henkel innovation process or ideas that have already been evaluated or even rewarded in previous rounds. There is only one winner or winning team every month, but the prize is declared empty if in a given month no concept shows an average rating of 4.0 or above. The winner or winning team is asked to present his/her or their idea in the executive committee of the business unit, regardless of the part of the world he/she or they come from. All winners are additionally announced in the quarterly newsletter from the CEO.

## 4  Additional Measures

One year after Henkel's InnoGate had been launched in the business unit Laundry and Home Care, results were rather satisfying.

The number of ideas and new concepts submitted during the first 12 months had accumulated to almost 1,000, and their average monthly number stabilized at a constant level. Nevertheless, statistics showed that there was room for improvement. Over 90 % of all submitted ideas came from only five countries, many countries and many employees within the country organizations did not participate. The good news was the trend towards team submissions, which had increased from 21 % to 31 % during that first year. Moreover, the distribution of new concepts between the different product categories of the business unit resembled the actual product portfolio and included almost 10 % of new business ideas. In order to find ways to improve the usability of the system, a feedback from the employees who used InnoLounge appeared necessary. In an MBA-thesis (Briehs 2011), a questionnaire was developed and sent to 150 employees, and the results were evaluated. The findings confirmed that one target which had been considered very important when developing the tool: InnoLounge is seen as simple, easy to navigate, and user-friendly. The launch of the whole process had resulted in much higher awareness and interest regarding all issues surrounding innovation. But only 40 % of users confessed to perceiving their idea generation skills as good, and almost 30 % judged their *concept writing skills* to be poor. Comments also centered on the fact that employees did not have enough time for idea generation or did not make enough time for this purpose in their daily routine.

A plan for improving InnoLounge was decided on and implemented quickly. InnoLounge should be linked more strongly to the operations side of the business through specific contests limited in time, the first one being 'Let's Futurize', which directly addressed the Persil brand. A symbol was created in the form of a robot, which was half light bulb, representing creativity, and half washing machine, linking it to the laundry care product category. The contest explicitly solicited new idea concepts specifically for the Persil brand and based on relevant consumer insights and market needs, the duration of the campaign was only 1 month. Unprecedented internal marketing activities were started, including posters, t-shirts, ties, video information screens, door signs, an announcement e-mail with a video, and more, which were sent to all employees. The result was overwhelming, the number of submissions increased by a factor of 5.

## 5 Lessons Learned

The creation of one hub of idea concepts worldwide in an organization in the *consumer business* like Henkel, and the implementation of a process starting from consumer insights and their problems results in an important increase of new product and concept ideas, both in terms of quantity and quality. The *evaluation process* works best if evaluators do not know the submitters, if there is an ever changing set of evaluators, and if they also evaluate concept ideas outside their own product categories. Additional measures are always necessary to maintain the momentum: specific campaigns, creativity skills training, information on trends, and additional activities.

## 6 Checklist

Based on the findings learned in the process of developing and running InnoLounge at Henkel, the following checklist for an idea creation process has been compiled:
- Submitters: Is the process of idea creation simple and clear to everybody? Have all employees worldwide been instructed about targets and the process?
  Have they generated the skills to understand consumer needs? Have they generated the skills to formulate concept ideas?
- Evaluators: Is it guaranteed that submitters stay anonymous? Have evaluators been instructed about the evaluation criteria? Is the changing set of evaluators guaranteed?
- Measures: Are additional accompanying campaigns under way? Do these campaigns reach the relevant employees? Are measures in place which delivers a feedback from employees about problems or issues which should be improved?

# Crowdsourcing: How Social Media and the Wisdom of the Crowd Change Future Companies

Johann Füller, Sandra Lemmer, and Katja Hutter

> *No matter who you are, most of the smartest people work for someone else. (Bill Jo, co-founder of Sun Microsystems, 1990)*

How is it possible to measure the weight of a live pig in a farm setting from a distance with a portable device? This is an interesting question, and the person who has the best answer will get a reward of $50,000 (InnoCentive 2012b). Such a question belongs to a crowdsourcing competition conducted on *InnoCentive*. InnoCentive is an open innovation and crowdsourcing platform on the World Wide Web where commercial, governmental and humanitarian organizations have the opportunity to solve key problems by giving them to a network of millions of people searching for the right solution (InnoCentive 2012a). Howe (2008) points out that a large enough number of amateurs may compete with professionals in various fields – whether astrophysics, meteorology, marketing, or journalism. This *shadow labor force*, as he calls it, may become one of the driving sources of innovation. Apart from InnoCentive, companies like *HYVE* (www.innovation-community.net), *One Billion Minds* (www.onebillionminds.com) or *NineSigma* (www.ninesigma.com) developed virtual spaces where interested and capable individuals unleash their minds to generate creative solutions.

Instead of the pure consumption of information, today's users of the internet have become active producers of content. The formerly passive crowd of consumers nowadays accomplishes a variety of impressive tasks. One such outstanding example is *Wikipedia*, an online encyclopedia that comprises information

---

J. Füller (✉) • S. Lemmer
Hyve AG, Schellingstr. 45, 80799 München, Germany
e-mail: johann.fueller@hyve.de; sandra.lemmer@hyve.de

K. Hutter
University of Innsbruck, Institute for Strategic Management, Marketing and Tourism, Universitätsstr. 15, 4. Stock West, 6020 Innsbruck, Austria
e-mail: katja.hutter@uibk.ac.at

on pretty much anything you can think of. It currently contains 4,133,184 articles offered in more than 250 languages (Wikipedia 2012). Not only is the information used by a huge crowd of people, the content is also created by the crowd. This success would never have been possible without the many-faceted masses, with various cultural and educational backgrounds, and their willingness to share existing, and collaboratively generate new, knowledge. The incredible number of approximately 2.4 billion internet users (Stats 2012) nowadays are interlinked and form online communities to discuss their problems, generate knowledge, evaluate products, articulate needs, design and analyze innovations, recommend and promote their favorite products and brands, create stars, collect donations, canvas and elicit support for politicians, show public resistance, and become antagonists.

How can companies benefit from the crowd? This article shows how organizations can outsource various activities to the crowd and sheds light on a number of underlying principles.

Companies may rely on the crowd especially for the development of successful innovations. They may open up their rather closed innovation processes and invite the unknown crowd of co-developers to generate ideas or provide solutions, for example through the application of innovation competitions (Terwiesch and Xu 2008), ideagoras (Tapscott and Williams 2006), or problem broadcasting platforms (Lakhani and Jeppesen 2007). While many crowdsourcing platforms only allow final contributions from single or team efforts and do not disclose these ideas to other participants, more and more crowdsourcing environments offer full community functionality. They allow joint idea development as found in open source software communities, such as *Apache* and *Firefox*. Participants can now interact, collaborate, vote for their favorite ideas, discuss various topics by leaving comments on other users' pin boards, and learn from the aggregated knowledge and feedback of others while still competing for prizes (Hutter et al. 2011).

The German-based fragrance brand 4711, for example, launched an 8 week online design contest on a company-branded microsite. The contest was timed to match their July 2011 market introduction of the brand new 'Nouveau Cologne' fragrance (see Fig. 55). Designers, students, professionals, creative's as well as other involved individuals were invited to submit their designs for attractive merchandising products and to compete for €10,000 in prize money and the actual realization of their submission. In addition to eliciting freely created designs, the contest facilitated non-designers to participate by using an online configuration tool, which allowed them to configure and individualize their favorite 4711 city bag, laptop or smartphone case. With 932 registered community members, a total number of 1,707 submitted designs (764 freely created; 943 configured), and 18,357 comments and messages, the contest by far exceeded 4711s expectations and turned out to be highly successful in terms of numbers and especially community interaction.

*Siemens* provides another example of a company relying on the idea of crowdsourcing. Siemens conducted an internal contest to increase its employees' awareness of social responsibility and sustainability and to get access to successful innovations (see Fig. 55). It asked for innovative ideas that extend its environmentally, economically and socially sustainable business approaches. Employees from

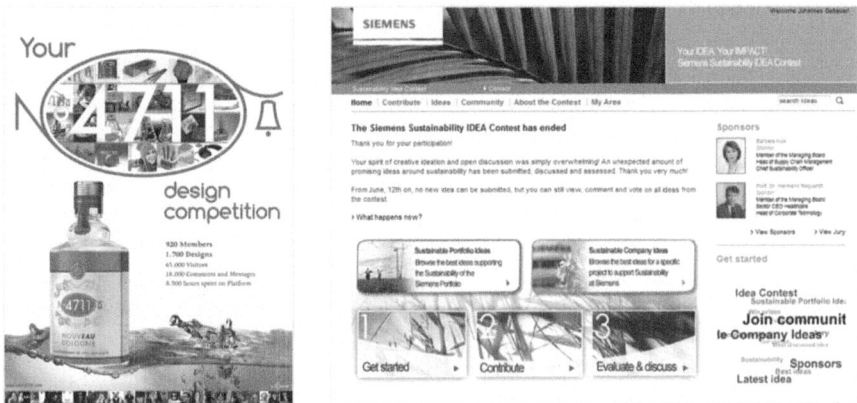

**Fig. 55** 4711 design contest flyer (*left*), Siemens sustainability contest platform (*right*) (Source: www.innovation-community.net)

all fields and locations submitted their ideas and further improved the submissions of others. 30,000 Siemens employees, which equates roughly 7.5 % of all Siemens employees worldwide, visited the website. More than 3,100 employees registered for the contest and created more than 850 ideas of impressive quality. The contest also supported cross-organizational collaboration. Colleagues from all business fields of Siemens and 43 different countries engaged in the collective idea generation. In total, the contest website generated 1.3 million views and almost 5,000 detailed evaluations of submitted ideas. This contest truly addressed all employees including both blue and white collar workers, executive managers as well as top level engineers.

While the two examples described above illustrate how the crowd can contribute to a company's innovation process, nowadays cases for almost every function along a company's value chain can be found in areas including marketing, logistics, finance, purchasing, production, or after sales service, as well as for any kind of industry, such as health, oil & gas, insurance, banking, and IT. The crowds help to accomplish complex challenges and come up with valuable results or attractive alternatives. Consider *Dell*'s support forum where customers post their challenges and solutions; they support each other, and fix problems, and ultimately save Dell the cost of customer support calls (Bernoff and Li 2008). Since 2001, *IBM* has been using online brainstorming sessions to involve its more than 300,000 employees worldwide in discussing and redefining the core IBM values, looking for new business ideas or sharing strategies and experiences that help to establish a new approach to sustainability in business (IBM 2008).

# 1 Crowdsourcing Principles

How does crowdsourcing work? Crowdsourcing is based on several principles that provide a rough understanding why and when collaborative approaches may work.

## 1.1 First Principle: Evolution

> Given a large enough beta-tester and co-developer base, almost every problem will be characterized quickly and the fix is obvious to someone. Or, less formally, "given enough eyeballs, all bugs are shallow." I dub this: "Linus's Law". (Raymond 1999, p. 41).

Linus's law already implies that innovations of higher quality may emerge if they are created in collaboration with a large enough mass of contributors pooled into an innovation community. *Open source* software development projects, such as the Apache web server, show that concurrence in design and the testing of software modules enables an efficient use of distributed resources connected by the internet. In contrast to the evolutionary, trial and error approach found in innovation communities, companies tend to standardize their innovation process by decomposing it into standard modules. However, Glass (1995) points out that creative ventures do not follow a fully structured approach that consists of routine actions. He further states: "Methodologies that convert design into a disciplined activity are not suited to addressing new problems to be solved" (Glass 1995, p. 41). Such structured step-by-step innovation approaches follow the logic of the *weak link chain* where the least productive element in the process determines the quality of the output (Becker and Murphy 1992). By contrast, crowdsourcing initiatives and innovation communities follow an evolutionary logic where the output is as good as its most productive member. It is only the talent and time available within the community that constrain the quality of the achievable output, not the structure of the approach.

## 1.2 Second Principle: Swarm Intelligence and Analog Knowledge

The knowledge relevant for the generation of a successful innovation may be widely dispersed within a society. It may be that it rests neither in the academic community nor in corporate boardrooms (Hayek 1945). Hayek states: "It is because the circumstances in which the different individuals find themselves at a given moment are different, and because many of these particular circumstances are known only to them, that there arises the opportunity for the utilization of so much diverse knowledge" (Hayek 1978, p. 9). Knowledge scattered among experts, amateurs, producers and users may be necessary to come up with superior innovations that do not reside in any individual, but are embedded in the intelligence of the crowd. In a computer model, Page (2007) discovered that teams incorporating a range of perspectives generally outperform groups of like-minded experts and also find better solutions than brilliant individuals who work alone. The model showed that "diversity trumped ability" (Page 2007, pp. XIX–XX). These findings contradicted the common understanding by showing that it is not the best and brightest who come up with the most promising solutions, but rather a randomly selected collection of (at least on average) 'less gifted' individuals. Innovation communities are seen as an effective organizational form of promoting innovation because they allow tapping into the distributed intelligence of their participants (Kogut and Metiu 2001).

Innovative solutions often stem from a (re)combination of existing knowledge from different domains. Hence, in addition to the swarm intelligence, it is the knowledge from other fields – not considered central by companies and by analog markets – embedded in the crowd which helps to generate superior innovations. In studying the InnoCentive problem broadcasting platform, Lakhani and Jeppesen (2007) found that complete outsiders to the field of the broadcasted problem had a probability of winning which is 10 % higher compared to insiders. Outsiders, with their 'fresh eyes', play an important role as they consider problems from different perspectives and apply solutions that are uncommon or entirely novel to the problem domain. New knowledge emerges through the constant exchange of information within innovation networks. Participants learn from the community and complement their knowledge. The disseminated knowledge inspires other community members to build on their ideas. Finally, through intense interactions, new solutions emerge that are superior to those which would have been created by a single user and are superior to the sum of the individual outputs.

## 1.3 Third Principle: Self-Selection

While in theory every community member can contribute and provide substantial improvements, in reality only a small number of contributors provide high-quality inputs. However, less skilled members are not of less value. They can comment on and test provided solutions. In their role as evaluators and testers, they contribute to lean processes and less deficient solutions (*Mockus* et al. 2000). As innovation communities provide coordination between task and competence, they are capable of exploiting the intelligence of their members. Members who are highly qualified tend to engage in more complex activities while less skilled members tend to participate in less complicated tasks. This kind of self-selection is certainly welcomed in crowdsourcing as it makes participants choose tasks they are interested in and think they are also capable of accomplishing. Qualified participants engage in crowdsourcing and innovation communities because they are naturally interested in the topic and because they can choose the activity which matches their ability at an intensity which is appropriate for the intended level of engagement. In contrast to the traditional division of tasks practiced in companies, where employees accomplish the tasks which have to be done, participants in crowdsourcing initiatives can select the activities they like to work on and also consider interesting themselves. Research shows that humans tend to perform at their peak levels when they consider the assigned task intrinsically rewarding and this task is neither too complex nor too easy for them (Füller 2006).

## 1.4 Fourth Principle: Task Distribution and Aggregation

In order to crowdsource an activity, it has to be distributable in form of an open call. This means that the task accomplished by the crowd has to be comprehensible and clearly described so that the undefined public can work on it. Before an activity is announced to the public, it often has to be divided into sub-tasks and sub-modules which can be completed individually by participants with reasonable effort and which also may require different skill sets. *Amazon*'s Mechanical Turk[1] presents an interesting example of how to solve tasks such as book translations, product reviews, photo descriptions, or database entries in short periods of time by dividing them into micro tasks and outsourcing them to a large enough community. The *TopCoder* (www.topcoder.com) software developer community presents a further example where firms can even crowdsource the development of entire software systems (Lakhani et al. 2011). At TopCoder, complex software solutions are realized by slicing the undertaking into various modules, phases, and task types such as planning, designing, architecture, algorithms, programming, testing, assembling, and bug fixing, and then exposing them to its 450,000 community members via different contest formats. Each sub-task has to be achievable within a compliant timeframe of about 2 h. In this case, crowdsourcing takes the form of *peer production* where contributors work on different sub-tasks which are put together once they have been accomplished. Modularity is essential in order to not only decompose an activity into sub-tasks, but also to aggregate the outputs of the distributed sub-tasks into one final solution. Modularity allows relying upon different sources and benefiting from the distributed intelligence of suppliers, producers, and users (Baldwin and Clark 2000). Task distribution is often complex, as the adjacent aggregation of the individual outputs has to be thought of as early as in the distribution phase. Certain standards, templates, formats, processes, guidelines, and norms as well as the provision of interfaces and assembling intelligence may be needed in order to recombine and aggregate individual outputs. The design and organization of crowdsourcing platforms that allow collaboration and concurrent work on different tasks often proves difficult. Besides allowing for the correct aggregation of contributions and information, the platform design and organization

---

[1] "The Mechanical Turk is a crowdsourcing Internet marketplace that enables computer programmers (known as requesters) to co-ordinate the use of human intelligence to perform tasks that computers are currently unable to do. The Requesters are able to post tasks known as HITs (Human Intelligence Tasks), such as choosing the best photo of a store front from among several photographs, writing product descriptions, or identifying performers on music CDs. Workers (called providers in Mechanical Turk's Terms of Service, or, more colloquially, turkers) can then browse among existing tasks and complete them for a monetary payment set by the Requester." http://en.wikipedia.org/wiki/Amazon_Mechanical_Turk

also has to empower interaction and collaboration, spur diversity in solutions, ensure independence of contributions (Surowiecki 2004), and at the same time to provide a rewarding and beneficial experience to its engaged contributors (Füller 2010; Kohler et al. 2011).

## Conclusion

Although crowdsourcing has become popular in the context of innovation, it is not limited to it. Indeed, many tasks – if not anything! – can be outsourced to the crowd. Instead of referring to crowdsourcing merely as a new method that can help foster a business, one should see it as a new way of structuring and organizing work by relying on the principles of evolution, swarm intelligence, self-selection, and task distribution. More and more companies may try to leverage the distributed intelligence of their customers, suppliers, employees, and internet users by setting up crowdsourcing platforms and broadcasting their problems and tasks to the internal and external crowd. However, a sound understanding of crowdsourcing and its underlying principles is needed in order to decide whether crowdsourcing is appropriate. Although crowdsourcing has become quite popular, it also faces some difficulties that must be overcome in order to provide value. Problems with crowdsourcing initiatives often emerge because its initiators lack the sound knowledge of crowdsourcing discussed above. The absence of a clear task description and problem explanation, failure to provide an appealing platform design, solid terms and constraints of participation, or neglecting to offer a fair prize structure may create difficulties for companies that apply crowdsourcing. Once they have gained the appropriate knowledge, companies can truly benefit from crowdsourcing and generate superior innovations.

# Building a Bridge from Research to the Market: IBM's Industry Solutions Labs

Matthias Kaiserswerth

## 1 Innovation at IBM

IBM is all about innovation. Innovation makes us essential to our clients and partners. It means helping them be prepared for the challenges ahead, and it means being creative and forward-looking in everything we do – in services, software, and hardware. This dedication to innovation is reflected in one of our three core values: 'Innovation that matters – for our clients and the world'. In other words: making the world work better.

A cornerstone of IBM's innovative capacity is, and always has been, its research organization – ever since the first corporate research lab was opened in 1945. *IBM Research* – 3,000 world-class scientists and a network of 12 research centers spanning the globe – paves the way for radical innovation, while delivering value to today's clients. Its mission is to pursue a bold research agenda to push the boundaries of science, technology and business.

Innovation is no longer merely a synonym for a brilliant idea, an invention or a patent. Instead, it is defined by its impact, the opening-up of new spaces (e.g., technical products, new markets, or organizational transformations). Its 'wow effect', according to Nowotny (2008), does not result from the idea, but from how it influences our lives and work, our way of seeing and thinking. Increasingly, innovation is no longer created by individuals in secluded laboratories. It has become multidisciplinary, global, and multidimensional. It no longer revolves only around product innovation: Innovation thrives in services, business processes, business models, management, corporate culture, and in politics and society.

Last but not least, innovation is collaborative. Open and joint research have become an imperative. They will be crucial for success. In the 2012 IBM Global CEO study, 70 % of the 1,700 CEOs interviewed said they were aiming to partner

M. Kaiserswerth (✉)
IBM Research – Zurich, Säumerstr. 4, 8803 Rueschlikon, Switzerland
e-mail: kai@zurich.ibm.com

extensively. Business leaders around the world are recognizing the value of *collaborative innovation* to sustain competitive advantages in a world of relentless change, economic pressure, and increased global competition. And, looking ahead, collaboration is a prerequisite when tackling some of the world's greatest challenges – in infrastructure, transportation, healthcare or energy – to name just a few.

In recent years, another key change has occurred in how IT innovations are brought from the lab to the market. The business model has evolved from a solely product-based or software-licensing model to software as a service and to joint pilot projects and joint development with the client.

IBM Research has embraced this notion of innovation and it has become part of its DNA. The world has become IBM's laboratory where it works with clients and partners to pilot new solutions and to deliver value. To support this effort, IBM Research has created several instruments to build a bridge from research to the market, foster real-world innovation, and establish innovation ecosystems. The *client centers* at IBM's research sites, the so-called *Industry Solutions Labs* (ISL), are a crucial piece of the puzzle.

## 2   IBM Research and the Market

In 1997, IBM established the first client centers at research sites. It did so with the vision of creating synergy between the future needs of its customers and emerging technologies from IBM Research. The opening of the first research client centers represented a historic step towards opening up IBM Research – the company's crown jewel and one of its best-kept secrets – to the external world. This opening was one of the results of IBM's so-called 'near death experience' at the beginning of the 1990s. At that time, IBM was very close to bankruptcy. And the new CEO, Lou Gerstner, who had previously been with *American Express*, soon identified one of the main reasons for that development. He realized that somehow IBM had lost contact with its clients. A strong belief in the high level of their own technical expertise had lured many employees into a pure 'talking mode' when meeting with customers: the art of listening had vanished at IBM. One of Gerstner's first activities was, therefore, to launch an initiative called 'bear hug'. He asked the 'IBMers' to go out and get into a true dialog with their clients. And he opened up IBM's research labs to foster this kind of active exchange right from the beginning. Today, close interaction with the external world as well as open and collaborative innovation are imperative for IBM Research to drive innovation.

Currently, there are four industry solutions labs – located in New York, Zurich, New Delhi, and Beijing. Their role in IBM's innovation process is manifold:
- **Forums:** The Research Client Centers promote active dialog and exchange among IBM's researchers and business experts, on the one hand, and industry, government and representatives of academia, on the other hand. Here, decision makers can meet and brainstorm with world-leading scientists and business experts and discuss their complex business challenges. IBM researchers, in

turn, obtain a market's-eye view of the ways in which their technologies can be applied to real-world problems.
- **Knowledge hubs:** Research Client Centers also provide subject-matter expertise on technology trends, knowledge and strategic innovation management, and industry-specific insights.
- **Enablers:** The Research Client Centers prepare the ground for joint projects and *innovation partnerships*.

Each visit is tailored to the client's specific needs, with thought-provoking discussions, brainstorming sessions on demanding business issues, and demonstrations of key solutions. Each center offers demonstrations of emerging technologies – ranging from computer systems, to cloud computing applications, to advanced security solutions, to data analytics and demos designed for specific industries, such as healthcare, energy, and finance.

## 3  Innovation at Work

The industry solution lab's goal is to pose the right questions to stimulate fruitful discussions and generate insightful answers to complex challenges. Decision makers who visit the lab on a regular basis appreciate its open atmosphere and creative environment. They enjoy being challenged with questions about their vision of the future, questions that 'unfreeze' their established view of reality, questions that provoke outside-the-box thinking.

The right questions are the key to profound insights, and IBM's R&D personnel is among those best suited to pose them. In this way, a relationship based on inspiration emerges between scientists, innovation experts, and business leaders. Looking at things from different perspectives and in a systemic and holistic way often stimulates innovation impulses. What is discussed often stretches the imagination, but is not science fiction: It is about scrutinizing the work of leading scientists, who are building prototypes of technologies that will change the future of business and society. And it is based on solid science being conducted in IBM's labs today.

**Discovering IBM Research's Look into the Future.** One of the core elements within industry solution lab workshops is the discussion of key technology trends based on the *Global Technology Outlook* (GTO) – IBM Research's look into the future. The GTO constitutes an important element in shaping IBM's corporate strategy. It is IBM's annual vision of the future of *information and communication technology* (ICT) and its impact on business and society at large. At the Research Client Centers, the implications of these developments are explored and shared with the participants in the workshops. The global technology outlook is not designed to benefit solely IBM. IBM invites clients, partners and other interested parties to discover this vision of the future and to leverage the insights for their own organization's benefit.

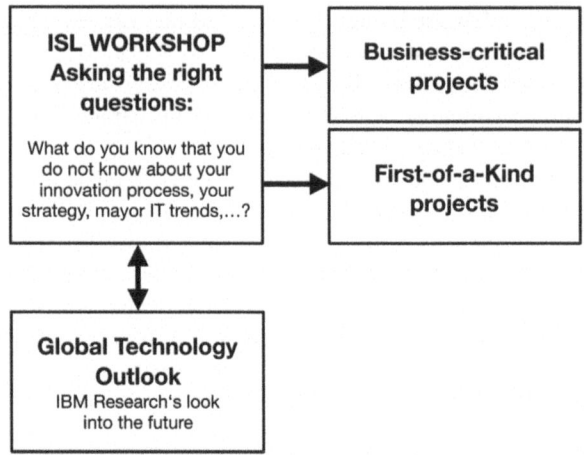

**Fig. 56** The IBM Industry Solution Labs (ISL) as the starting point for innovative partnership projects

While the global technology outlook is used for industry solution labs, information that is created within industry solution labs may also flow into the generation of GTOs (see Fig. 56). The development of the technology outlook is a true bottom-up process. Every IBM scientist is encouraged to engage in the GTO process and propose what he or she considers transformative technology trends. It is a way of tapping into the minds of scientists at the cutting edge of research. This annual activity encompasses generating ideas, gathering data, and rigorously debating the proposals, including their possible impact on business and society. The key trends identified are presented to IBM's CEO and used within the company to define areas of focus and of future investment. In the past, the technology outlook has predicted such emerging trends as the Internet of Things, virtualization and cloud computing, and the growing importance of data and analytics. The current GTO reports on six key findings that share a common denominator: analytics. The explosion of unstructured – and increasingly uncertain – data will amplify the need for new models and new classes of computing systems that can handle the unique demands of big data analytics.

**Creating Innovation Partnerships.** Often industry solution labs workshops are the starting point for a project with IBM Research that is critical to a particular business. The Flexlast project, for example, resulted from a series of industry solution lab workshop sessions. In the Flexlast project, IBM scientists work together with Swiss energy utilities and the largest Swiss supermarket chain to pilot a smart energy grid using refrigerated warehouses as buffer to help balance fluctuations in the availability of sun and wind energy. In the same vein, a client from the automotive industry decided to engage in energy-related science projects with IBM Research and sums up the company's industry solution lab experience as follows: "Visits to IBM Research have always helped us reflect on current issues and given us the opportunity to see how these issues are judged from a different perspective. Discussions on such subjects have been rewarding and have benefited both partners. In the future, I hope we shall find further areas of common ground that will allow our companies to continue their relationship of trust."

Workshops at IBM Research also help initiate so-called *first-of-a-kind (FOAK) projects*. The FOAK program brings together IBM scientists and clients to test a new technology on a real business problem for the first time, thus leapfrogging the traditional development cycle and helping guide research efforts toward strategic markets. Costs and risks are shared between IBM and the client, and the partner benefits from the competitive advantage gained by being the first to use the new technology. A prominent example is Aquasar. IBM scientists in Zurich have explored innovative hot-water cooling technologies for computer systems. Using nature-inspired concepts, they demonstrated that it is possible to cool a computer with water that is 60 °C hot, which not only decreases the energy consumption of the system by 40 %, but also facilitates direct re-use of the removed heat, e.g., for heating buildings. In 2009, the Aquasar FOAK project with *ETH Zurich* was kicked off, and the pilot computer system was put into operation in 2010. The novel heat-removal concept now heats one of the main ETH buildings. Delivering this proof of concept led to the commercialization of the technology in 2012 and its use in one of Europe's most powerful supercomputers, the SuperMUC at the *Leibniz Supercomputing Center* near Munich in Germany.

## 4 Success Factors

Of course, not one single aspect of an ISL experience drives innovation through thought-provoking workshops, but rather it is always a combination of factors. Being co-located with an IBM Research lab, the centers have direct access to the scientists and benefit from the inspiring and creative atmosphere in the labs, an ideal framework for an open *think tank*. The ability to demonstrate emerging technologies is also important: Clients can explore and experience potential future solutions not only 'minds-on' – but also hands-on. Seeing research prototypes at work can make all the difference, generate excitement about a particular technology, and set the stage for FOAK projects or other forms of innovation partnerships.

The tailor-made design of the workshops is another critical success factor: It addresses an organization's specific challenges with the help of the right experts in a mind-opening setting. Significant innovation often starts out as a 'fuzzy' and 'crazy' idea. Turning such an idea into game-changing innovations requires a special kind of person. As Gerd Binnig, one of IBM's Nobel laureates, puts it, grounded dreamers are the key to success. They succeed in managing, and exploiting, the fertile tension that arises from having the head in the sky and the feet firmly rooted on the ground, and they do so in a highly creative and productive way. This is the kind of person who will be brought to the table at an industry solution lab workshop.

# The MINI Countryman: Successful Management of the Early Stage in a Cooperative Product Development Environment

Markus Seidel, Patrick Oberdellmann, and Antony Clayton

## 1  Initial Situation and Objective

In 2001, the *MINI* brand was successfully re-launched by its new owner, the *BMW group* with headquarters in Munich, Germany. For the MINI brand, BMW decided to pursue a unique strategy in the automotive industry: For the first time a small car segment was combined with a dedicated premium approach – BMW's core competence. Today, MINI is one of the fastest growing brands in the industry. One of the most important growth factors is the consistent expansion of the MINI brand into different market segments without the brand losing its DNA. The MINI Countryman was launched in 2010 and immediately became a best-seller, reaching an annual sales volume of approximately 100,000 units. The Countryman was the first small premium Sports Activity Vehicle (SAV) and has been copied intensively since its market introduction.

The product's initial 'birth' was extremely difficult since several challenges had to be overcome. With a very iconographic appeal, the MINI Brand was strongly associated with the unique MINI vehicle concept that had not been changed since its initial market introduction in 1959. Market research indicated that customers could not imagine changing this concept or even altering it slightly. Various automotive specialists and historians argued that it would not be possible to create a larger MINI in a modern market segment. Therefore, the development of the MINI Countryman was not only a matter of management excellence in product development. In addition, it was very much a matter of brand, market and sales channel development and professional *change management* along the whole value chain.

In the following article, the key challenges and problems during the product development process and the specific solutions that have been developed will be

M. Seidel (✉) • P. Oberdellmann • A. Clayton
BMW AG, 80788 München, Germany
e-mail: markus.seidel@bmw.de; patrick.oberdellmann@bmw.de; antony.clayton@bmw.de

described. The focus lies on the early stage of the process, since in this phase most process disturbances occurred that sometimes almost led to project termination. Based on this description, general interdependencies and causal nexuses will be described.

## 2 The Early Stage of the Countryman's Product Development Process

### 2.1 BMW's Vehicle Development Process

Basically, the BMW vehicle development process has two very distinct phases: the so-called 'early stage', and 'series development'. While the early stage is designed to find the right product concept that will meet all corporate targets (sales volume, production costs, fuel consumption, quality, profitability, etc.), series development aims to launch the vehicle successfully within a given time and budget framework. The early stage itself is subdivided into a strategy-/initial phase and the concept development phase. At the end of each phase, the project has to report to the BMW board. The board then evaluates the project status and authorizes any activities to be carried out in the following phase (Fig. 57). The early stage ends with the milestone 'target agreement' that defines specific targets for a wide set of different parameters intrinsically linked to the company's balanced scorecard.

If the project passes the milestone 'target agreement', costs start to increase exponentially since huge investments in machines, supplier contracts, prototypes, testing, etc. have to be authorized. For a single vehicle project, the overall development budget can exceed 1 billion EUR. Contracts with hundreds of different suppliers on several tiers have to be made, and dozens of teams inside and outside the organization have to be staffed and managed professionally.

After the target agreement has been reached, the project's targets cannot be changed easily since profitability is very sensitive to changes in parameters and interdependencies between the various vehicle components and modules are huge. It is very unusual, not to say impossible, to stop a project after target setting since sunk costs would substantially harm overall company profits. Therefore, the basic target of the early stage is to create a firm organizational commitment towards a specific vehicle project for a decade, with total production costs sometimes amounting to more than 10 billion EUR.

### 2.2 Strategy and Initial Phase

Since the re-launch of the MINI brand in 2001 had been very successful and initial sales targets had outperformed significantly, in 2005 the central product planning department of BMW started to consider further *product portfolio* expansion to stimulate additional growth. As part of a strategic framework, several targets were defined initially:

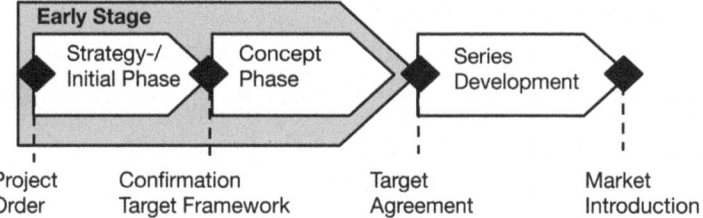

**Fig. 57** Simplified development process (Source: BMW Group)

- A significant contribution to further growth through a larger and popular vehicle concept (no niche model).
- Seizing the global automotive trend towards SAVs that have a four-wheel drive powertrain, a higher and comfortable driver seating position in the cabin, which provides a better outside view, and many other use cases that support leisure activities such as mountain biking.
- Ensuring product authenticity regarding styling and driving characteristics (in particular the MINI 'go-cart' feeling).
- Achievement of the BMW group profitability targets.

This plan meant a radical innovation for MINI with a high degree of uncertainty, many chances and risks, and a large extent of organizational turmoil. Since its market introduction in 1959, 'the' MINI had become an icon of the automotive industry with a very specific vehicle concept. At no time in the brand's history had the expansion of the brand into totally different market segments been successful.

For these reasons, the initial discussions and considerations in the organization were very controversial. But soon it became clear: There was no alternative to extending the brand in order to reach the critical mass necessary to manage an automotive brand successfully and to cushion the volatility of the sales volume. Moreover, the functional managers involved came to the conclusion that the project would be able to reach the strategic targets – but with one exception: Since internal development and production capacity was insufficient, the project would have to be outsourced completely to an external partner. Market introduction was scheduled for 2010 without decision-makers knowing what the project would look like and where it would be developed and produced. In 2006, the BMW board authorized the start of the concept phase and allocated the required resources to the project.

## 2.3 Concept Phase

During the concept phase, four major challenges had to be overcome that had significantly increased complexity in comparison to regular vehicle projects:

**Management of the Cooperation.** After long and strenuous negotiations, *Magna Steyr Fahrzeugtechnik (MSF)* in Graz, Austria, was selected to be the partner for development and production. The cooperation model was designed in such a way

**Fig. 58** Simplified cooperation model (Source: BMW Group)

that Magna was phased in during the concept phase, but the MINI team had overall responsibility until target agreement. After that milestone had been reached, MSF was in charge of the whole vehicle project including production, and MINI assumed a supervisory role (Fig. 58).

Such a cooperation model places very high demands on the management on both sides, since the complexity is huge. Therefore, the BMW/MINI purchasing department developed a unique business model with a dedicated incentive system to make the project manageable.

MINI and MSF signed a combined development and delivery contract. To steer such a highly complex structure, a highly skilled team and a specifically defined committee structure are required.

Especially the MINI project management had to be qualified to control this structure in cooperation with MSF. It had to be ensured that the interests of the technical project leader and the BMW purchase departments were adequately enforced.

**Target Setting Process.** On the other hand, the optimum of the project targets had to be identified and refined to achieve the maturity level required for 'target agreement'. This resulted in a time-consuming convergence process of technical, design, sales and economic requirements.

The frequent changes of the project targets during this process were the result of the insecurity of the organization as well as the degree of innovation of the project.

As neither a preceding model nor a direct competitor existed, there were no reference data available to validate the targets. Therefore, the parameters of the business case fluctuated greatly in this phase. This was also communicated in frequent reports to the top management of the organization. If the MINI Countryman was not going to be successful in this new vehicle segment, the MINI long-term strategy of the BMW group would have to be redefined. The global economic crisis from 2008 to 2010, i.e., during the concept phase of the project, put additional stress on the final decisions of the project.

**Organizational Development.** Prior to the development of the MINI Countryman, MINI had already developed the existing vehicle projects together with a cooperation partner (*IDG* in Turin, Italy) in a complex and international engineering structure. This experience provided a good basis for qualifying the organization to develop the MINI Countryman as a second product line in parallel with a new cooperation partner.

Normally the development activities of the BMW group are focused in one location only, since this facilitates optimized information exchange and superior problem-solving capabilities. Moreover, it guarantees smooth know-how transfer between the different vehicle projects and departments. Therefore, working in the complex and culturally diverse MINI development network was a real challenge for all employees and executive managers involved. This required a special process of employee selection and training.

Furthermore, the communication infrastructure needed to be improved. Most of the meetings were held per videoconferencing. At one single point in time, up to seven videoconferencing systems were used at MINI simultaneously to minimize travel costs and time. This led to a new way of working together for all employees involved and was a challenge which is not to be underestimated: For many employees it had been normal to have face-to-face communication with internal colleagues in Germany before joining the Countryman team. After starting to work for the project, the engineers had to work with a technical peer group from a different company, which added intensive negotiation processes of commercially driven topics to their workload. Not everyone found this development environment easy.

**Concurrent Engineering.** Quickly it became apparent that in the concept phase, the development and the cooperation processes were interwoven. The use of significant management resources in top management subsequently led to a considerable delay in the process. As a consequence, series development had to be started before successful completion of the concept phase and in the absence of an appropriate target agreement, since the launch date had already been set.

Both phases overlapped for about 6 months. In this period, the top management was involved intensely on both sides of the project and was constantly striving to ensure reliable and close co-operation between the participating teams and to resolve conflicts as quickly as possible. Firstly, far more conceptual and technical changes were necessary in this later stage than had originally been anticipated. Secondly, substantial costs had to be triggered without a final target agreement. In this phase, a task force-like structure was established within MINI and MSF, which was seen as a shared 'company within a company' and acted accordingly in a straightforward and goal-oriented manner.

## 3 Summary and Outlook

From the example of the MINI Countryman, the following findings can be derived that help overcome the special challenges of the early phase of product development in a cooperation model:

The more innovative and risky a development project is, the more turbulent the course it takes – especially in the early stage. With an increasing degree of innovation comes a decrease in predictability and structuring of the project. Often questions or problems occur which could not be foreseen at the beginning of the project: delays, unplanned increase in costs, unexpected interactions between different components of the product and technologies: these lead to a considerable increase in change intensity.

If we judge the product at this stage only on rational criteria and lose sight of the strategic considerations, there is a high risk that this will prevent fundamental change or that the project will be stopped. This can lead to serious conflicts.

The early phase of the MINI Countryman had a pronounced entrepreneurial character, in which not only analytical criteria, but also subjective 'gut feelings', expert assessments and strategic visions were considered.

The findings outlined above are particularly relevant for innovative projects in the automotive industry: Based on the average of 4–6 years of product development, a 6–7-year period of marketing and subsequent multi-year spare parts supply, very long-term forecasts have to be made and consistently reviewed. Due to the intensity of the investment in the automotive industry, even relatively small variations in sales volume or production costs can lead to large fluctuations in the company's economic business case.

Co-operations in the automotive industry are becoming increasingly important. Only in this way is it possible to compensate for the constantly rising development costs and capacity needs due to constant product portfolio extensions, and to quickly develop and realize market opportunities. Against this background, the ability to manage partnerships is a major strategic key competence for a car manufacturer.

# Controlling the Early Innovation Phase at Autoneum

Javier Perez-Freije

## 1 The Need for Innovation and Related Controls at Autoneum

Effective and efficient research and development (R&D) is increasingly held accountable for company success. At the same time, the management at *Autoneum* is confronted with the greater complexity associated with a wide range of often interrelated technological, market and process options from which to choose in constraint conditions.

Competition in the automotive supply industry is fierce, and technology is and will become an even more important differentiator. With the rise of global platforms and many OEM (Original Equipment Manufacturer) preferred supplier programs, Autoneum needs to achieve global scale and competitiveness. This can be achieved through delivering advanced technologies, especially around fuel efficiency, weight reduction, and managing complexity. Strong, innovative suppliers are likely to swallow up competitors who can't support global programs across all important regions or can't deliver technology or cost leadership to global OEMs.

However, to control innovation efforts, Autoneum addressed a severe dilemma: although some degree of freedom and flexibility is an essential ingredient to productive R&D teams, management must still institute effective control mechanisms, move projects in the right strategic direction, monitor progress towards project goals, and allow for adjustments if necessary. Current controls facilitate an environment conducive to innovation, which reduces specific barriers and drivers that constrain or foster creative performance. Autoneum regards performance as innovation efforts successfully transformed into new products, processes or materials.

J. Perez-Freije (✉)
Autoneum Management AG, Schlosstalstr. 43, 8406 Winterthur, Switzerland
e-mail: javier.perez@autoneum.com

## 2 Controlling Concept

Traditionally, Autoneum has established formal controls for the development phase of the innovation process. However, to compete on the basis of innovation, formalizing the early innovation phases was clearly necessary and, according to an emerging principle in R&D management (Cooper et al. 2001), competence in controlling the early phases, also called the fuzzy front-end (FFE), is crucial for new product success. The early innovation phases describe that period between the first consideration of an opportunity and when it is judged ready for development (Kim and Wilemon 2002). Literature on the management of R&D is not clear whether the early innovation phase should be managed differently depending on the project characteristics such as risks, ambiguity, or lack of routineness (Perez-Freije 2008). However, controls have increased recognition as a vital instrument for coordinating activities and their results. Accordingly, the related gain in effectiveness and efficiency of innovation activities has been crucial for Autoneum.

Autoneum differentiates the FFE activities and related controls into (1) invention and (2) innovation. Invention activities relate to the idea generation and evaluation phase and, therefore, are characterized by a high level of uncertainty and ambiguity. Accordingly, the knowledge about cause-and-effect relationships and outputs is vague. Autoneum uses various input controls that can be considered as a form of resource allocation. Inputs represent raw materials or stimuli that a system receives and processes. In addition to people resources, inputs, therefore, include equipment, facilities, and funds needed to complete various activities. Input controls aid in general the exposure to and the acquisition of new knowledge. Outputs are also measured while they are not known beforehand.

Innovation activities relate to Autoneum's core competencies and summarize the efforts to generate new products, processes or materials. These activities lead to a well-defined written and visual description of a product that includes its primary features and customer benefits, combined with a broad understanding of the technology needed. The controlling approach at Autoneum aims at fulfilling various aspects: first, potential customers have difficulty articulating the needs that the new technology might fulfill. Thus, managers confront uncertainty about where the most fruitful market opportunities lie. Second, new product managers are uncertain about how to turn new technologies into tangible new products to meet the needs of prospective customers. Often managers are uncertain which product features and benefits can be delivered, at what cost, and with which pace of technological advancements. Third, higher-level managers suffer uncertainty also in terms of how much capital to deploy and when, in pursuit of new markets. Extensive investments usually are necessary to turn new technologies into tangible products and then successfully take those products to market.

The basic premises of ambiguity and uncertainty inherent to invention and innovation activities have been considered when designing the controlling concept of the early innovation phase at Autoneum. Inventions are intrinsically non-routine and dynamic, and typically involve ad-hoc decisions and loose management of activities. However, for effective innovation, a singleness of purpose is required,

# Controlling the Early Innovation Phase at Autoneum

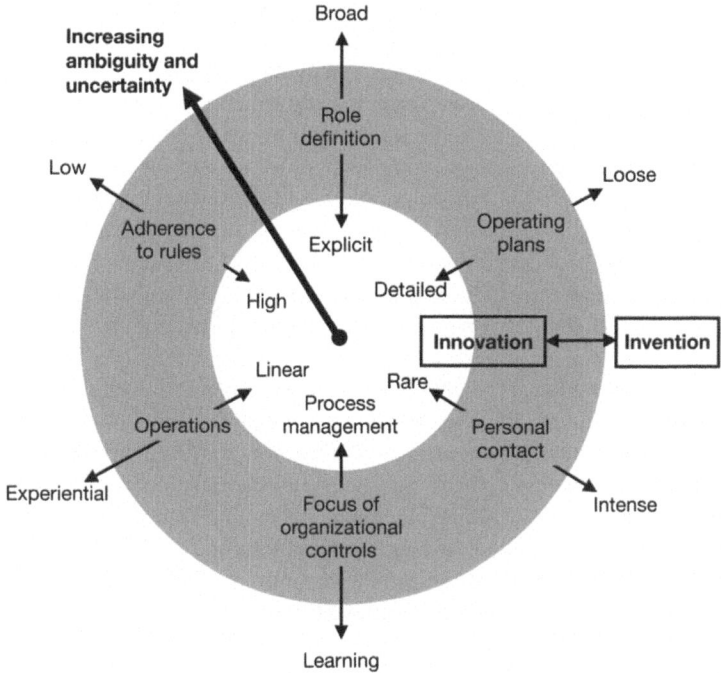

**Fig. 59** Basic premises of Autoneum's control system

which can be achieved through structuring processes and activities. These momentous assertions suggest that certain structures might be better suited for particular organizational tasks. Control helps manage creativity throughout the FFE and prevents the circumstances in which too much creativity reduces R&D performance because of a loss of focus. Accordingly, managing the early innovation phase through boundary-spanning, communication, feedback, and information sharing facilitates a climate of active participation and minimal dysfunctional conflict. The generic differences among control practices of the later (innovation) and earlier stages (invention) are indicated in the control characteristics (see Fig. 59).

## 3 Stage-Gate Process

Autoneum differentiates three types of innovation:
- Structural innovation to maintain current market position (substitution of existing product technologies and processes)
- Sustainable innovation to enhance current market position (new technologies and products)
- Innovation to increase company profitability (cost leadership).

All innovation types are handled in the central Research and Technology (R&T) department and require a strategically and organizationally embedded form of innovation management. To manage innovation properly, Autoneum has formalized the early innovation phase by implementing a *stage-gate process* (process controls). Even though the knowledge of cause-and-effect relationships for all innovation types is not complete, process controls are important in Autoneum's control efforts. Autoneum addressed the task of creating procedures or routines to prioritize promising opportunities or combine related ideas in new ways. Accordingly, process controls are emphasized when specific processes to be followed can be characterized by less complexity. This allows bottom-up opportunity generation and top-down prioritization playing a key role in ensuring R&D effectiveness.

The consistency of the stage-gate process suggests that innovation opportunities are likely to be compared. However, though individual R&T professionals focus on their particular work, management, with a broader overview, deliberates on multiple options. Even if process controls may not be appropriate for specific innovation, these controls facilitate evaluation. Frequent interactions and procedures, among other things the attributes associated with process controls, enhance the detection of an option's value. Accordingly, the primary benefit in using process controls is to reduce the variance in the methods employed during the early innovation phases. Consequently, close adherence to the prescribed methods and repetition reduce errors and enhance precision of cause-and-effect relationships and desirable outputs.

The stage-gate process has two main phases: (1) Ideation & Screening and (2) Product, Process, and Material Development. During ideation and screening Autoneum links its technological capabilities with the customer and, ultimately, market opportunities, because higher-impact applications defined in respect of current products, processes, or materials can lead to long-term competitive advantages. Phase (2) of the process involves making strategic choices about which markets and products to invest in, as well as resource allocations that define how to spend scarce resources. A structured, formalized process avoids pressure from interest groups, justifies decisions, and communicates decisions. Condition precedent to fulfill all related R&T tasks is the embedment of market and competitor analysis, customer needs, and Autoneum's strategic planning into a closed loop with a regular review of the controlling instruments in the early innovation phase (see Fig. 60). The closed loop ensures that so-called *Innovation Essentials* are defined: Innovation and technology needs by product family depending on their life cycles and market analysis, demand for incremental or radical innovation, competitor differentiation, and definition of core competencies to ensure high effectiveness. Subsequent *Fact Finding Missions* indicate the relevant products, market trends and technologies as well as the related characteristics and innovation capabilities. As a result, the stage-gate process bears innovation that can be transferred to the development departments, where customized parts are created.

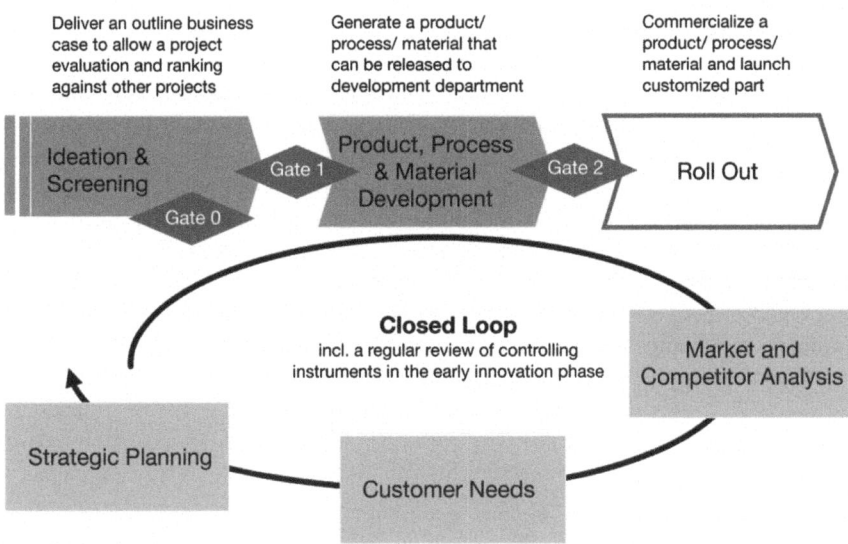

**Fig. 60** Stage-gate process at Autoneum

## 4 Checklist

The use and impact of process controls in the early innovation phase, including better coherence and relevance of R&T initiatives and efforts, as well as facilitating *R&T effectiveness and efficiency*, have been validated and specified with the example of Autoneum. The following checklist provides support when creating a controlling concept for the early innovation phase:

- Operation of controls: What is the intention of the controlling system (e.g. evaluation of innovation projects for their prioritization)? What are the goals and what is controlled? Control approaches depend on (1) nature of cause-effect relations, (2) nature of interdependency, complexity, and integration, (3) degree of regularity and continuity of control analysis, and (4) degree of judgments.
- Purpose of controls: Controls can refer to projects or the R&D function. When related to projects, controls have to (1) facilitate learning, (2) indicate corrective actions at the right moment in running efforts, and (3) prioritize promising and relevant projects for development.
- Controls and their results impact organizations and their behavior: (1) Increase knowledge of cause-and-effect relationships and outputs, (2) meet information requirements, (3) facilitate well-balanced arbitration and decision-making, and (4) enhance motivation.

## 5 Lessons Learned

Process controls (e.g. stage-gate process) guarantee that decision makers consider all relevant and available information. In this regard, controls can be brought to the forefront of the innovation process without endangering innovation or creativity, while taking a facilitative role. Regular assessments of innovation projects facilitate discussion, arbitration, decision making, and the management of priorities. In addition, control improves information access and contributes to the development of common understanding. However, while controls aim to improve R&D efficiency and effectiveness, innovation success is also driven by the availability of resources, core competencies, and the support of the management.

# SAP: Bringing Economic Viability to the Front End of Innovation

Uli Eisert

## 1 Business Models as a Complement to Design Thinking

The pioneers of design thinking postulate that innovations should start with a focus on desirability, but in the end should satisfy three perspectives: human desirability, technical feasibility, and *economic viability* (Brown 2008; IDEO 2012). With its proven and 'tech savvy' development organization, technical feasibility has never been an issue for *SAP*. Over the past few years, the development organization has increasingly been influenced by the design thinking approach, and first analyses of innovation projects using this approach have indicated that design thinking is very effective at addressing human desirability. However, economic viability is equally important, but less in the focus of design thinking (Vianna et al. 2012). Therefore, SAP looked closely into *business model innovation*. After carrying out various business model innovation (BMI) projects, including the example described below, SAP considers BMI a possible method to complement design thinking, which is deeply rooted in SAP's philosophy.

A business model is a model that abstracts the complexity of a company by reducing it to its core elements and their interrelations. It specifies the core business logic of the firm, in particular those aspects that are relevant for building its competitive advantage. It has to be developed according to the firm's strategy and can be seen as an instantiation of the strategy (Afuah and Tucci 2000; Morris et al. 2005; Linder and Cantrell 2000). While in practice the focus is often exclusively on the enterprise view or canvas (e.g. Osterwalder et al. 2005), the network view helps to fully understand and capture the relationships between all relevant business partners, to analyze the value flow (in particular in multi-sided business models), and to compare the position of the company relative to the competition.

---

U. Eisert (✉)
SAP (Schweiz) AG, Blumenbergplatz 9, 9000 St. Gallen, Switzerland
e-mail: uli.eisert@sap.com

Business model innovation can be defined as an iterative process resulting in a qualitatively new and value-adding business model (Bucherer et al. 2012). To support BMI systematically, certain process phases are essential, i.e., analysis, design, validation, implementation planning and implementation. While it is possible to develop best practices for the first phases (see example below), the implementation itself is rather specific to the individual project and a matter of change management. The combination of business model innovation and design thinking could be intriguing because both procedures are very similar and a combined approach allows incorporating the strengths of the BMI approach with regard to economic viability into design thinking with its focus on capturing human needs and desires. In the end, the objective is to facilitate the creation of new business models with the same professionalism that is common in the area of product innovation. Indeed, in most companies there is a striking discrepancy between the common acknowledgement of the importance of business model innovation and its poor implementation (Bucherer et al. 2012; Chesbrough 2009).

## 2 Business Model Innovation in Practice

Our research team in Switzerland carried out a project that aimed to find suitable business models to integrate all kinds of services from SAP and its current and potential future partners into our commercial platform that had been focusing solely on software applications up to that point. We leveraged our close partnership with the Institute of Technology Management of the University of St. Gallen to jointly explore platform-based business models to commercialize all kinds of service offerings and to investigate the potential of BMI.

In the analysis phase, we started by reflecting on the triggers for the envisioned BMI. In this case we wanted to seize an opportunity: why not leverage an existing commercial platform beyond software applications for all kind of services? For this purpose we had to investigate which types of business-related services could be offered via the platform and how these could be clustered. In addition, we documented the current business model (for applications) as a baseline, as well as the models of the competition. Besides an analysis of changes in the environment (e.g. technology, eco-system, and industry), another important step was a detailed assessment of customers' needs. For this end, all (potential) customer groups had to be identified. Customers included internal entities and external partners that were needed to make the business model successful and that demanded an individual value proposition. In a last step, the objectives for the design phase were derived from the insights gained. In this project, we had to find suitable business models for all service clusters identified, and we had to gain a detailed understanding about their overall attractiveness for the SAP Store.

In the design phase, we created a large number of new business model options. The crucial steps were developing ideas in a systematic manner and using methods that were adequate for the formulated objective. Consequently, a combination of methods proven for ideation were used that allowed both for a systematic variation of potential options (e.g., morphological analysis of all relevant elements of a business model (Schief and Buxmann 2012)) and for creative invention of previously unknown possibilities (e.g., Blue Ocean approach (Kim and Mauborgne 2005)). Some of the methods leveraged existing business models (e.g., pattern recombination (Gassmann et al. 2012)). Since the description of our business model was consistently used as the basis for all methods, all options created could be clustered easily.

During the validation phase, the various options were evaluated to determine the best business model for each service cluster using a reproducible process that could be executed very quickly and that laid the foundation for a broad acceptance of the new ideas. All options had to be discussed with all relevant internal and external experts and stakeholders. In addition, a framework for evaluation and basic business cases were created and included in the discussions with the experts. What was most important for the business cases were transparent and reliable assumptions. The framework for evaluation illustrated the impact of the different business models versus their ease of implementation at a glance and allowed for combining qualitative criteria, such as customer acceptance, and quantitative criteria, like revenue potential. Finally, we developed a generic framework of platform-based models and factors that influence the choice of the platform provider (Weiblen et al. 2012). It turned out that service standardization and the level of desired control are the most prominent drivers that determine the applicability of the different models.

In the implementation planning phase, suggestions for various pilots representing the most attractive service clusters were made, and a roadmap as well as a timeline for overall implementation were drawn up. Driven by this project, services from SAP and its partner eco-system are now being included step by step in the SAP Store.

## 3 Economic Viability

Many people think that in innovation projects economic viability can be addressed simply by calculating business cases early on and by creating detailed business plans at a later stage. These elements are necessary; however, this is far too little. Economic viability requires an approach like business model innovation that changes the mind-set and influences all activities.

Throughout the entire process, BMI puts economic viability at the very core of innovation. The focus on the business model forces the team involved to center their thoughts and ideas, from the analysis to the implementation, on value creation for the customer groups identified and even more on value capture. As soon as the (potential) customer groups have been identified, it is most crucial to (1) deeply understand the customers, (2) derive a convincing value proposition (taking into account what the competition is able to offer), (3) analyze and quantify the value for the customer(s), and (4) determine the most appropriate and effective mechanism of capturing the value for the company (and, if required, for the partners one depends on). In addition, the team (5) has to work out the most efficient value chain including all partners that could contribute to enabling the company to offer the value propositions defined at the lowest possible costs as well as in an agile and responsive manner.

By carrying out these steps based on a solid understanding of the market and the competition, the team focuses on the core logic of the firm, orchestrating the contributions of various internal and external resources for optimum market success in a sustainable fashion. This is what BMI is all about, and this is its key contribution: bringing economic viability to the front end of innovation.

## 4 Benefits of a Combined Approach

The design thinking and business model innovation approaches can benefit from each other by integrating fitting elements from one into the other. We performed this exercise from a BMI perspective and found that BMI can benefit from design thinking in various areas, e.g.:
- By leveraging the human-centered approach for the analysis of customer needs to derive promising value propositions. The 'persona' approach can be applied to customer groups, both for B2C (persons as 'persona') and B2B (companies as 'persona').
- By leveraging the rapid experimentation and prototyping approach. We adapted it for BMI under the name of 'Rapid Feedback Loops'. The objective remains the same: 'act rough and rapid, to fail early and cheap'. Only by learning and through iterations the optimum solution will evolve.
- By leveraging the workshop formats and the focus on creativity. In our approach, there is a constant switch between workshops including creative elements and work in small teams to prepare or elaborate on certain aspects.

The investigation of a possible combination of design thinking and our BMI approach indicates that both approaches have many similarities that facilitate a close integration: similar process steps, a phased and iterative approach, and a compatible mind-set with a focus on creativity, diverse teams, and a balance between speed and reliability. The main benefits of a tight integration are:
- Parallel consideration of desirability and viability aspects
- A mind-set that is customer-centric and business-centric at the same time
- Creative process steps that focus on solutions and business models simultaneously
- New solutions and business models that are in synch at any time in the process

A combined approach delivers comprehensive results step by step as illustrated in Fig. 61. There is a reduced risk that the team focuses too much on a solution that is great for customers, but hardly economical, or, vice versa, that they create a great business model, but do not find a solution that is convincing enough to provide the required value proposition. Given these obvious advantages, we have recently started test driving a combined approach to establish if it has an edge over previous attempts.

# SAP: Bringing Economic Viability to the Front End of Innovation

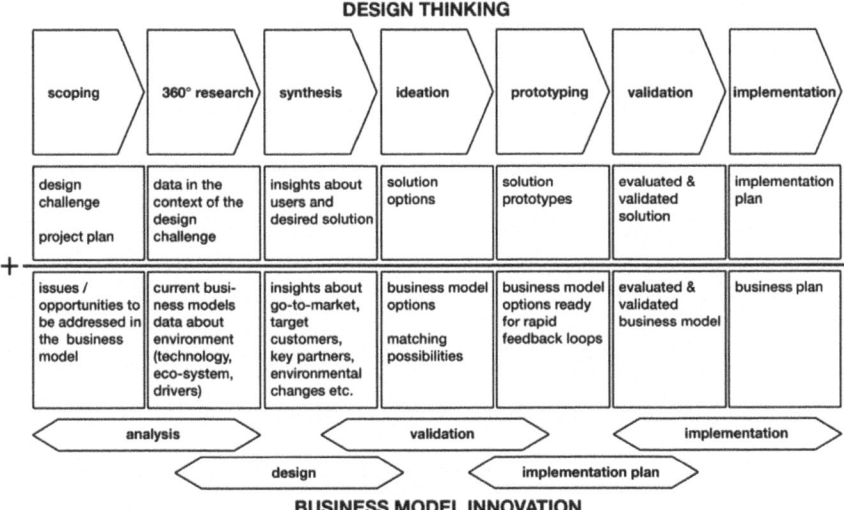

**Fig. 61** The approach which combines design thinking and business model innovation

## 5 Checklist

Checklist for bringing economic viability to the front end of innovation:
- Approach: Do you only think of business cases or do you have a broader view? Are business models part of your analysis and design efforts?
- Attitude: Do people only focus on the next products and services? Or do they understand that customers have to desire new solutions and, even more importantly, that they have to be willing to pay for them?
- Team: Are people with different skill-sets involved in your innovation projects? Do the teams include team members with a solid business background and deep knowledge about the market and the competition?

# Sprint Radar: Community-Based Trend Identification

Denis Eser, Kurt Gaubinger, and Michael Rabl

## 1   Project Setting & Focus

The early stages of innovation processes have a central leverage effect on the success of an innovation project. To maximize the performance of the front end of innovation, a goal-oriented and appropriate systematic approach is needed. Therefore the faculty of Engineering and Environmental Sciences at the University of Applied Sciences Upper Austria has founded the research and transfer center *sprint* (systematic product innovation transfer center). *sprint* focuses on the development and the utilization of advanced methods and tools to enhance the effectiveness and efficiency of the front end of innovation (FEI). The center maps the areas of expertise along the entire front-end process and reflects the following competences:

- *sprint>research*: methods of market research and innovation research,
- *sprint>lab*: ideation and creativity workshops,
- *sprint>design*: design thinking and design concepts,
- *sprint>tec*: engineering methods for the early phases.

Within the area sprint>lab, a new methodical approach was evolved to identify trends and developments in pre-defined clusters like mechatronics, life sciences, information and communication technologies or logistics.

---

D. Eser (✉)
HYVE AG, Schellingstr. 45, 80799 München, Germany
e-mail: denis.eser@hyve.de

K. Gaubinger • M. Rabl
University of Applied Sciences Upper Austria, Innovation and Product Management, Stelzhamerstr. 23, 4600 Wels, Austria
e-mail: kurt.gaubinger@fh-wels.at; michael.rabl@fh-wels.at

**Fig. 62** Three-step procedure 'sprint>radar'

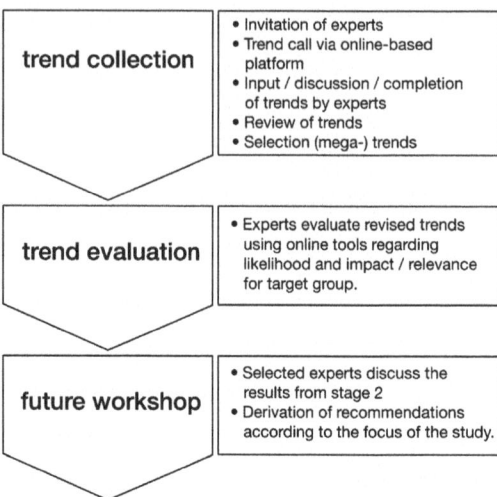

## 2    Method

To support the approach mentioned, a three-step procedure was developed following the *Delphi method*. The Delphi method is a multi-stage survey method, which collects the opinion of experts from various disciplines. The goal is to identify future technological trends and their timeline by merging and analyzing opinions of experts. The method is based on the individual and intuitive judgment of the experts. It is assumed that experts have an in-depth knowledge in their field and therefore can produce very good estimates of possible developments. Therefore the selection of experts according to their area of expertise is very important since the quality of the forecast depends on it. Generally, the Delphi method begins with a written survey of expert opinion on what the future key issues will be, and the likelihood of the developments. The result is then reflected back to the experts until a consensus emerges. Learning processes that arise from the results presented are an integral part of the concept. The aim of the Delphi method is to obtain a stable opinion amongst the experts. The number of survey rounds can vary. The experts remain anonymous and have no contact with each other.

In the case of the sprint>lab, the first written survey was substituted with an online-based *collaboration platform*. This platform also represents the heart of the three-step procedure used (see Fig. 62) and focuses on the collection of trends from experts. In the next step the analytical results of the *trend-collection* phase are reflected back to the experts. The method applied in this stage is a quantitative online survey. Here the experts are asked to evaluate the trends regarding their probability of occurrence and relevance. Based on the results of this survey, *strategy-workshops*, the so-called *future workshops*, are conducted, where experts from industry and academia discuss the impact of the identified trends on the field of interest (e.g. mechatronics).

In the following section, each step of the three-step procedure will be described.

## 2.1 Trend Collection

To develop the online-based trend identification tool, the sprint team cooperated with *HYVE*, a specialist for online customer integration. Based on their established web-based *crowdsourcing* tool *myIdeaNet*, the *sprint>radar* was developed. The idea management tool myIdeaNet is typically used to support company-internal idea management processes and crowdsourcing projects such as idea contests. As the platform is hosted externally (software as a service), members can log into the community from any device with internet connection and html browser capability.

As all crowdsourcing campaigns, also *sprint>radar* is divided into three phases, a preparation phase, a campaign phase and an evaluation phase. In the *preparation phase* the online platform is adapted to meet the project requirements. The design of the interfaces is adapted to reflect the project hosts' corporate identity, including their logo and corporate color. Then a welcome text, a precise description of the campaign's goal and the legal terms and conditions are added to the platform. Together, the topic visualization and the description of the campaign's goal form a 'trend call'. Each trend call asks a specific group of experts to submit trends in their relevant clusters. For the *sprint>radar* each of the project clusters is addressed to discuss a relevant trend in a predefined period of time. Once this timeframe is over, a new call for another cluster starts.

The first campaign aims to collect trends in the field of mechatronics (Fig. 63).

The final part of the preparation phase is the training of a moderator, also called community manager. The software tool itself can support the trend-collection process only to a limited extent. The main value of such a platform is created through a continuously active network of cluster experts. It is the community manager's role to guide and support this network. This will help keep the level of participation and involvement high.

The *campaign phase* starts with the invitation of cluster experts. The community manager starts the campaign by sending out invitation e-mails to all pre-registered mechatronic experts. This invitation contains information on the goals of the initiative and the benefits for participants. In this particular case, the main benefit for cluster experts is the knowledge and experience exchange (Füller et al. 2008). In addition, members receive continuous reporting on latest trend discoveries provided by the sprint team. During the campaign phase, cluster experts can submit their opinions about current and future trends in their relevant cluster.

The collaborative mechanisms of the platform encourage experts to not only submit their own trends but also discuss other members' submissions (Füller et al. 2007). The sprint>radar platform incites new participants to explore the existing trends on the platform. Members can see what other experts have already discussed and submitted. Generally, all content on the platform is open to every participant, so that a free discussion can evolve. All submitted trends are displayed in a central trend pool. Once the community have understood the ongoing trend call and know which trends are currently discussed, they can submit their own trends and comment on existing trends. Moreover the members of the community can *evaluate the trends* based on four criteria, occurrence probability, chance for the industry, risk

**Fig. 63** Campaign 'Trends in Mechatronics'

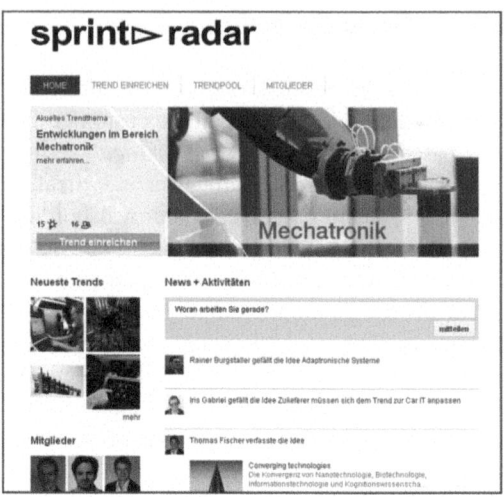

for the industry and timeframe. Through these measures the participants are transformed from passive into active members. In this phase it is also important that the community manager appreciates the activity of the members regularly and gives them feedback. The community manager supervises all discussions like a moderator to avoid unfairness and inappropriate content (Mühlbacher et al. 2011).

## 2.2 Trend Evaluation

In the second process stage the results of the trend-collection phase are reflected back to a selected group of experts. Especially experts who were outstandingly active during the campaign phase are targeted in this stage. Following the Delphi method, this phase aims to initiate anonymous interaction between the participants (Tidd and Bessant 2009). The method applied is an online survey. Here the experts are asked to evaluate the trends regarding their probability of occurrence and relevance and are also asked for their explanatory comments. The data are analyzed again and fed back to the experts, who are asked for their evaluation. This sequence should be repeated until a consensus is reached (Ahmed and Shepherd 2010).

## 2.3 Future Workshop

The term *future workshop* goes back to the ideas of Jungk and Müllert (1987), who saw an application of this method in all fields of society and concentrated in their work mostly on communities and political action groups. In the context of the sprint>radar, this method is modified with regard to three groups of stakeholders, namely firms, universities and public authorities. The basic steps of a future workshop are:

- Depending on the targeted group of stakeholders, generate visions about how the identified trends influence the industry, tertiary education or public funding programs.
- Discuss and analyze the feasibility of the visions and ideas that emerged during the first step.
- Develop a future strategy based on various future visions, barriers, and constraints.

## 3 Lessons Learned

Using an *online community* of selected experts for structured *trend collection* and evaluation has some similarities to offline focus groups. But it also offers certain advantages. The asynchronous mode of communication gives every expert the possibility to take part at the time of their convenience. The easy access to the online tool allows even larger group of experts (>150) to join the discussion no matter where they are. Offline discussions in workshop formats are limited to smaller numbers of participants. The use of modern *social media* gives researchers the chance to form a much larger expert panel that can deliver constant input over a defined period of time.

Another aspect is the identification of 'lead experts'. Similar to Von Hippel's lead user (1986), these individuals stand out as key players in the online community due to their frequent and high-quality contributions. Often these highly involved participants have an in-depth knowledge of the subject and function as thought leaders in the specific area. By identifying them in the community they can be leveraged to keep the community active and drive further discussions.

## 4 Checklist

Checklist for the implementation of an online-based trend-collection system:
- All members who use the platform for the first time have to be welcomed by a short welcome message that describes the intention of the platform and the goal of the project. All members can submit their own trend proposal and comment on existing trends.
- All trends submitted have to be displayed in a central trend pool.
- Constant motivation of community members to log into the platform regularly is very important.
- A 'recent activities wall' supports the community manager to leave messages to the community. An enclosed notification system sends emails to members if new content is on the platform that is directly addressed to them.
- Reporting metrics help to identify key experts by counting the amount and quality of submissions, comments and also how often users log into the community.

- A central success factor of a web-based trend collection system is the community manager. He supervises all discussions.
- The identification and mobilization of experts is another key aspect for the project's success.

# Landis+Gyr: Designing and Analyzing Business Models in Value Networks

Amir Bonakdar, Branko Bjelajac, and Alexander Strunz

## 1 Project Setting and Focus

The setting for this contribution was a joint research project of *Landis+Gyr* and the University of St. Gallen. The goal of the project was to develop a standardized approach for the design, analysis and comparison of business model ideas that follow the logic of *value networks*. In value networks, value is created through the cooperation of several companies that jointly form a business network that adds value to the end customer as well as to all participants.

The global energy market, characterized by deregulation and steered by governmental stimuli for the use of alternative energy, lends itself as a perfect application area for the new approach as its evolution, triggered by persistent climate change concerns paired with rising energy prices, opens up a tremendous potential for new business opportunities.

Landis+Gyr as the global industry leader in total metering solutions made a big effort to gather various new business model ideas considering their future role in the electric power value network. As future energy value networks are still evolving, with roles that partly do not exist yet, it was quite challenging for Landis+Gyr to decide which ideas are the most promising ones and hence worth focusing on. Much like other companies, Landis+Gyr has limited resources and therefore needs to focus on the financially and strategically most reasonable business model opportunities.

---

A. Bonakdar (✉) • A. Strunz
University of St. Gallen, Institute of Technology Management, Dufourstr. 40a, 9000 St. Gallen, Switzerland
e-mail: amir.bonakdar@unisg.ch; alexander.strunz@student.unisg.ch

B. Bjelajac
Landis+Gyr AG, Theilerstr. 1, 6301 Zug, Switzerland
e-mail: branko.bjelajac@landisgyr.com

Therefore an approach has been developed that entails the most relevant decision dimensions to evaluate the attractiveness of value networks as well as the attractiveness for each participant. There are several benefits of such an approach. First, a standardized approach ensures that no important aspect is overseen. Second, it enables all stakeholders to efficiently and effectively discuss the business model and reach a common understanding about its potential. Third, a standardized approach allows comparing two or more value networks and enables decision makers to select the most promising alternative(s).

## 2 Method

The University of St. Gallen, Institute of Technology Management, was commissioned to design the new approach for the design and analysis of business models in value networks. In a 4-month research project four workshops were conducted. Participants included researchers from the University of St. Gallen and several of Landis+Gyr's subject matter experts and managers with strategy, product management and technology background. In the first workshop the background and challenges of the project were discussed. Subsequently the requirements of the approach and deliverables of the project were defined. In the second workshop a first version of the approach developed was proposed. In a joint discussion the improvement potential was identified and subsequently tested during the third workshop, where also some minor issues were resolved and the final steps were defined. The focus of the fourth and final workshop was on the so-called *Management Cockpit* (Fig. 65).

## 3 Results

The result of the undertaking is an eight-step approach for the design and analysis of business models that follows the logic of value networks (Fig. 64).

**Step 1: High Level Description of the Business Model Idea and Definition of Value Propositions.** This helps all stakeholders to gain a basic understanding about the proposed business model. It describes the benefits delivered in the form of products and/or services and the customers receiving them (Bieger and Reinhold 2011).

The business model idea developed and further elaborated in this project is founded on the German government's initiative to encourage investments in photovoltaic (PV) energy generation. By providing high feed-in-tariffs (FIT) and high own-use-tariffs (OUT) for electricity produced by PV plants, investors are incentivized to produce electricity with their own PV plants and thereby increase the overall ratio of renewable energy in total energy production in Germany. Here the investor has two different options for amortization – either feeding electricity into the public electricity network and receiving a FIT of €0.28/kWh or consuming

**Fig. 64** Eight step approach for the design and analysis of value network business models

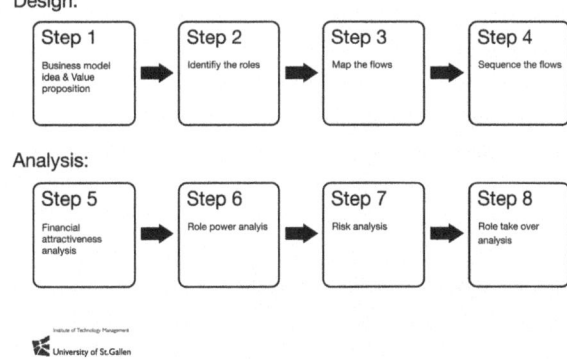

the electricity produced and receiving an OUT of €0.16/kWh. Both the FIT and the OUT scenario can be used for the amortization of the PV plants. Assuming a market price of electricity of €0.22/kWh, the most profitable way for a PV energy producer to manage electricity would be to keep 100 % of the electricity produced and either consume it or sell it to third parties for a price at or below market price. In order to have an understanding of the business model idea that is adequate for further analysis, a high-level description of key roles and their responsibilities is provided in.

**Step 2: Identification of Network Roles.** In this step, key roles necessary to realize the value proposition and their key resources/capabilities are identified. A role can be any kind of actor that initiates action, adds value, engages in interactions and makes decisions such as small groups or individuals, business units, organizations or even states (Allee 2011, 2008).

In the context of the selected business model idea, nine different roles were identified and are illustrated in Fig. 65. For illustration purposes, examples of companies that have the capabilities and resources required to carry out the respective roles are provided.

**Step 3: Identification and Mapping of Value Flows.** A value flow in a value network is composed of a transaction and a deliverable. A transaction shows how a particular value moves between two roles and is illustrated by a one-directional arrow. A deliverable is the actual object, the value that moves between two roles. It can be physical like a product or intangible like the exchange of expertise, knowledge or information. Mapping the value flows helps to clarify what each role contributes and how value is created in the network (Allee 2011, 2008).

**Step 4: Sequencing of Value Flows.** Sequencing identifies the order in which value exchanges are taking place, e.g. normally first a bill for a certain product is sent out before the payment is made. Sequencing is important because it makes sure that all relevant roles and value flows are considered (Allee 2011, 2008).

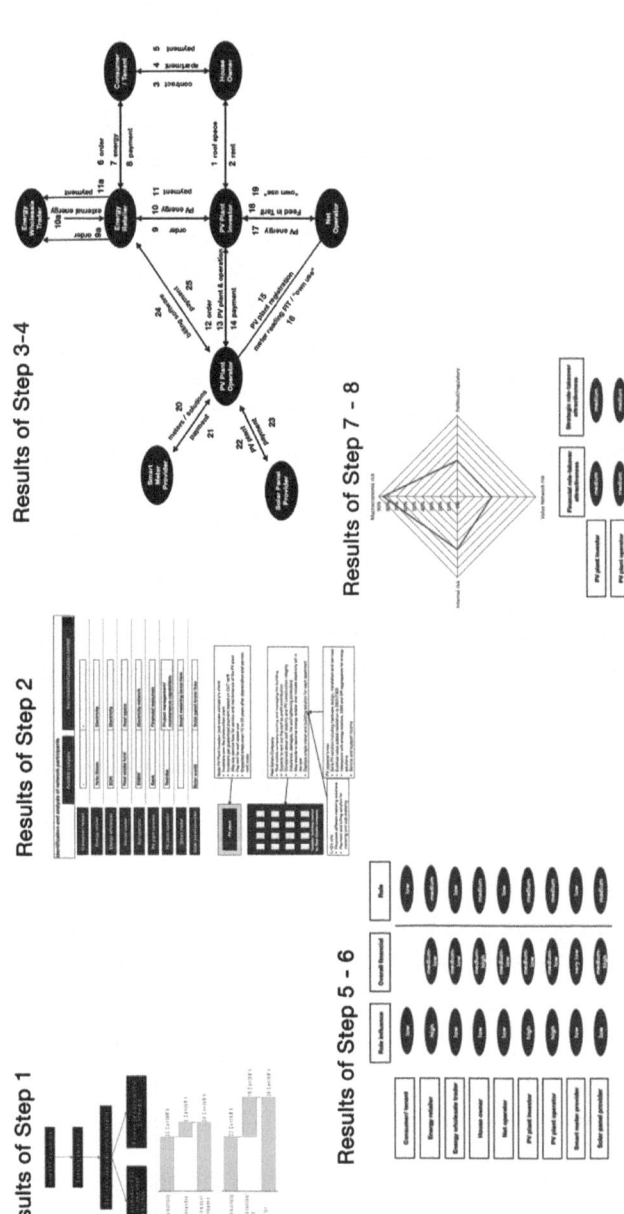

Fig. 65 Management Cockpit

The result of steps 3 and 4 is displayed in Fig. 65 in the form of a value network map. In most instances, the value exchange between the interacting roles is either a service/product or of monetary nature.

**Step 5: Financial Attractiveness Analysis.** The goal of this analysis is to evaluate how financially attractive each role of the network is. The analysis is conducted in two steps. First, the additional revenue streams for each role are identified. In a second step the upfront investments as well as the additional operating costs for participating in the network are identified per role. The results of financial attractiveness analysis will provide an indication about the roles that are potentially attractive and if the value network needs to be aligned to make certain roles more attractive. This is important particularly due to the sustainability of the network.

The result of step 5 is the so-called role attractiveness analysis shown in Fig. 65. The analysis shows that for most players the financial benefit is relatively low and thus would not motivate them to enter the value network. Only two out of eight roles (PV plant operator and PV plant investor) provide medium-high financial incentives to enter the market.

**Step 6: Role Power Analysis.** In order to understand how power is distributed in the network and which roles are most powerful, three centrality measures should be considered: degree centrality, betweenness centrality and closeness centrality. Degree centrality is defined as the number of ties to other roles a role has (Wassermann and Faust 1994). The higher the number, the more central the role is and the more power it has. Betweenness centrality is the number of times a role acts as a bridge for two other roles along their shortest path (Wassermann and Faust 1994). The higher the betweenness centrality, the higher the amount of power the role has. Closeness centrality measures how many steps on average it takes for one role to reach every other role in the network (Costenbader and Valente 2003). Closeness centrality is measured by taking the reciprocal of the sum of the distances between one role and any other role in the network (Freeman 1979). The higher the closeness centrality, the higher is the role's power in the network. We used the plain assumption that the more central a role in a network is, the more opportunities the role has to choose from value-creating activities and the more it is able to distribute information according to its own interests.

The result of the role power analysis reveals that there are three powerful roles: energy retailer, PV plant investor, and PV plant operator. These roles influence many other roles and are essential for the network to function.

**Step 7: Risk Analysis.** Not only should the results of the earlier steps be included, but also other external and internal factors that have an influence on the probability of success of the network. On a more granular level, the external risks contain macroeconomic risks (e.g. impact of negative economic trends, risk of lower market prices), political/regulatory risks (e.g. risk of negative regulatory changes/ negative political changes), value network risks (e.g. network is financially viable for all roles from step 5, power of one's own company in the network from step 6, and level of network development). The internal risks on the other hand comprise e.g. the risk of new technologies included, the financial attractiveness for one's own company from step 5, the complexity of the project, and the new capabilities needed). All these risk dimensions should be evaluated if the probability of their occurrence is high, medium or low. In order to receive a consolidated final risk, the single risk dimensions should be weighted by a predefined calculation key. Then the overall probability of success can be derived.

An overall probability of success of 48 % was calculated for this network. There are two main reasons why the probability of success is so low: First, the value network is highly dependent on the FIT and the OUT values arbitrarily set by the government. A decrease in these tariffs would have massive negative implications on the success of the overall value network. Second, the overall financial attractiveness of the value network is low. Most players have low incentives to join and/or to stay in the value network.

**Step 8: Role Take-Over Analysis.** The objective of this part of the analysis is to make it feasible for the focal firm of the analysis to evaluate if a forward or backward integration along the value chain/value network has a sufficient probability of success. The guiding principle behind this analysis is that a forward or backward integration or a role take-over is attractive if it makes strategic and financial sense. Therefore the analysis consists of a financial and a strategic analysis. The analysis follows a decision-tree approach in which first a strategic analysis is conducted. Only if its output is positive, implying that a role take-over makes strategic sense, a financial analysis is conducted. If this analysis also yields a positive result, implying that a role take-over would make financial sense, a deep-dive on both dimensions should be conducted.

The results yielded that a role take-over only makes strategic sense for two out of seven analyzed choices. For the two roles identified, a financial role take-over analysis was conducted from the perspective of Landis+Gyr. It showed that for both roles the overall financial attractiveness is medium.

The results of the approach are illustrated in the Management Cockpit as shown in Fig. 65, which provides stakeholders a first, but still comprehensive overview of a business model idea and basic indicators for investment decisions.

## 4 Lessons Learned and Checklist

In order to apply the described method successfully, a multi-disciplinary project team of experienced and cooperative individuals with various backgrounds, skills and capabilities is required:
- Steps 1 through 3: Vision, creativity, 'big picture' thinking and experience in describing business models is required in order to properly describe a not-yet-existing ecosystem and identify all relevant roles (while disregarding irrelevant ones), their interdependencies as well as the value added for each role.
- Step 4: The capability to think through and model dynamic systems is needed.

- Step 5: Knowledge of business finance fundamentals and understanding of financial KPIs, as well as experience in writing and analyzing business cases, is essential.
- Step 6 and 7: Analytical skills and practical business experience to understand the concept of power in a business ecosystem, as well as the business risk of a role in a given business context (ecosystem), is required.
- Step 8: The capability and neutrality to judge rationally over the potential of the focal company's chances to successfully play the envisaged role is essential.

The ecosystem described in Steps 1 through 4 and analyzed in Steps 5 through 8 is supposed to be in 'steady-state'. It is important to think through what has to happen in order to get to such a state. In case that non-rational or too risky moves are required by one or more stakeholders (roles) in the start-up phase, the ecosystem might never come to life without flanking measures. These must be identified and their financial (and other) impact must be properly modeled and reflected in Steps 5 through 8.

The model is useful as it provides a structured approach to describe new, complex business models and ecosystems. Its prediction power has yet to be validated by applying it to existing business models and ecosystems and benchmarking the predicted performance of the ecosystem as well as selected roles against real-life data. Only after this has been done with a positive outcome, the model will have a chance to be endorsed by academia and the business community as a tool to support business decision-making.

# Voestalpine Anarbeitung: Commercialization Framework for Technology Development Projects

Kurt Gaubinger, Fiona Schweitzer, and Hans-Jörg Kirchweger

## 1 Project Setting & Focus

In the *automotive supply industry*, sustainable success increasingly depends not only on systematic development of technologies, but also on early *commercialization* activities.

In keeping with the company's slogan 'one step ahead', *voestalpine Anarbeitung GmbH (voestalpine)* designed and implemented a process-oriented framework for commercializing technology for its strategic business field *automotive*. This framework complements the existing *technology development* (TD) process and integrates specific activities of commercialization into every stage of the TD process to increase its effectiveness and efficiency. The key objectives of this project were, firstly, to identify the crucial process steps for market-orientated TD, and, secondly, to define and adapt proper tools and methods for the individual process steps in a market-orientated TD process.

## 2 Preparatory Work

In order to gather insights into success factors and challenges in early technology commercialization, voestalpine commissioned the University of Applied Sciences Upper Austria with a benchmark study. The study combined qualitative exploratory research with quantitative insights. In the qualitative benchmark project, a two-step

---

K. Gaubinger (✉) • F. Schweitzer
University of Applied Sciences Upper Austria, Innovation and Product Management, Stelzhamerstr. 23, 4600 Wels, Austria
e-mail: kurt.gaubinger@fh-wels.at; fiona.schweitzer@fh-wels.at

H.-J. Kirchweger
Voestalpine Anarbeitung GmbH, Voestalpine-Str. 3, 4020 Linz, Austria
e-mail: hans-joerg.kirchweger@voestalpine.com

probability sampling process was used to select 14 persons in charge of TD and pre-development (e.g., head of R&D, managing director, project team manager) from 11 Upper Austrian companies in the automotive supply industry with outstanding records in TD. These persons were selected for face-to-face semi-structured interviews. All interviews were recorded with prior permission and analyzed using the four-step procedure suggested by Lamnek (2005). This research revealed successful process structures and appropriate management tools for commercialization activities within TD.

The quantitative study was used to identify how different variables of technology management and technology commercialization influence the success of innovations in TD processes. The corresponding questionnaire was pre-tested and used a benchmarking tool composed of 120 individual questions. This benchmarking tool allowed respondents to rank their companies' TD activities on a five-point scale. Selected managers of 360 companies were invited to use the benchmarking tool, and detailed insights into the TD framework of 12 % of the companies invited to respond were gathered.

**Framework for Technology Commercialization.** Based on theory and the empirical findings obtained as well as according to the existing voestalpine technology development process, a process-oriented framework for commercialization was developed (see Fig. 66). It demonstrates how the commercialization activities within TD processes should be typically structured and which management tools should be applied in the particular process steps.

The framework for commercialization is classified into a preliminary phase and three main phases (strategic marketing planning, customer integration & co-development, broad marketing), which in turn can be divided into several sub-phases.

## 2.1 Preliminary Phase

The implementation of effective commercialization activities starts with an assessment of the internal corporate base of technological competence and an analysis of the macro and micro environments focusing on the systematic search for weak signals (Diller 2007). PESTEL analysis (Hungenberg 2011) and an extensive structural analysis of company specific industries (Porter 1980) are suitable for the holistic analysis of the environment. Based on the results of this analysis, the SWOT analysis finally permits deriving options for strategic action (Müller-Stewens and Lechner 2005), which determine concrete search fields for the following phase of idea generation. In the phase of idea generation, many technology ideas compliant with the search fields identified should be generated drawing on different internal and external sources of information. A structured suggestion scheme and the integration of lead experts and lead users ensure the appropriate direction of the idea-finding process. Finally, the individual, alternative ideas for new technologies are evaluated by interdisciplinary teams that use checklists and value benefit analysis, which consider a certain market orientation at an early stage.

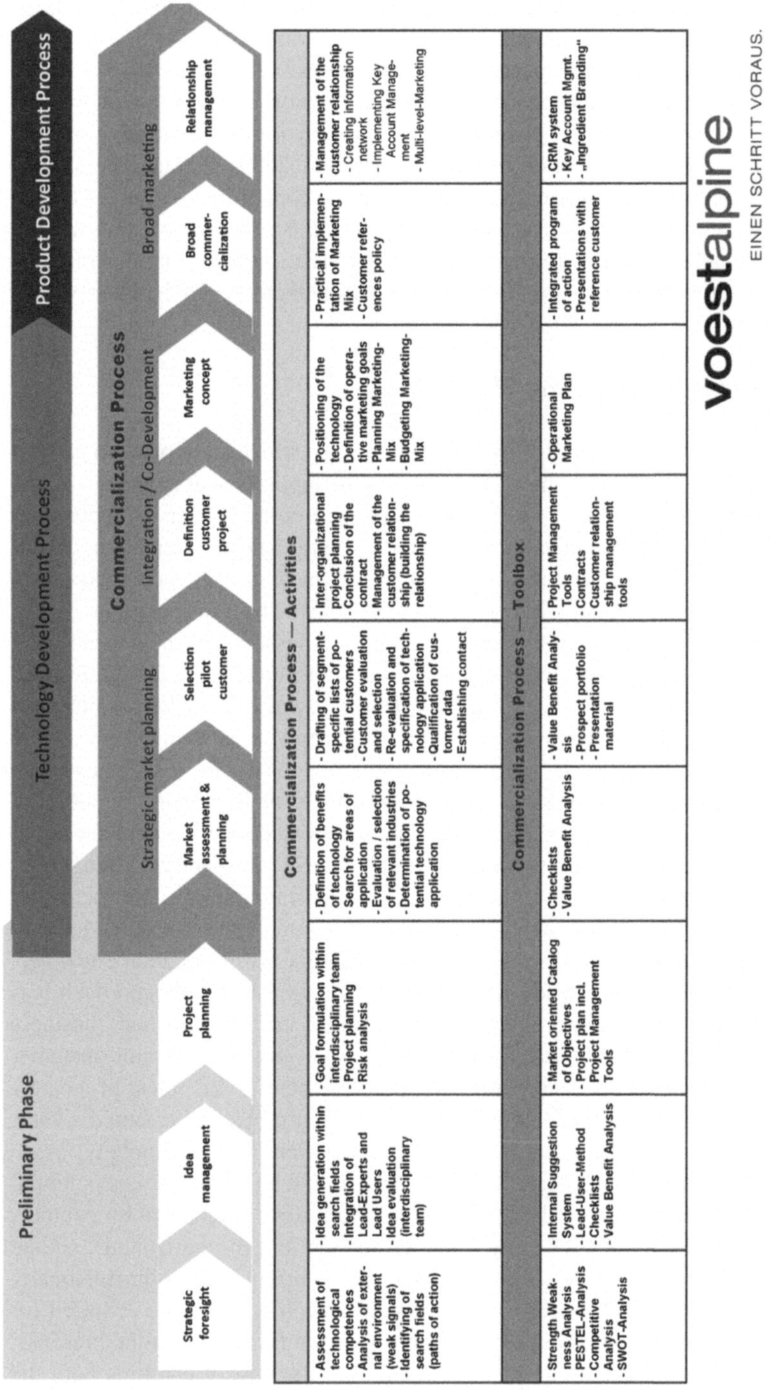

**Fig. 66** Framework for TD commercialization (own illustration)

Besides technicians and members of the R&D department, voestalpine integrated selected members of the marketing and sales department in this stage, which accelerated this and ensuing process steps significantly.

In gate 1, the decision is made to transform a technology idea into a concrete TD project. In the phase of project planning, it is necessary for the whole development team to create a shared state of knowledge (Slama et al. 2006). A catalogue of clear objectives is an important instrument for planning and controlling the entire process. Building on this, an initial project plan is prepared. Since TD projects are usually based on vaguely defined information at project start, project planning must be gradually specified further as the level of information increases in the later phases.

## 2.2 Strategic Marketing Planning

In the first main phase of *strategic marketing planning*, which takes place in parallel to technology concept development, the potential of application associated with the new technology concept has to be evaluated. Because possible areas of application of technologically induced innovation ideas are often unknown (Herstatt and Lettl 2000), promising areas of application and target segments for the new technology have to be identified (Bower and Christensen 1995). The starting point of this activity is the determination of the strengths and weaknesses of the new technology (Schwery and Raurich 2004) and the subsequent translation of these features into utility functions (Kotler et al. 2003). Based on these results, a list of potential industries can be narrowed down by means of a stepwise assessment procedure (Meffert and Bruhn 2003). In such a procedure, checklists are used to evaluate industries with potential application fields according to their strategic fit and their attractiveness. An even more detailed analysis serves to determine relevant target industries and target market segments. In the course of specifying the target segment, voestalpine decided to pursue a single-segment strategy. The focus was on the industry that had been identified as most attractive in a cost-benefit analysis. For this target industry identified, voestalpine created a list of potential customers with whom the company already had a business relationship. This approach was chosen because it is especially already known and established business relationships that offer potential for pilot projects. In this way, a pilot customer was identified, who would ensure the application-oriented development of the new technology from the beginning. Furthermore, this pilot project represented a valuable reference for supporting the later commercialization of the technology.

In addition to a utility analysis, a prospect portfolio can help to assess potential pilot customers (Homburg et al. 2008). The vertical axis of this portfolio captures the expected attractiveness of prospective customers and the horizontal axis the probability of acquisition, with both axes being measured multi-dimensionally (Homburg and Daum 1998). Finally, a potential customer has to be acquired for the application-oriented development project. It should be noted that the presentation of simulation results strongly supports this step because it confirms both the technological competence of the company and the feasibility of the new technology.

## 2.3 Customer Integration & Co-development

In the second major phase, *customer integration & co-development*, the basic conditions and customer-specific technological requirements are identified and evaluated in co-operation with the acquired pilot customer(s). Inter-organizational planning of the project has to be carried out using various project management tools. To secure the relationship between supplier and customer, the project-specific investments and activities to be undertaken by both partners have to be determined, and the distribution of project-specific costs and revenue has to be negotiated. In general, these aspects are settled with contracts (Backhaus and Voeth 2007). Careful management of the business relationship is of major importance, because voluntary customer retention is desirable from the perspective of the supplier firm. In this context, particular attention should be paid to an interdisciplinary key-account management team building and consolidating customer relations (Silber 2007). The subsequent planning of the operational marketing concept for the technology is the focus of the following sub-phase and is aimed at establishing business relations with other potential customers. In line with the value-based positioning of the new technology, the operational marketing plan includes a definition of operational marketing objectives and details specific marketing activities including the corresponding budgeting. Especially a B2B-specific communication mix is essential to advancing awareness of the technology in the target segment.

## 2.4 Broad Marketing

The third main phase focuses on the *broad marketing*, which includes the action-oriented implementation of the operational marketing plan. In this context, voestalpine took the following factors into account to ensure a successful introduction of the innovation: An integrated action program with clearly allocated responsibilities, a clearly defined project team, adequate incentive systems, long-term personnel planning, and the integration of marketing strategy and marketing plan in the corporate culture. In addition to the continuous monitoring of project-specific aspects of the TD project, measuring innovation success is necessary, too. Appropriate key figures are customer acceptance, customer satisfaction, and the profit margin. Finally, the reference policy is of high importance in this phase.

Because of the complexity of the product range, a systematic reference policy was crucial for voestalpine. In this context, references reduce the risk perceived by potential customers and thereby directly promote widespread marketing. Maintaining customer relations is the central focus of the last sub-phase. For this purpose, voestalpine installed key-account management, which allows for the effective coordination of customer contacts and customer information. Finally, a multi-level marketing approach combined with an ingredient branding strategy can support customer retention. However, this strategy is based on the prerequisite that the technology is significant and identifiable for the end user.

## 3 Lessons Learned

Our developed framework shows how a commercialization process can be structured and which tools and methods should be applied in the particular process steps to turn technological ideas and inventions into effective action. *voestalpine Anarbeitung GmbH* benefits from the established framework, because it includes *a manageable number of phase-specific practicable management tools* that assist the company in increasing the *effectiveness* of their TD. Furthermore, implementing the framework enhances the *efficiency* of the TD activities due to the strategy-orientated and systematic procedure resulting in reduced time to market and a higher return on TD activities. A project-specific validation showed that the model complements the existing TD process with a valuable market-pull perspective.

## 4 Checklist

The following checklist serves to assist companies in the automotive supply industry in implementing a framework for commercialization:
- **Preliminary Phase:** (1) assessment of the internal technological corporate competence base; (2) analysis of the macro and micro environments; (3) derivation of options for strategic action and identification of search fields; (4) generation of technology ideas; (5) idea evaluation; (6) transformation of ideas into a concrete TD project; (7) preparation of an initial project plan.
- **Strategic Marketing Planning:** (1) determination of strengths and weaknesses of the new technology; (2) identification of areas of application, target segments, target industries and target market segments for the new technology; (3) identifying and winning potential pilot customers for the application-oriented development project
- **Customer Integration & Co-Development:** (1) evaluation of the customer-specific technological requirements; (2) inter-organizational planning of the project; (3) determination of the project-specific investments, activities, costs and revenue of both suppliers and customers; (4) building customer relations through an interdisciplinary key-account management team; (5) planning of the operational marketing concept for the technology (objectives, activities, budgeting).
- **Broad Marketing:** (1) action-oriented implementation of the operational marketing plan; (2) continuous monitoring of project-specific aspects of the TD project; (3) evaluating innovation success; (4) installation of key-account management; (5) support of customer retention through a multi-level marketing approach combined with an ingredient branding strategy.

# Volkswagen: Open Foresight at the Front End of Research Innovation

Caroline V. Rudzinski and Gereon Uerz

## 1   Project Setting and Focus

The goal of this project was (1) to support the strategy process of one department within Volkswagen Group Research and to (2) foster the acceptance of innovative foresight and innovation management tools in the long-range strategic planning processes. The project combined an *open innovation*, or *information market*, approach and strategy development based on scenarios that had been developed beforehand. It was conducted as a joint project of two departments at Group Research of the *Volkswagen Group*, Wolfsburg. The project combined two methods, i.e., the information market approach and the systematic deduction of strategic implications by *wind tunneling*, which was based on four alternative global scenarios that had been developed by Future Affairs, the internal foresight unit at Group Research.

The alternative scenarios were used to challenge the current strategy of the department. By providing alternative scenarios as 'wind tunnels' the current strategy was tested for its robustness and 'best fit' strategies for the alternative scenarios were lined out. An information market pools effectively the *'wisdom of the crowd'* through a 'virtual stock exchange' and was used in this project to gain new insights by embedding a larger group to the strategy process.

As methodologically sound participatory foresight and strategy development are not widely used within large organizations and *open foresight* often faces rejection, the focus was on (1) engaging top management and experts in an open dialog on strategy and product innovation, on (2) supporting 'future proving' the mid- to

---

C.V. Rudzinski (✉) • G. Uerz
Volkswagen Aktiengesellschaft, Konzernforschung Zukunftsforschung & Trendtransfer (K-EFZ), P.O. Box 1895, 38436 Wolfsburg, Germany
e-mail: caroline.rudzinski@volkswagen.de; gereon.uerz@volkswagen.de

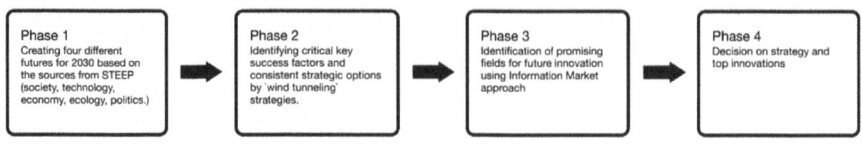

**Fig. 67** Overview of the different project stages

long-term strategy of the respective department, and (3) on getting strategic insights by integrating the knowledge of the whole Group Research organization via the information market approach.

## 2    Method

The project comprised different stages, which are described in Fig. 67.

Four alternative global scenarios that were based on megatrends and sketched four different futures for 2030 had been developed by the foresight experts of Future Affairs. Scenario development (e.g., Steinmüller 1997; Kosow and Gaßner 2008) was conducted relying on the outcomes of horizon scanning, which used a wide and diverse range of scientific sources from STEEP (society, technology, economy, ecology, politics). The relevant drivers were condensed into key factors that were used for developing alternative lines of development for these factors. Relying on a scenario software tool, cross impact analysis was used for calculating the most consistent scenarios.

The resulting four scenarios were discussed with top management and the experts of the department and were in a first step used as a framework for wind tunneling the current innovation strategy. Critical success factors were identified for the department, and subsequently a set of projections was defined for each of these key factors. Consistency evaluation of the projections was performed by using a software tool, a scenario and strategy software that had been used for the generation of scenarios as well. A set of highly consistent scenarios that were highly diverse from each other was identified and used as an input for wind tunneling of scenarios and strategic options. For each of the four scenarios, the best fitting (most consistent) strategic options were identified, and the current strategy of the department was tested against the scenarios.

In a second step, the white spots of the current strategy, as detected by wind tunneling, served as a starting point for the information market, which aimed to identify promising fields for future innovation by drawing on the expertise of the whole of Group Research. An information market is an online-based real-time idea trading market (Fig. 68). Experts from all departments of Volkswagen Group Research were invited to take part in the information market. On the market they could share their knowledge on pre-defined fields of innovation. They were asked either to contribute ideas, to comment on ideas, or to trade ideas.

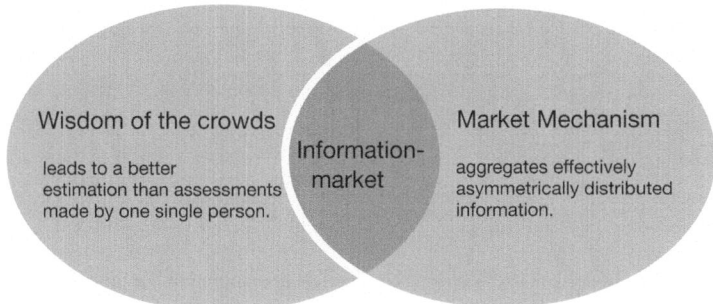

**Fig. 68** The basic information market mechanism (Rudzinski 2009)

The insights from the wind tunneling process and the output of the information market were used by top management to re-discuss strategic issues and top innovations.

## 3 Results

Combining the strategic foresight and the open innovation/crowdsourcing approaches enabled a structured and transparent, bottom-up and top-down analysis as well as a qualified discussion of different, consistent possible future strategies and future innovations within the department. It also provided the basis for identifying white spots within the current research agenda.

The process was a permanent balancing act between focusing on the issues relevant for the department and opening up to new (innovation) topics (Fig. 69).

By starting with four possible futures and narrowing them down to the relevant key factors for the department, consistent strategic options were identified which matched the possible future challenges.

Based on this insight, relevant fields were deduced, and the process was opened up by inviting all employees of Volkswagen Group Research to participate via the information market and to bring their expertise to the strategy and innovation process for the respective department. The information market was run for 2 weeks and generated 136 ideas, with 250 comments posted and 2,642 opinions traded. 41 % of the ideas, 39 % of the comments, and 30 % of the trades originated from other Group Research departments than the (internal customers) department. The fact that five out of the top ten ideas came from other departments showed that the approach allowed to consider knowledge which otherwise would not have been taken account of. The ease of contributing ideas and the joy of exchanging them was mentioned as one factor critical for success by several participants.

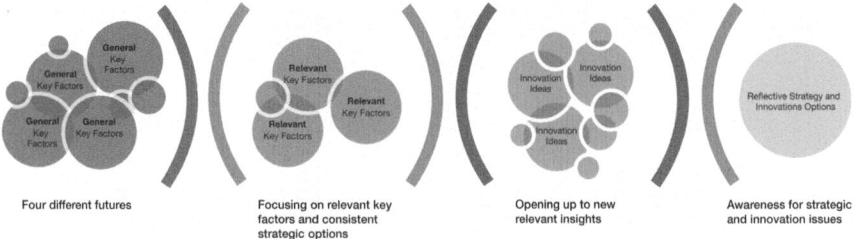

**Fig. 69** Overview of the balance between focusing and opening up in the process

## 4 Lessons Learned

One has to consider that the process cannot be completed 'along the way'. Employees participating in the wind tunneling process have to be officially assigned to it.

Running an information market, one could see that the employees appreciated the process and were highly motivated to be part of it. They also liked the simple and non-bureaucratic way of participating due to fast and easy access to the market via the Volkswagen intranet. However, it needs to be mentioned that the quality of the ideas for innovation generated via the information market can vary. One cannot expect to get ready-to-use ideas that have already been analyzed completely. The approach facilitates the aggregation of different knowledge and ideas in a very early innovation (exploration) stage. It is the responsibility of the top management to identify the potential of an idea and to provide the infrastructure necessary to bring the idea to a more concrete level.

The project showed that the combination of two established methods from the innovation management repertoire – the *scenario thinking* approach and information markets – is able to generate strategic insights/value and to establish a long-term perspective on strategy, embedded in an iterative learning cycle on strategy and innovation. Thus, the chosen approach of combining open innovation and foresight is able to serve both the long-term strategic requirements of a large organization and the more short-term need to foster innovation at the same time.

## 5 Checklist

For implementing a wind tunneling and an information market approach, the following aspects have to be considered:

**Organizational enablers:** futures thinking must be owned and adequately supported
- Supported by top management.
- Allocated sufficient resources and capacity.

- Cross-functional teams.
- Embedded in business units.

**Process enablers:** the process must be pluralistic, engaging and focused.
- Clear objectives.
- Sufficiently broad to generate new thinking.
- Accepting of multiple futures.
- Consultative: Internally and externally

**Implementation enablers:** implementation must be strategic and sufficiently embedded.
- Make the process transparent.
- Outputs relevant to corporate strategy.
- Iterative.
- Integrated into planning process.
- Established audience.

# Fuzzy Front End of Innovation: Quo Vadis?

Oliver Gassmann and Fiona Schweitzer

## 1 Fuzziness as Opportunity for Innovation

Throughout this book, several approaches, tools and principles have been presented that allow managing the front end of innovation professionally. The management task in the front-end phase involves capturing vague opportunities and ideas as well as their professional transformation into clear concepts that can be transferred into the next stage of the innovation process. It is also the task of the manager responsible for the early innovation phase to reduce uncertainty and prepare the technological and market basics for the ensuing product or process development. Fuzziness is therefore a key concept that is incorporated at this phase.

The overall innovation output of a company will only be exceptional if an active search for latent needs, weak signals and emerging trends is undertaken, instead of just following the evident and palpable. For this reason, innovation managers should enjoy embracing the fuzziness of the front end. A set of concepts and processes along with innovative, new tools support them in bringing yet faint ideas down to earth (see Fig. 70). The criteria for evaluating rather disruptive ideas or visionary concepts are different from those appropriate to more incremental concepts. But even more importantly, the mindsets and attitudes of those decision makers who evaluate such ideas should also be different in order to prevent them from selecting incremental ideas at the cost of radical ideas owing to a preference for less risky, known and proven solutions (Levinthal and March 1993).

---

O. Gassmann (✉)
University of St. Gallen, Institute for Technology Management, Dufourstr. 40a, 9000 St. Gallen, Switzerland
e-mail: oliver.gassmann@unisg.ch

F. Schweitzer
University of Applied Sciences Upper Austria, Innovation and Product Management, Stelzhamerstr. 23, 4600 Wels, Austria
e-mail: fiona.schweitzer@fh-wels.at

**Fig. 70** Grasping opportunities at the front end of innovation

In this sense, the front end of innovation will stay fuzzy. Managers not only need to have holistic assessment abilities incorporating market, technological and system perspectives, but they also need to have the personality and standing within the company to make outstanding decisions. Risk-averse managers are rarely dismissed; they seem to be wise and responsible. Visionary leaders fail in their careers more often, because the typical path to radical innovation is paved with failures, and clear ways to success do not exist. Too often an attitude that is willing to take risks and empowers imagination of creative minds is punished by the corporate thinking of large firms.

Successful innovation leaders help turn great ideas from chimeras into tangible concepts. They navigate those ideas through the front end by constantly striving to reduce market and technological uncertainties associated with these ideas, where possible, and at the same time allowing fuzziness – in the sense of creative input and vague suggestions that help to develop and amend the original idea – at any stage of the front end.

## 2 New Tools: New Ways

In the last few years, progress in ICT-technologies, increasing global interconnectedness, and cost reductions for *virtual solutions* have led to new ways of opening up the front end of innovation. Software for exploring and tracking technological trends, netnographic procedures to observe user behavior and collect user ideas online, technical advancements to increase the validity of virtual prototyping, and new visualization tools to immerse customers in virtual realities offering new product experiences are currently explored to improve efficient and effective decision-making at the front end of innovation.

Analysis of 'big data' offers new opportunities for identifying relevant trends in consumer behavior. Judging from the number of initiatives and start-ups that are currently appearing with ever new and more elaborated tools and methods of using online data, it is very likely that ICT-technology will again revolutionize the way information is gathered and ideas are produced and materialize at the front end of innovation. While gathering information was a key challenge a few decades ago, the key challenge today is how to make sense of the host of ideas and data. The focus will be more and more on the creation of relevant knowledge and on evaluating the huge amount of available ideas.

Crowdsourcing ideas from the external world is a strong trend. In 2007, *Cisco* was among the first companies to initiate a crowdsourcing competition. The company offered a prize of $250,000 for the winner. The result was very convincing: within 5 weeks, 2,500 ideas from 104 countries had been blogged. From among these ideas, 450 pitches were conducted on *webex,* and 12 out of these had the opportunity to present their ideas in front of the Cisco top management. The winner, Anna Gossen, an IT student of TU Karlsruhe, won the prize and initiated an investment of 10 million US dollars at Cisco in the area of energy efficiency. Integrating external sources into the innovation process via web-based crowdsourcing initiatives and other *open innovation* tools is facilitated primarily through current ICT. Many more initiatives from several firms are expected. Yet, the large numbers of potential ideas that are generated through crowdsourcing initiatives challenge company decision-makers. New approaches in testing market acceptance of large numbers of new product ideas, such as the securities trading of concepts (Soukhoroukova et al. 2012), are still in their infancy.

Some managers contemplate outsourcing screening processes. But true innovation is only successful, if the company itself is innovative. Typically, a pure shopping mentality does not work for innovation. A company needs to understand the outside world in order to evaluate and absorb outside knowledge and to translate it into its own products and services. In other words, companies need to innovate themselves in order to develop an absorptive capacity which enables successful implementation of crowdsourced ideas.

As the quality of contributions from different external sources can vary considerably, another approach to handling external information input lies in limiting participation to only such individuals who might contribute valuable ideas and information. For example, *BMW* restricts access to its corporate-wide idea platform Red Square and invites only selected individuals inside BMW to participate. Once inside the system, everyone is treated as equal, and everyone can choose a nickname to disguise origin and position. In this selective open innovation approach that cultivates the sense of belonging, members are extremely active, provide top-notch ideas and intensively discuss and refine posted ideas. The idea of creating an exclusive club has a strong impact in many companies. Sony introduced so-called 'gold badge' members, who are more closely involved in idea creation and selection than other employees. Finding and motivating the right people to participate in the innovation process has been and will be the key issue of innovation success.

Technological advancement and new forms of interacting with external sources might not only improve the effectiveness and speed of front-end processes and allow the inclusion or exclusion of selected individuals, but even lead to entirely new business models. An example is *Quirky*, a social product development platform where registered members do not only post and refine ideas and develop product names and slogans for them, but also select which ideas will be produced and finally participate in product sales according to their contribution.

Another example is 3-D printing, which does not only allow speeding up prototyping in the early stages of product innovation, but might have the potential of changing the whole manufacturing process. At the moment, additive manufacturing through three-dimensional printing is limited to certain materials (plastics, resins and metals) and still expensive compared to mass manufacturing. Yet, its appeal lies in its high precision, resource efficiency, waste reduction, and the fact that basically any shape that can be produced on the screen can be printed. Slight differentiations between several parts can be achieved relatively easily and at no extra cost, which allows new flexibility and new possibilities as compared to traditional manufacturing processes. These benefits make it attractive for many applications, such as medical engineering, where it allows producing artificial limbs, dental braces, or other highly individual single items precisely and quickly. Given wider availability and decreasing prices, more industries might join in, and the technology may lead to completely new business models.

## 3    From Product Innovation to System Innovation

Nowadays, opening up a company's own innovation process at the front end of innovation mainly focuses on outside-in approaches of leveraging external knowledge, and only at a smaller scale on inside-out initiatives, e.g. by licensing patents out to other companies. In the future, the front end of innovation could be a place where radically new ways of innovating take place via coupled processes with diverse network partners to enable system innovations. Innovation management would profit from shifting the current focus on mere technological innovation more towards business model innovation. To live up to expectations, future innovation has to be enabled by technology, but has to move beyond mere technological solutions to more complex systemic solutions. For example, Deepak Phatak from the Indian Institute of Technology – one of the 50 most influential Indians according to Bloomberg Businessweek – has just developed the Aakash. Aakash is Indian for 'heaven' and can be described as an iPad for the poor, which claims to revolutionize education. India has millions of young people, but not enough schools to educate them, and 25 % of the population are illiterate. Together with the Indian government, Pathek is planning to distribute the tablet PC, which comes at the cheap price of approximately €40, to pupils and students in India for it to function as a virtual classroom.

For several decades, detachment of economic growth from resource usage has been discussed by green economists. Moreover, a move from product innovations to service innovations, often IT-enabled services, has been recognized in many industries. The diffusion of open innovation throughout companies, the integration of diverse stakeholder groups, and the increasing collaboration of whole *networks* of different partners may provide new impetus for this development, if sustainability issues are considered in a serious way at the front end of innovation. Open innovation could work as a leverage of individual efforts and enable a move from product innovation to system innovation. For example, the diffusion of electric cars powered by renewable energy is not only a question of technically advanced batteries, but the whole system of e-mobility has to evolve covering smart grid solutions, renewable energy supply, and intelligent battery switching or charging services at filling stations. Such system innovations cannot be tackled by an individual company, but only through concerted actions of whole networks of companies, individuals and governmental institutions. In this way, future endeavors at the front end of innovation may not only lead to strengthened competitive advantage of individual companies, but have the power to create more ecologically, economically and socially sustainable innovations.

## 4  The 20 Rules for Successful Innovation

The fuzzy front end of innovation has certainly gained more attention among practitioners and academics in the past few years. The importance of the early phase has been acknowledged, and 'fuzziness' has been recognized not only as a disturbing factor in planning, but also as a source of creativity and new opportunities. In a broader view, the mechanistic view of the world has been replaced by a more dynamic and sensitive one: the wing beat of the butterfly and the theoretical strings around chaos, fractals and uncertainty have also influenced management theory and practice. Systemic management and cybernetic perspectives on enterprises as living organisms have emerged; interestingly, managers in the real world have discovered the value of these theoretical approaches more rapidly and eagerly than could be said of the academic world.

If we allow ourselves to open up to the fuzziness of the front end of innovation and see the opportunities it offers, there are many chances of creating better products by exploiting new technologies for superior value for the users. In order to manage the unmanageable better, some success factors of the fuzzy front end have to be considered:

1. **Balance exploration and exploitation**
   Too often, there is a strong emphasis on exploitation. Companies like *AEG*, *Nixdorf*, *Nakamichi* and *Schlecker* had been great companies in exploiting their technologies, capabilities and brands. But they failed in exploring new opportunities. Today's cash cows can be the nails to the company's coffin

tomorrow. One of the strongest obstacles to taking risk is today's success. But companies have to care about innovation while they are successful, and not when they are in a crisis.

2. **Energize the project start**
   There is an old saying in project management: "Tell me how the project is started, and I will tell you how it will end." The early phase, and therefore the management of the fuzzy front end of innovation, is extremely important for project success. Many innovations fail because the objectives and directions were not clearly set in the beginning and team members did not truly buy in. Requirement engineering in the early phase includes collecting facts about the users and their needs. The fuzzy front end is often misunderstood to be fuzzy overall; instead, a clear mission, vision, and objectives are crucial for successful projects. The Roman philosopher Seneca summarized this when he stated that "no wind is favorable if one does not know to which port one is sailing".

3. **Stay on discovery mode**
   The fuzzy front end of an innovation project is an adventurous discovery journey. While heading to India, it could be that one ends up in America. The value of unplanned outcomes and deviations from goals has to be appreciated more. *Pfizer* never planned Viagra as a lifestyle drug; it had been intended as a medicine for cardiovascular diseases. Today, it is the most sold and most imitated drug worldwide.

4. **Deviations are always two-sided.**
   When the famous *3M* experiment with a new glue failed, other companies would have stopped the project. At *3M*, people are trained to always seek upside potentials of a failure. There might be new applications, such as the Post-It note, which can build on the failed experiment. A strong user perspective is most important.

5. **Innovate around users**
   Generations of MBAs have unlearned to talk about humans as only targeting market segments. But customers are humans with individual values, characteristics and requirements. Design thinking has helped to regain focus on the user as a person. User-centric innovation is rarely wrong. *Virgin* founder Richard Branson has the mantra, "build the business around people and users". Most of the time, he has been right.

6. **Use freedom of the fuzzy front end of innovation to discover new needs**
   *BMW*'s innovation management strives to develop something the user never knew he was seeking, but which he says he always wanted as soon as he gets it. This means, at BMW, innovators do not ask the customer what he wants, but they observe the customer and seek to understand his or her latent needs. In the automotive industry, only 10 % of all innovations are inspired directly by customer statements. At BMW, engineers try to understand the customer's latent requirements; they try to understand the customer better then he knows himself. This is a fine line between 'exceeding customer expectations' and 'happy engineering'. Steve Jobs once said: "How can the consumers know what they want; it is our job to identify that."

7. **Open up the innovation process**
   Too many engineers still see their lab as the world. By doing so, they underutilize the value of customers, users, suppliers and outside inventors as powerful engines for innovation. *Procter&Gamble* changed their research & development department to the new connect & develop department. Today more than 50 % of all innovations are contributed by this open innovation department, and P&G plans to triple current revenue through open innovation. This is a strong message.
8. **Fail earlier to succeed sooner**
   Michael Dell always pushed his team towards producing failures in the fuzzy front end of innovation. The more failures in the early phase, the better the team knows where the dead ends of a technology or project are. A deeper analysis of change request protocols offers interesting insights. Good and fast innovators request/perform many changes in the early phase; ineffective and slow innovators start with high stability and introduce changes when the project has already reached a late development stage – with the result of running overtime and over-budget.
9. **Lead people, rather than administer people**
   The fuzzy front end needs strong leadership in the sense of providing value and meaning, inspiration, intellectual stimulation, and individual coaching. Steve Jobs was one of these famous leaders. Today, *Apple*´s managers still remember him fondly, although he was also described as egocentric and difficult to work with by some. Many small companies and hidden champions are led by such strong leaders.
10. **Use agile development environments to speed up innovation**
    In the software world, a revolution happened at the end of the 1990s: traditional waterfall models, which had been taught as good practice for decades, were replaced by agile programming. This involves less documentation, more iterations, and much more interaction between the user and the programming team, which ultimately leads to overall more rapid prototyping. A picture is worth a 1,000 words, but a prototype is worth a 1,000 pictures; therefore rapid prototyping in agile development proved to be very successful. The internet company *Xing* relies solely on agile environments, as do most of the software start-ups. However, linear processes in stable environments can still have their value.
11. **Only start when resources are available**
    Too often, new projects are started when not enough resources are available. This leads to too many projects being pursued in parallel with not enough focus on any of them. As a result, the queuing effect ensues: many projects, but no outcome. From experience we know that two projects per engineer constitutes the optimal R&D capacity. Strategic 'star projects' should warrant a full assignment of team members to the project in a new venture mode – without line tasks or involvement in other projects. Discipline in starting a project supports project success and the overall innovation rate.

12. **Use project stop as the hidden success factor of innovation**
    Too many projects are walking zombies – too alive to die, but no hope of success. By evaluating against clear criteria, dead ends should be acknowledged early, if the team itself does no longer believe in the technology. This should not be confused with killing concepts and ideas too early. Instead, the company should develop a culture where project teams can stop their projects and are praised for this. At *Phonak*, a world market leader in hearing devices, project teams are praised in the in-house magazine of the company for their foresight capabilities and responsibility if they stop their own project. What is important is a corporate culture that emphasizes that project failures are not wrong. *Google* has a strong culture of encouraging experimentation, and project failures are understood as a natural result in Google's experimental culture.
13. **Don't restrict innovation to products and technologies**
    The supreme discipline is business model innovation. The most innovative firms of the world – e.g., those named in the yearly BCG of Fortune 500 ranking – have all been business model innovators, with *Apple* and *Google* leading this list. However, business models do not have to be disruptive. The construction technology company *Hilti* developed fleet management for its customers, where the customer no longer has to buy drilling devices, but can also rent them. In the future, new methods of developing business model innovation will arise and gain a level of proficiency similar to that we know in the engineering disciplines today.
14. **Use the power of effective tools without becoming their slave**
    The biggest innovation is still the method of innovation. In the front end of innovation, many methods have been developed with tremendous impact in the last 15 years. They range from hundreds of creativity techniques for each situation and problem, over dozens of computer aided innovation methods to innovation and business model navigators. Companies which use these methods create strong competitive advantages; however, the usage of tools in itself is no guarantee for success, as 'a fool with a tool is still a fool'.
15. **Exploit the flat world**
    Pulitzer Prize winner Thomas Friedman wrote 'the world is flat' where competition is global. Companies which use the fact that our competitor, partner, colleague and customer sits a mouse click away in Bangalore or elsewhere are no longer limited by geographic limitations on speed, insight, and costs. Modern ICT and English as the Lingua franca of engineers and scientists enable global innovation processes.
16. **Use the power of vision**
    A vision is a dream with a deadline. But dreams alone do not move the organization, and milestones alone weaken the direction and leave busy people in Hamster wheels. Especially in the early fuzzy front end of innovation, the normative power of a shared dream is too often neglected. A strong, shared vision can energize teams and align their efforts with an invisible hand. There are two major strategies: 'Killing the dragon', or 'winning the princess'. If you define and interpret an external threat – the dragon, e.g., 'the Chinese

competitors will overrun us with low-cost products' – and communicate the threat. This will enable the company to focus its activities. If you develop and interpret a strong, positive vision – the princess, e.g., *Google*'s 'make a better world with information' – the company aligns its energy to follow the vision. Important in both cases is that the team is encouraged and believes that the goal can be reached. *Bombardier* has developed the vision of the 'zero maintenance train'; this has had a huge, energizing effect on the organization.

17. **Take crisis as an opportunity to innovate**
  In a recession such as in 2009, teams can be mobilized to seek new ways. Many second-tier automotive suppliers in high-precision areas have leveraged their core competencies into new fields, such as the medical industry. In times of economic boom, companies have the challenge to deliver and implement. Therefore, often the power and resources to innovate are lacking. In serious crises, the energy to innovate can be generated, and the willingness to break with the past is higher. This is a prerequisite for innovation.

18. **Learn how to unlearn**
  One of the biggest barriers in thinking is the dominant logic of a business, technology, product, or customer. Even if 'something always used to be this way', it might change. Sociologists call the dominant logic orthodoxy; HR people call it corporate identity. If these common, shared values and beliefs are too strong, the mental barriers will block new opportunities. This is often the reason why leading firms are overtaken by newcomers to the industry. It was not *Nokia* or *Motorola* that revolutionized the mobile phone industry, but the outsider *Apple*. What is interesting in this context: Apple will have to fight hard in order to reinvent itself if it wants to survive. Hard to imagine today, but Apple might one day have to unlearn that fancy devices make the difference. In the future, it might be the cloud or some other innovation in this fuzzy world.

19. **Take calculated risks**
  Empirical research shows that managers are risk-averse, and the higher they move up in their careers, the more risk-averse they become. This is a major barrier to innovation. Instead of trying new ways, managers defend the old ways and try to find reasons why the new way might be very risky and could fail. Of course innovative projects can fail, in fact failure is the very nature of innovation projects. But in dynamic industries, the biggest risk of all can be not taking risks at all.

20. **Develop a culture of innovation**
  The core of innovative companies is their innovation culture. According to Harvard colleagues Robinson and Stern, companies have an innovative culture when they align their activities, encourage self-initiated activities, allow and encourage unofficial activities, encourage serendipity, support diversity in teams, and promote informal communication. The culture of innovation lived by firms such as *3M* might be their strongest ability. If there is one thing to focus on, it is the right culture.

If the rules outlined above are followed, there is no guarantee for success in the next new product development, but the probability of high innovation rates and successful projects increases. No guarantee, only probability – but in our new world we have to live with this fuzziness. This is a serious game we are in; companies which do not innovate become extinct like dinosaurs. If we do not challenge existing concepts, a garage industry somewhere in the world might start with an idea – and become a serious competitor like *Microsoft*, *Apple*, *Dell*, but also *Daimler* and other established firms. But we should not look at innovation like the rabbit at the snake. Instead, we have to take the sportive approach: let's innovate and challenge the way we are working today, challenge the products and services we are providing to the customer today, challenge the supply chain architecture we take for granted today, challenge the business logic we are in today. Even more, we have to enjoy the fuzziness and exploit it to open up great opportunities for more value and a better world. The winners of tomorrow's competition will certainly be the agile and innovative companies, not the large, slow ones.

# About the Authors

*Martin A. Bader* is managing partner of the innovation and intellectual property management advisory group BGW AG and head of the Competence Centre for Intellectual Property Management at the Institute of Technology Management at the University of St. Gallen, Switzerland. He is a European and Swiss Patent Attorney with almost 20 years of professional experience.

*Branko Bjelajac* is Chief Technology Officer and Executive Vice President for Landis+Gyr, Switzerland. After graduating from the University of Belgrade, Serbia, he gained a Ph.D. in electrical engineering and information technology from Aachen University of Technology (RWTH Aachen) in Germany as well as executive qualifications from INSEAD, Fontainebleau, France, and the Stanford University Graduate School of Business, Stanford, CA, USA. He is member of the Board of Directors of Electrosuisse CES, member of the Research Committee of the Münchner Kreis, a non-profit supra-national association dedicated to communication research, as well as member of the IEEE and the VDE.

*Amir Bonakdar* is Ph.D. student and research associate at the Institute of Technology Management, University of St.Gallen, Switzerland. After obtaining his diploma in business administration from the University of Erlangen-Nuremberg, Germany, in 2008, he worked as a consultant for SAP, supporting SAP clients in process analysis and optimization. His research interests include business model innovation as well as social capital and ambidexterity. He worked in various research projects with renowned clients like SAP, Siemens, Hilti, Swisscom, Landis+Gyr, Holcim, ABB, MTU, and Bühler.

*Walter Buchinger* is head of marketing for Austria and Switzerland at Emporia Telecom. Besides communication, his daily work comprises the conceptualization and interpretation of usability studies and other market research for the generation 50+. After completing his business administration studies, he worked in sales, marketing, and general management for different consumer goods companies in Austria for many years. In addition, he is a trainer of, and lecturer on, communication, team building, and social skills.

*Antony Clayton* is responsible for committee management and target management within the MINI project management function.

*Michaela Csik* is a Ph.D. student and research associate at the Institute of Technology Management, University of St. Gallen, Switzerland. Her research interests include business model innovation, business model patterns, and innovation management. She holds a diploma in business administration from the University of Mannheim and was a visiting researcher at Stanford University.

*Michael Daiber* is business development manager at ABB Turbo Systems in Baden, Switzerland. Before his association with ABB, he earned his Ph.D. in Innovation Management at the Institute of Technology Management at the University of St. Gallen. During his Ph.D. studies, he spent 1 year as visiting researcher at Tongji University in Shanghai. His main research areas included the integration of external actors into the innovation process and the internationalization of R&D. He studied virtual idea platforms and open innovation initiatives of industrial companies in depth. Besides his background in business, he holds an M.Sc. in Mechanical Engineering from ETH Zurich, Switzerland.

*Uli Eisert* is Research Manager in the 'New Assets & Business Model Innovation' program within the SAP Next Business and Technology organization. Uli has been working for SAP since 1995 in various roles in consulting and solution management. In 2006, he joined the research unit and built up the Swiss research labs. He earned a doctorate in economics from the University of St. Gallen and a diploma in mechanical engineering from TU Darmstadt and in industrial engineering from FH Mainz.

*Ellen Enkel* is professor for innovation management and head of the Dr. Manfred Bischoff Institute for Innovation Management of EADS at Zeppelin University. Her research comprises the topics of open innovation and cross-industry innovation, business model innovation, innovation networks, innovation metric systems and strategic communication of innovativeness as well as entrepreneurial culture. She also possesses broad experience from extensive collaboration with well-known enterprises across a myriad of industries.

*Denis Eser* is project manager at HYVE Innovation Community GmbH and supports companies and organizations in building software-based communities and in using open innovation techniques and crowdsourcing tools. As project manager for the sprint> radar platform he is responsible for the technical set-up of the community and the training of the community manager.

*Karolin Frankenberger* is assistant professor at the Institute of Technology Management, University of St. Gallen, Switzerland. Previously, she had been working as a consultant for several years. Her current research interests include business model innovation, strategic networks, imitation, new ventures, strategic initiatives, and strategic renewal. Her research has been published in outlets such as the Academy of Management Journal.

*Sascha Friesike* works as postdoctoral researcher at the Alexander von Humboldt Institute for Internet and Society in Berlin, Germany. There, he is in charge of the research group on internet-enabled innovation. He holds an engineering degree from the Technical University in Berlin and a Ph.D. from the University of St. Gallen. Prior to his engagement in Berlin, he worked as a researcher at the Center for Design Research at Stanford University.

*Johann Füller* has just taken office at the University of Innsbruck's faculty for Business Administration, where he holds the recently founded chair for Innovation and Entrepreneurship in the Department of Strategic Management, Marketing and

Tourism. He is Fellow at the NASA Tournament Lab at Harvard University and CEO of HYVE AG, an innovation and community company based in Munich.

*Oliver Gassmann* is Professor of Technology Management at the University of St. Gallen, Switzerland, and Director of the Institute of Technology Management. At the core of his research is the pervading question of how companies innovate and profit from innovation. Thus, he is dedicated to discovering new approaches to the management of technology and innovation that contribute to firms' competitive advantage.

*Kurt Gaubinger* is head of the degree program 'Mechatronics and Business Management' at the Faculty of Engineering and Environmental Sciences of the University of Applied Sciences Upper Austria. His research interests include industrial marketing and innovation management in industrial markets with a special focus on the front end of innovation.

*Sebastian Heil* is currently a research fellow and Ph.D. candidate at the Chair for Innovation Management at Zeppelin University. His research focuses on the antecedents and consequences of collaborative innovation, especially beyond established industry boundaries (cross-industry innovation). Furthermore, he is involved in cooperation projects with firms across industries to develop new theory-based concepts and gain new insights from managerial practice.

*Jürgen Heubach* is head of Corporate Development–Innovation at Bayer. He is a board certified Pharmacologist and Toxicologist. After his medical studies at Ulm University he spent several years in cardiovascular research at different universities before joining the pharmaceutical industry in 2003. At Bayer HealthCare he was responsible for pharmacogenetics and clinical biomarker application in the context of Personalized Medicine.

*Katja Hutter* is research assistant at the Innsbruck University School of Management. Katja holds a doctoral degree in Social and Economic Sciences from the University of Innsbruck. She is a Fellow at the NASA Tournament Lab at Harvard University. Her research interests are in the fields of innovation management, online innovation communities, and co-creation.

*Matthias Kaiserswerth* has been director of IBM Research – Zurich since 2006. Before, he was managing director of an IBM Integrated Account. Matthias Kaiserswerth joined IBM Research in 1988. He received his M.Sc. and Ph.D. in Computer Science from McGill University in Montreal, Canada, and from Friedrich-Alexander-University in Erlangen-Nuremberg, Germany, respectively. He regularly shares thoughts and ideas on his blog: http://ibmzrl.wordpress.com/.

*Hans-Jörg Kirchweger*, MBA is managing director of voestalpine Anarbeitung and voestalpine Steel Service Center. He is responsible for production, maintenance, IT, quality- & process management and R&D.

*Petr Korba* is head of the electric power systems and smart grid lab at the Zurich University of Applied Sciences. His responsibilities and research interests include lecturing and consulting in the field of advanced power and control systems and their industrial applications. Dr. Korba has worked with ABB Switzerland Ltd. in various positions for more than 12 years. He was nominated by the European Patent Office for the Best European Inventor Award in 2011.

*Markus Kretschmer* is Professor of Product Design and Design Management at the University of Applied Sciences Upper Austria. He teaches design strategy, design management, and theory of design and also supervises design projects. His research interests focus on strategic design, the management of design in innovation processes, and design as a driver of sustainable innovation.

*Larry J. Leifer*, Ph.D. is a Professor of Mechanical Engineering Design and founding Director of the Center for Design Research (CDR) at Stanford University. A member of the faculty since 1976, he teaches the industry sponsored master's course ME310, 'Global Project-Based Engineering Design, Innovation, and Development'; a thesis seminar, 'Design Theory and Methodology Forum'; and a freshman seminar 'Designing the Human Experience.' Research themes include: (1) creating collaborative engineering design environments for distributed product innovation teams; (2) instrumentation of that environment for design knowledge capture, indexing, re-use, and performance assessment; and (3), design-for-sustainable well-being.

*Sandra Lemmer* is executive assistant to Johann Füller at HYVE. During her studies of German philology and psycholinguistics (M.A.) and her graduation at the LMU Munich, she worked as graduate assistant at the LMU. After finishing her studies, she worked in quality management in the market research sector for 8 years and, within the scope of her honorary post at the D.A.H., obtained a certification as online-counselor.

*Stephan Michael Meier* is assistant to the member of the Board of Management of Bayer AG being responsible for Technology, Innovation and Sustainability. He was born in Neuss, Germany in 1978 and started his career at Bayer in 2006 after a Ph.D. in Chemical- and Bioengineering from Friedrich-Alexander-University Erlangen-Nurnberg. He is a qualified European Patent Attorney and a visiting Professor at Dongseo University Busan, Korea.

*Thomas Müller-Kirschbaum* is responsible for global research and development as well as production and supply chain of the Laundry and Home Care division at Henkel. He is board member of the Research Committee for the Association of the German Chemical Industry (VCI) and honorary professor for Innovation & Sustainability Management at the University of Applied Sciences in Krefeld.

*Patrick Oberdellmann* was responsible for the project management of the MINI Countryman from the early phase until start of production. Currently he is responsible for the purchasing of the BMW i8.

*Georg Oenbrink* holds a Ph.D. in chemistry and started his professional career at Dynamit Nobel AG in the application technology for thermoplastics in 1987. In 1989, he joined the former Hüls AG, and acted as Senior Vice President Innovation Management of the High Performance Polymers business line from 1998 to 2009. In 2010, he took over the responsibility as Senior Vice President Innovation Networks & Communications within the Corporate Innovation Strategy & Management organization at Evonik Industries AG. He is member of several industrial and scientific advisory boards in Germany and presently heads the Macromolecular Division of the Gesellschaft Deutscher Chemiker (GDCh).

# About the Authors

*Javier Perez-Freije* heads the Corporate Controlling department at Autoneum in Winterthur/ Switzerland. Before, he worked at BMW Group in Munich. Javier graduated in Business Administration and Engineering and received his Ph.D. from the University of St. Gallen. He was visiting research scholar at New York University focusing on controls in the early innovation phase. He has published several articles highlighting the interaction between controls and innovation.

*Wolfgang Plischke* has been a member of the Board of Management of Bayer AG since March 1, 2006. He is responsible for Technology, Innovation and Sustainability, as well as for the Asia/Pacific region. He was born in Stuttgart, Germany in 1951 and studied biology at Hohenheim University before starting his career in 1980 with Bayer's subsidiary Miles Diagnostics. Since July 2011 Wolfgang Plischke has been an honorary professor for Business Chemistry at the Ludwig Maximilian University, Munich.

*Michael Rabl* is head of the department of Innovation and Product Management at the Faculty of Engineering and Environmental Sciences of the University of Applied Sciences Upper Austria. His special interests in teaching, research and consultancy are entrepreneurship & innovation and process-oriented aspects in product development and management.

*Stephan Rahn*, born in 1963, studied business management at the University of Cologne. After working in business consulting, he joined 3M Germany, located in the city of Neuss in the Rhineland area of Germany near Dusseldorf, in 1997. Today, he is responsible for 3M's innovation marketing activities.

*René Rohrbeck* is Associate Professor of Strategy at the Aarhus University in Denmark. He teaches MBA and master's courses on corporate foresight, strategy, and innovation management. His research is focused on corporate foresight, technology management, business model design, and strategic innovation management.

*Caroline V. Rudzinski* works at the future affairs department at Volkswagen Group Research. For seven years, she has applied the open innovation approach and worked on the interface between crowdsourcing, strategy and organization. Besides, she is writing her Ph.D. on 'Open Strategy' and also holds a lecture at the University of Witten/Herdecke in Germany.

*Patricia Sandmeier Kahmen* is the business development manager for the Swiss market at ABB Switzerland Ltd. Her responsibilities comprise organizational development as well as the identification and development of growth opportunities. She has an academic background in innovation management from the University of St. Gallen, University of New South Wales, and UC Berkeley. Her research focuses on customer integration in industrial new product development processes during the innovation front end phase.

*Fiona Schweitzer* is Professor of Marketing and Market Research at the department of Innovation and Product Management of the University of Applied Sciences Upper Austria. She teaches diverse courses of marketing, innovation management, and market research. Her academic research focuses on open innovation, customer integration into the innovation process, the front end of innovation, and technology adoption.

*Markus Seidel* is head of MINI project management within the BMW Group.

*Martin Steinert* is Acting Assistant Professor ME, and is Deputy Director of the Center for Design Research (CDR) and of the Hasso Plattner Design Thinking Research Program (HPDTRP) at Stanford University. His research focuses on optimizing the fuzzy front end, the intersection between engineering, new product development and the design process. Dr. Steinert teaches a seminar on 'Design Theory and Methodology' and a project based course on 'Design Theory and Methodology' and a project-based course on 'Designing Emotion Interactive Interfaces'. Three of his four current, sponsored research projects combine engineering design behavior and real-time physiological measurements.

*Alexander Strunz* is a bachelor student of business administration at the University of St. Gallen. He accompanied the research project and wrote his bachelor thesis in the area of business models and value networks at the Institute of Technology Management. During his studies he gained working experience in various companies like McKinsey, T-Systems and UniCredit.

*Stefan Thomke*, an authority on the management of technology and product innovation, is the William Barclay Harding Professor of Business Administration at Harvard Business School. Having authored or co-authored more than fifty articles, books, and case studies, Thomke has focused his research primarily on the process, economics, and management of business experimentation in innovation.

*Gereon Uerz* is project director at Volkswagen Group Research. He lectures at the University of Hanover, Germany, and is a member of the Planning Committee of the German Node of the Millennium Project of the World Federation of United Nations Organizations (WFUNA), a global think tank on the future.

*Christoph H. Wecht*, MBA, is co-founder and managing partner of the BGW Management Advisory Group, a strategy consulting company focusing on innovation management. He also heads the Competence Center for Open Innovation at the Institute of Technology Management at the University of St. Gallen. Previously, he worked in the automotive supply industry in Austria, Germany, and the US.

*Juan Carlos Wuhrmann* is responsible for international R&D Management in the R&D/Technology department of the Laundry and Home Care division at Henkel. In recent years he has worked on different approaches to innovation and put these into practice. He has successfully organized benchmarking projects on innovation processes with other industries, consultants, and academia.

*Marco Zeschky* is Assistant Professor of Technology Management at the University of St. Gallen, Switzerland. His current research focuses on the question how firms effectively set up their international R&D organization and how product architecture influences the design of international R&D organization. Major parts of his research are conducted in the light of the rise of emerging markets.

*Nicole Ziegler* is research associate at the Institute of Technology Management at the University of St. Gallen, Switzerland. Her research topic is intellectual property management with focus on the commercialization of intellectual property. Before Nicole came to St. Gallen, she studied industrial engineering at the Karlsruhe Institute of Technology, Germany, and the Grenoble Institute of Technology, France.

# About the Institutes

## Excellence from St. Gallen

### Managing Innovation

#### Research

The **Institute of Technology Management at the University of St. Gallen** is the intellectual core of our 'managing innovation' approach. With its four international professors and over 60 employees the institute focuses on close collaboration with companies and organizations by means of major research and consulting projects. Research results from publications flow directly into our courses as well as into our spin-offs. The final test for thought leadership is the implementation into practice. www.item.unisg.ch

| Institute of Technology Management<br> University of St.Gallen | Contact:<br>Prof. Dr. Oliver Gassmann<br>Managing Director<br>oliver.gassmann@unisg.ch<br>phone +41 71 224 7220 |
|---|---|

#### Consulting

The **BGW Management Advisory Group** is a spin-off of the University of St. Gallen which is specialized on coaching companies in issues around management of innovation and intellectual property. Companies such as BASF, Bayer, Holcim, IBM, Nestlé, Siemens or Syngenta, as well as many SMEs have been coached in innovation strategy, R&D strategy, open innovation, innovation processes and methods of the fuzzy front end of innovation. www.bgw-sg.com

|  Management Advisory Group<br>St. Gallen – Wien | Contact:<br>Dr. Martin Bader<br>Managing Partner<br>martin.bader@bgw-sg.com<br>phone: +41 71 840 0831 |
|---|---|

## Business Model Innovation

The **BMI-lab** is a recent spin-off of the University of St. Gallen which is specialized on executive teaching and action based learning around the new methodology how to develop business models. Many companies such as BASF, Hilti, Holcim, Landis&Gyr, PWC, Sennheiser, Siemens, Swisscom have successfully developed new business models according to our construction methodology. The process helps to overcome mental barriers of the existing business logic. www.bmi-lab.ch

Contact:
Karolin Frankenberger
CEO
karolin.frankenberger@unisg.ch
phone: +41 71 224 7302

## Department of Innovation and Product Management (IPM)

### Shaping Innovation

Studying innovation and product management at IPM provides a comprehensive interdisciplinary view on innovation from Mechatronics, to Business Economics and Marketing, complemented by Industrial Design and Development Process know-how. At IPM experienced experts from industry and academia teach latest trends and developments, ensuring a modern and future-orientated course and practical learning through real-life cases. The three cornerstones of IPM are:

**Engineering**
**Design**
**Management**

ipm▷

| Bachelor's degree | Master's degree |
|---|---|
| IPM's fulltime B.Sc. degree program offers insights into the principles of innovation management with an emphasis on high-tech management in a six semesters' course held in German | IPM's fulltime M.Sc. provides comprehensive skills in strategic innovation management with specialization in Product Concept Design or Development Process Engineering and is taught in English for four semesters |

For details on the degree programmes visit www.fh-ooe.at/ipm

### Research
**sprint>** Systematic Product Innovation Transfercenter is IPM's research center that specializes on the front-end of innovation. **sprint>** is relying on a systematic utilization of advanced methods and tools to enhance the effectiveness and efficiency of the early phases of innovation. Researchers of **sprint>** have received international awards for their research, transfer the knowledge into daily teaching practices @IPM, and consult B2B-companies and high tech companies. Methods used in consulting and research projects comprise multimedia ideation workshops, crowdsourcing platforms, various quantitative and qualitative market research techniques, emphatic design and design thinking. www.fh-ooe.at/sprint

Contact:
Prof. (FH) Dr. Michael Rabl
Head of Department for Innovation and Product Management
michael.rabl@fh-wels.at, phone + 43 (0)50804-43355
University of Applied Sciences Upper Austria
Stelzhamerstr.23, A-4600 Wels

# References

Achtenhagen L, Müller-Lietzkow J, zu Knyphausen-Aufseß D (2003) Das open source-Dilemma: open source-software zwischen freier Verfügbarkeit und Kommerzialisierung. Schmalenbachs Zeitung für betriebswirtschaftliche Forschung 55:455–481
Afuah A, Tucci C (2000) Internet business models and strategies: text and cases. McGraw-Hill, New York
Ahmed PK, Shepherd CD (2010) Innovation management: context, strategies, systems and processes. Pearson, Harlow
Ahuja G, Katila R (2004) Where do resources come from? The role of idiosyncratic situations. Strateg Manage J 25(8–9):887–907
Alexander C (1964) Notes on the synthesis of form. Harvard University Press, Cambridge
Allee V (2003) The future of knowledge: increasing prosperity through value networks. Butterworth-Heinemann, Boston
Allee V (2008) Value network analysis and value conversion of tangible and intangible assets. J Intellect Cap 9(1):5–24
Allee V (2011) Value Networks and the True Nature of Collaboration. ValueNet Works and Verna Allee Associates. http://www.valuenetworksandcollaboration.com/home.html
Allen TJ (1966) Studies of the problem-solving process in engineering design. IEEE Trans Eng Manage 13(2):72–83
Andreasen MM (2005) Vorgehensmodelle und Prozesse für die Entwicklung von Produkten und Dienstleistungen. In: Schäppi B, Andreasen MM, Kirchgeorg M, Radermacher F-J (eds) Handbuch Produktentwicklung. Hanser, München, pp 247–264
Ansoff IH (1975) Managing strategic surprise by response to weak signals. Calif Manage Rev 18(2):21–33
Aquino LM, Steinert M, Leifer L (2011) Designing to maximize value for multiple stakeholders: evidence of the challenge to the medical-technology innovation. In: Proceedings of the international conference on design engineering. Presented at the ICED 2011, Copenhagen
Argyris C, Schon DA (1978) Organizational learning: a theory of action perspective. Addison-Wesley, Reading
Ayers D, Dahlstrom R, Skinner SJ (1997) An exploratory investigation of organizational antecedents to new product success. J Market Res 34(1):107–116
Back A, von Krogh G, Seufert A, Enkel E (2005) Putting knowledge networks into action. Springer, Berlin
Backhaus K, Voeth M (2007) Industriegütermarketing. Vahlen, München
Backman M, Borjesson S, Setterberg S (2007) Working with concepts in the fuzzy front end: exploring the context for innovation for different types of concepts at Volvo Cars. R&D Manage 37(1):17–28
Baldwin C, Clark K (2000) Design rules: vol. 1. The power of modularity. MIT Press, Cambridge
Banathy BH (1996) Designing social systems in a changing world. Plenum, New York

Barczak G, Griffin A, Kahn KB (2009) Perspective: trends and drivers of success in NPD practices: results of the 2003 PDMA best practices study. J Product Innovat Manage 26(1):3–23
Barker JR (1993) Tightening the iron cage: concertive control in self-managing teams. Adm Sci Q 38:408–437
Bartol KM, Srivastava A (2002) Encouraging knowledge sharing: the role of organizational reward systems. J Leadersh Organ Stud 91(1):64–76
Bayer Annual Report (2010) http://www.annualreport2010.bayer.com/
Bayer Annual Report (2011) http://www.annualreport2011.bayer.com/
BCG (2008) BusinessWeek/BCG innovation survey
Beck E, Obrist M, Bernhaupt R, Tscheligi M (2008) Instant card technique: how and why to apply in user-centered design. In: Proceedings of the tenth anniversary conference on participatory design 2008, Indiana University, Indianapolis, 30 Sept to 04 Oct 2008, pp. 162–165
Becker GS, Murphy KM (1992) The division of labor, coordination costs and knowledge. Quart J Econ CVII:1137–1160
Bergmann I, Butzke D, Walter L, Fürste J-P, Möhrle MG, Erdmann VA (2008) Evaluating the risk of patent infringement by means of semantic patent analysis: the case of DNA-chips. R&D Manage 38(5):550–562
Bernoff J, Li C (2008) Harnessing the power of the Oh-So-Social web. MIT Sloan Manage Rev 49 (3):36–42
Bieger T, Reinhold S (2011) Das wertbasierte Geschäftsmodell – Ein aktualisierter Strukturierungsansatz. In: Bieger T, Krys C, zuKnyphausen-Aufseß D (eds) Innovative Geschäftsmodelle, Konzeptionelle Grundlagen, Gestaltungsfelder und unternehmerische Praxis. Springer Verlag, Berlin, pp 13–70
Bjelland OM, Wood RC (2008) An inside view of IBM's 'Innovation Jam'. MITSloan Manage Rev 50(1):32–40
Boeddrich H-J (2004) Ideas in the workplace: a new approach towards organizing the fuzzy front end of the innovation process. Creativity Innov Manage 13(4):274–285
Bohn RE (1995) Noise and learning in semiconductor manufacturing. Manage Sci 41:31–42. doi:10.1287/mnsc.41.1.31
Bonsiepe G (1992) Die sieben Säulen des design: design braucht keine Manifeste, sondern Fundamente. Form+Zweck, 6
Bower JL, Christensen CM (1995) Disruptive technologies: catching the wave. Harvard Business Review 73(1):43–53
Box G, Hunter W, Hunter S (1978) Statistics for experimenters. Wiley, New York
Braungart M, McDonough W (2008) Die nächste industrielle Revolution, 141. Europäische Verlagsanstalt, Hamburg
Briehs S (2011) Employee integration in the idea generation of the NPD process at Henkel. Unpublished MBA-thesis, Henkel Center for Consumer Research of the WHU – Otto Beisheim School of Management, Vallendar
Brown T (2008) Design thinking. Harv Bus Rev 86(6):84–92
Brown SL, Eisenhardt K (1998) Competing on the edge, strategy as structured chaos. Harvard Business School Press, Boston
Bucherer E, Eisert U, Gassmann O (2012) Towards systematic business model innovation: lessons from product innovation management. Creativity Innov Manage 21:183–198
Burckhardt L (1995) Design ist unsichtbar. In: Höger H (ed) Design ist unsichtbar. Cantz Verlag, Ostfildern, pp 14–24
Burgelman RA (2002) Strategy as vector and the inertia of coevolutionary lock-in. Adm Sci Q 47 (2):325–357
Burt RS (1992) Structural holes: the social structure of competition. Harvard University Press, Cambridge
Cagan J, Vogel CM (2002) Creating breakthrough products: innovation from product planning to program approval. Prentice Hall, Upper Saddle River

Calantone R, Garcia R, Dröge C (2003) The effects of environmental turbulence on new product development strategy planning. J Product Innov Mange 20(2):90–103

Carbonaro S (2008) Radikales Umdenken erforderlich: an ökologischen und ethischen Fragestellungen kommt kein Unternehmen mehr vorbei. In: Nachtwey J, Mair J (eds) Design ecology! Neo-Grüne Markenstrategien. Verlag Hermann Schmidt, Mainz, pp 202–221

Chandy RK, Tellis GJ (2000) The incumbent's curse? Incumbency, size, and radical product innovation. J Market 64(3):1–17

Charter M, Chick A (1997) Editorial. The centre for sustainable design. J Sustain Prod Des 1(1): 5–6

Checkland P, Scholes J (1999) Soft systems methodology in action. Wiley, Hoboken

Chesbrough H (2003) Open innovation: the new imperative for creating and profiting from technology. Harvard Business Press, Boston

Chesbrough H (2009) Business model innovation: opportunities and barriers. Long Range Plann. doi:10.1016/j.lrp.2009.07.010

Chesbrough H, Rosenbloom RS (2002) The role of the business model in capturing value from innovation: evidence from Xerox corporation's technology spin-off companies. Ind Corp Change 11(3):529–555

Christensen CM (2000) The innovator's Dilemma. Harper Business, New York

Clark K, Fujimoto T (1991) Product development performance. Harvard Business School Press, Boston

Coan JA, Gottman JM (2007) The specific affect coding system (SPAFF). In: Handbook of emotion elicitation and assessment. Oxford University Press, New York, pp 267–285

Cockburn IM, Henderson RM (1998) Absorptive capacity, coauthoring behavior, and the organization of research in drug discovery. J Ind Econ 46(2):157–182

Coetzee F, Sandmeier P, Knöpfel M, Reinhardt P (2012) BESS – battery energy storage systems. In ABB Tower News 1/2012, pp 8–11

Cohen WM, Levinthal DA (1990) Absorptive capacity: a new perspective on learning and innovation. Adm Sci Q 35(1):128–152

Cooper RG (1993) Winning at new products: accelerating the process from idea to launch. Addison-Wesley, Cambridge

Cooper RG (1994) Third-generation new product processes. J Prod Innov Manage 11(1):3–14

Cooper RG (2001) Winning at new products: accelerating the process from idea to launch. Perseus, Cambridge

Cooper RG (2002) Top oder Flop in der Produktentwicklung. Erfolgsstrategien: Von der Idee zum launch. Wiley-VCH Verlag GmbH & Co. KGaA, Weinheim

Cooper RG (2006) Managing technology development projects. Res Technol Manage 49(6):23–30

Cooper RG (2008a) Perspective: the stage-gate idea –to-launch-process-update, what's new, and NexGenSystems. J Prod Innov Manage 25:213–232

Cooper RG (2008b) Maximizing productivity in product innovation. Res Technol Manage 51 (2):47–58

Cooper RG (2011) Winning at new products: creating value through innovation, 4th edn. Perseus, New York

Cooper RG, Kleinschmidt EJ (1986) An investigation into the new product process: steps, deficiencies, and impact. J Prod Innov Manage 3(2):71–85

Cooper RG, Kleinschmidt EJ (1990) New products: the key factors in success. American Marketing Association, Chicago

Cooper RG, Kleinschmidt EJ (1995) Benchmarking the firm's critical success factors in new product development. J Prod Innov Manage 12(5):374–391

Cooper RG, Kleinschmidt EJ (2007) Winning businesses in product development: the critical success factors. Res Technol Dev 50(3): 57ff

Cooper RG, Edgett SJ, Kleinschmidt EJ (2001) Portfolio management for new product development: results of an industry practices study. R&D Manage 31(4):361–380

Costenbader E, Valente TW (2003) The stability of centrality measures when networks are sampled. Soc Network 25:283–307
Cross N (2000) Engineering design methods: strategies for product design, vol 58. Wiley, Chichester
Cross N (2007) Designerly ways of knowing. Birkhäuser Verlag, Basel
Csikszentmihalyi M (2007) Kreativität: Wie Sie das Unmögliche schaffen und Ihre Grenzen überwinden. Klett-Cotta, Stuttgart, pp 41–79
Daft RL, Weick KE (1984) Toward a model of organizations as interpretation systems. Acad Manage Rev 9(2):284–295
Dahan E, Kim AJ, Lo AW, Poggio T, Chan N (2011) Securities trading of concepts (STOC). J Market Res 48(3):497–517. doi:10.1509/jmkr.48.3.497
Dahl DW, Moreau P (2002) The influence and value of analogical thinking during new product ideation. J Market Res (JMR) 39(1):47
Danneels E (2008) Organizational antecedents of second-order competences. Strateg Manage J 29(5):519–543
DARPA (2010) Defense Advanced Research Projects Agency. DARPA Mission. Retrieved from http://www.darpa.mil/mission.html
Davis JS, Harrison SS (2001) Edison in the boardroom: how leading companies realize value from their intellectual assets. Wiley, New York
Dawson R, Bynghall S (2011) Getting results from crowds: the definitive guide to using crowdsourcing to grow your business. Advanced Human Technologies, San Francisco
De Bono E (1990a) Lateral thinking for management. Penguin Books, London
De Bono E (1990b) Six thinking hats. Penguin, London
De Bono E (2013) Serious creativity. Vermilion, London
De Brentani U, Reid SE (2012) The fuzzy front-end of discontinuous innovation: insights for research and management. J Prod Innov Manage 29(1):70–87
De Luca LM, Atuahene-Gima K (2007) Market knowledge dimensions and cross-functional collaboration: examining the different routes to product innovation performance. J Market 71:95–112
Demil B, Lecocq X (2010) Business model evolution: in search of dynamic consistency. Long Range Plann 43(2/3):227–246
Deppe L, Kohn S (2002) The holistic view of the front end of innovation. Fraunhofer Technologie-Entwicklungsgruppe. In: Conference on IMTs and new product development, Mantova, 17–18 Oct 2002
Diller H (2007) Grundprinzipien des marketing. GIM, Nürnberg
Eco U (2002) Einführung in die Semiotik. Wilhelm Fink Verlag, Paderborn
Edelman J, Currano R (2011) Re-representation: affordances of shared models in team-based design. In: Plattner H, Meinel Ch, Leifer L (eds) Design Thinking. Spirnger Verlag, Berlin Heidelberg, pp 61–79
Edelman JA, Banerjee B, Jung M, Sonalkar N, Lande M (2009) Hidden in plain sight: affordances of shared models in team-based design. In: Proceedings of the international conference on engineering design. Presented at the ICED 2009, Stanford
Eisenhardt KM, Tabrizi BN (1995) Accelerating adaptive processes: product innovation in the global computer industry. Adm Sci Q 40(1):84–110
Enkel E (2010) Attributes required for profiting from open innovation in networks. Int J Technol Manage 52(3/4):344–371
Enkel E, Gassmann O (2010) Creative imitation: exploring the case of cross-industry innovation. R&D Manage 40(3):256–270
Enkel E, Heil S (2012) Assessing and managing cognitive distance in cross-industry innovation. In: Paper presented at the R&D management conference (RADMA), Grenoble
Enkel E, Mezger F (2013) Imitation processes and their application for business model innovation: an explorative study. Int J Innov Manage 17(1)

Enkel E, Gassmann O, Chesbrough H (2009) Open R&D and open innovation: exploring the phenomenon. R&D Manage 39(4):311–316

Eris O (2003) Asking generative design questions: a fundamental cognitive mechanism in design thinking. In: Proceedings of the international conference on engineering design, Stockholm, Sweden, pp 19–21

Eris O (2004) Effective inquiry for innovative engineering design. Kluwer Academic, Dordrecht

Eris O, Leifer L (2003) Facilitating product development knowledge acquisition: interaction between the expert and the team. Int J Eng Educ 19(1):142–152

Esslinger H (2012) A creative power shift. In: Esslinger H (ed) Design forward. Creative strategies for sustainable change. Arnoldsche Art Publishers, Stuttgart, pp 8–27

Fisher R (1921) Studies in crop variation: I. An examination of the yield of dressed grain from broadbalk. J Agric Sci 11:107–135

Fisher R (1923) Studies in crop variation: II. The manurial response of different potato varieties. J Agric Sci 13:311–320

Fisher R (1966) The design of experiments, 8th edn. Oliver and Boyd, Edinburgh

Fleming L, Sorenson O (2001) Technology as a complex adaptive system: evidence from patent data. Res Policy 30(7):1019–1039

Franke N, von Hippel E, Schreier M (2006) Finding commercially attractive user innovations: a test of lead-user theory. J Prod Innova Manage 23(4):301–315. doi:10.1111/j.1540-5885.2006.00203.x

Frederich MJ, Andrews P (2009) Innovation passport: the IBM First-of-a-Kind (FOAK) journey from research to reality. IBM Press, Upper Saddle River

Freeman LC (1979) Centrality in social networks: conceptual clarification. Soc Network 1:215–239

Friedman TL (2008) Die welt ist flach: Eine kurze Geschichte des 21. Jahrhunderts, vol 564. Suhrkamp Taschenbuch, Frankfurt/Main

Fuchs C, Schreier M (2011) Customer empowerment in new product development. J Prod Innov Manage 28(1):17–32. doi:10.1111/j.1540-5885.2010.00778.x

Füller J (2006) Why consumers engage in virtual new product developments initiated by producers. Adv Consum Res 33(1):639–646

Füller J (2010) Refining virtual co-creation from a consumer perspective. Calif Manage Rev 52(2):98–122

Füller J, Gregor J, Mühlbacher H (2007) Innovation creation by online basketball communities. J Bus Res 60(1):60–71

Füller J, Matzler K, Hoppe M (2008) Brand community members as a source of innovation. J Prod Innov Manage 25:608–619

Gagne CL, Shoben EJ (1997) Influence of thematic relations on the comprehension of modifier-noun combinations. J Exp Psychol Learn Mem Cogn 23(1):71–87

Gassmann O (2001) Multicultural teams: increasing creativity and innovation by diversity. Creativity Innov Manage 10(2):88–95. doi:10.1111/1467-8691.00206

Gassmann O (2012) Crowdsourcing: innovations management mit Schwarmintelligenz: – Interaktiv Ideen finden – Kollektives Wissen effektiv nutzen – Mit Fallbeispielen und Checklisten. Hanser, München

Gassmann O, Bader M (2010) Patentmanagement: innovationen erfolgreich nutzen und schützen. Springer, Heidelberg

Gassmann O, Bader MA (2011) Patentmanagement: innovationen erfolgreich nutzen und schützen, 3rd edn. Springer, Berlin/Heidelberg

Gassmann O, Zeschky M (2008) Opening up the solution space: the role of analogical thinking for breakthrough product innovation. Creativity Innov Manage 17(2):97–106

Gassmann O, Sandmeier P, Wecht CH (2006) Extreme customer innovation in the front-end: learning from a new software paradigm. Int J Technol Manage 33(1):46–66

Gassmann O, Kausch C, Enkel E (2010a) Negative side effects of customer integration. Int J Technol Manage 50(1):43–62

Gassmann O, Zeschky M, Wolff T, Stahl M (2010b) Crossing the industry-line: breakthrough innovation through cross-industry alliances with 'non-suppliers'. Long Range Plann 43 (5–6):639–654

Gassmann O, Csik M, Frankenberger K (2012) Aus alt mach neu: Ein Fahrplan für Innovationen. Harv Bus Manage 34:18–19, 06/2012

Gassmann O, Frankenberger K, Csik M (2013) Geschäftsmodelle innovieren. Hanser, München

Gaubinger K (2006) Grundlagen der Identifikation von Geschäftschancen: Situationsanalyse. In: Werani T, Gaubinger K, Kindermann H (eds) Praxisorientiertes business-to-business-marketing. Gabler, Wiesbaden, pp 57–69

Gaubinger K (2009) Prozessmodell des integrierten innovations- und Produktmanagements. In: Gaubinger K, Werani T, Rabl M (eds) Praxisorientiertes innovations- und produktmanagement. Gabler, Wiesbaden, pp 17–27

Gaubinger K, Schweitzer F, Zweimüller R (2012) A commercialization process model for technology innovations. Paper presented at the XXIII ISPIM conference in Barcelona, Spain on 17–20 June 2012

Gausemeier J, Fink A, Schlake O (1998) Scenario management: an approach to develop future potentials. Technol Forecast Soc Change 59(2):111–130

Gavetti G, Rivkin JW (2005) How strategists really think. Harv Bus Rev 83(4):152

Geim AK, Novoselov KS (2007) The rise of graphene. Nat Mater 6(3):183–191

Gemünden HG (2001) Die Entstehung von innovationen: Eine Diskussion theoretischer Ansätze. In: Hamel W, Gemünden H-G (eds) Außergewöhnliche Entscheidungen – Festschrift für Jürgen Hauschildt. Vahlen, München, pp 409–440

Gentner D, Rattermann MJ (1993) The roles of similarity in transfer: separating retrievability from inferential soundness. Cogn Psychol 25(4):524

Gershenfeld N (2005) Fab: the coming revolution on your desktop. Basic Books, New York

Geschka H (2006) Innovationsmanagement/Konzepte und Methoden des Innovationsmanagement. Technologie & Management, Schiele&Schön, Berlin, 9–10, pp 14–15

Geschka H, Schwarz-Geschka M (2007) Management von innovationsideen. In: Dold E, Gentsch P (eds) Innovation möglich machen. Symposion Publishing GmbH, Düsseldorf, pp 147–169

Gilsing V, Nooteboom B, Vanhaverbeke W, Duysters G, Van Den Oord A (2008) Network embeddedness and the exploration of novel technologies: technological distance, betweenness centrality and density. Res Policy 37(10):1717–1731

Glass RL (1995) Software creativity. Prentice Hall, Englewood Cliffs

Globocnik D (2011) Front end decision making: Das Entstehen hochgradig neuer Innovationsvorhaben in Unternehmen. Gabler Verlag, Wiesbaden

Gordon S, Tarafdar M, Cook R, Maksimoski R, Rogowitz B (2008) Improving the front end of innovation with information technology. Res Technol Manage 51(3):50–58

Gottman JM, Levenson RW (2000) The timing of divorce: predicting when a couple will divorce over a 14 year period. J Marriage Fam 62(3):737–745

Griffin A, Hoffmann N, Price RL, Vojak BA (2007) How serial innovators navigate the fuzzy front end of new product development. Paper presented at the PDMA research conference. Orlando, FL, USA

Gruner KE, Homburg C (2000) Does customer interaction enhance new product success? J Bus Res 49(1):1–14. doi:10.1016/S0148-2963(99)00013-2, 10.1016/S0148-2963%2899% 2900013-2

Hacklin F (2008) Management of convergence in innovation: strategies and capabilities for value creation beyond blurring industry boundaries. Physica-Verlag, Heidelberg

Hamel G (1991) Competition for competence and interpartner learning within international strategic alliances. Strateg Manage J 12(S1):83–103

Hamel G (2000) Leading the revolution. Harvard Business School Press, Boston

Hamel G (2002) Leading the revolution. Plume, New York

Hampton JA (1998) Conceptual combination: conjunction and negation of natural concepts. Mem Cognit 25(6):888–909

Hargadon A, Sutton RI (1997) Technology brokering and innovation in a product development firm. Adm Sci Q 42(4):716–749

Harison E, Koski H (2010) Applying open innovation in business strategies: evidence from Finnish software firms. Res Policy 39(3):351–359. doi:10.1016/j.respol.2010.01.008

Hauser MD (1996) The evolution of communication. MIT Press, Cambridge

Hayek FA (1945) The use of knowledge in society. Am Econ Rev 35:519–530

Hayek FA (1978) The mirage of social justice. University of Chicago Press, Chicago

Hedman J, Kalling T (2003) The business model concept: theoretical underpinnings and empirical illustrations. Eur J Inform Syst 12(1):49–59

Heger T, Rohrbeck R (2012) Strategic foresight for collaborative exploration of new business fields. Technol Forecast Soc Change 79(5):819–831

Hendersen RM, Cockburn J (1996) Scale, scope and spillovers: the determinants of research productivity in drug discovery. Rand J Econ 27(1):32–59

Henderson RM, Clark KB (1990) Architectural innovation: the reconfiguration of existing product technologies and the failure of established firms. Adm Sci Q 35(1):9–30

Herstatt C, Kalogerakis K (2005) How to use analogies for breakthrough innovations. Int J Innov Technol Manage 2(3):331–347

Herstatt C, Lettl C (2000) Management von technologiegetriebenen Entwicklungsprojekten. Arbeitspapier Nr. 5

Herstatt C, Verworn B (eds) (2007) Management der frühen Innovationsphasen, Grundlagen – Methoden – Neue Ansätze. Gabler Verlag, Wiesbaden

Hoffman DL, Kopalle PK, Novak TP (2010) The "right" consumers for better concepts: identifying consumers high in emergent nature to develop new product concepts. J Market Res 47(5):854–865. doi:10.1509/jmkr.47.5.854

Holtorf V (2011) Teams im front end: Steigerung des unternehmerischen Verhaltens durch strukturierte Teams. Gabler Verlag, Wiesbaden

Holyoak KJ, Thagard P (1995) Mental leaps: analogy in creative thought. MIT Press, Cambridge

Homburg C (2007) Betriebswirtschaftslehre als empirische Wissenschaft – Bestandsaufnahme und Empfehlungen. Zeitschrift für betriebswirtschaftliche Forschung 56(7):27–60

Homburg C, Daum D (1998) Das Management der Kundenstruktur als Controllingherausforderung. In: Reinecke S, Tomczak T, Dittrich S (eds) Marketing-controlling. St. Gallen, Thexis

Homburg C, Schäfer H, Schneider J (2008) Sales excellence. Vertriebsmanagement mit system. Gabler Verlag, Wiesbaden

Howe J (2008) Crowdsourcing: how the power of the crowd is driving the future of business. The Crown Publishing Group, New York

Howell JM, Shea CM (2001) Individual differences, environmental scanning, innovation framing, and champion behavior: key predictors of project performance. J Prod Innov Manage 18(1):15–27

Huber J (2001) Allgemeine Umweltsoziologie. Westdeutscher Verlag, Wiesbaden, pp 274–294

Hungenberg H (2011) Strategisches management in unternehmen. Ziele – Prozesse – Verfahren. Gabler Verlag, Wiesbaden

Huston L, Sakkab N (2006) Connect & develop – inside procter & Gamble's new model for innovation. Harv Bus Rev 84:58–67

Hutter K, Hautz J, Füller J, Mueller J, Matzler K (2011) Communitition: the tension between competition and collaboration in community-based design contests. Creativity Innov Manage 20:3–21

Iansiti M, MacCormack A (1997) Team New Zealand (A), (B) and (C). Harvard Business School Cases, No. 697–040, 697–041, and 697–042

IBM (2008) Welcome to the IBM Jam events page. https://www.collaborationjam.com/. Accessed 25 Nov 2011

IBM (2012) IBM global CEO study – the enterprise of the future. http://www-935.ibm.com/services/us/en/c-suite/ceostudy2012/

IBM 2012 Global technology outlook (2012). http://www.zurich.ibm.com/pdf/isl/infoportal/GTO_2012_Booklet.pdf

IDEO (2012) Human centered design, an introduction, 2nd edn. Available at: http://www.ideo.com/images/uploads/hcd_toolkit/HCD_INTRO_PDF_WEB_opt.pdf . Accessed 19 Dec 2012

Illich I (2012) Selbstbegrenzung: Eine politische Kritik der Technik. Verlag C. H. Beck, München, pp 17–74

Innocentive (2012a) Crowdsource your innovation challenge – improve the way you innovate & work – become an innoCentive solver. https://www.innocentive.com/. Accessed 15 Mar 2009

Innocentive (2012b) Measuring weight of live animals. https://www.innocentive.com/ar/challenge/9932863. Accessed 20 Jan 2012

Jaikumar R, Bohn R (1986) The development of intelligent systems for industrial use: a conceptual framework. Res Technol Innov Manage Policy 3:169–211

Jansen JJP, Van Den Bosch FAJ, Volberda HW (2005) Managing potential and realized absorptive capacity: how do organizational antecedents matter? Acad Manage J 48(6):999–1015

Jansen JJP, Van Den Bosch FAJ, Volberda HW (2006) Exploratory innovation, exploitative innovation, and performance: effects of organizational antecedents and environmental moderators. Manage Sci 52(11):1661–1674

Johnson MW, Christensen CM, Kagermann H (2008) Reinventing your business model. Harv Bus Rev 86(12):50–59

Ju WG (2008) The design of implicit interactions. Doctoral Dissertation, Stanford, Stanford University

Jung M, Chong J, Leifer L (2010) Pair programming performance: an emotional dynamicspoint of view from marital pair counseling. Electronic Colloquium on Computational Complexity ECDTR (2)

Jungk R, Müllert N (1987) Future workshops: how to create desirable futures. Institute for Social Inventions, London

Kalogerakis K, Lüthje C, Herstatt C (2010) Developing innovations based on analogies: experience from design and engineering consultants. J Prod Innov Manage 27(3):418–436

Katila R, Ahuja G (2002) Something old, something new: a longitudinal study of search behavior and new product introduction. Acad Manage J 45(6):1183–1194. doi:10.2307/3069433

Katz R, Allen TJ (1982) Investigating the not invented here (NIH) syndrome: a look at the performance, tenure, and communication patterns of 50 R & D project groups. R&D Manage 12(1):7–20. doi:10.1111/j.1467-9310.1982.tb00478.x

Kelley T, Littman J (2005) The ten faces of innovation: IDEO's strategies for defeating the devil's advocate and driving creativity throughout your organization. Doubleday, New York

Khurana A, Rosenthal SR (1997) Integrating the fuzzy front end of new product development. Sloan Manage Rev 2:103–120

Khurana A, Rosenthal SR (1998) Towards holistic "front ends" in new product development. J Prod Innov Manage 15(1):57–74

Kim WC, Mauborgne R (2005) Blue ocean strategy how to create uncontested market space and make the competition irrelevant. Harvard Business School Press, Boston

Kim J, Wilemon D (2002) Focusing the fuzzy front-end in new product development. R&D Manage 32(4):269–279

Klischewski R, Ukena S (2007) Designing semantic e-Government services driven by user requirements. In: Electronic government, 6th international EGOV conference, proceedings of ongoing research, project contributions and workshops. Trauner Verlag, Linz, pp 133–140

Koen P, Ajamian G, Burkart R, Clamen A, Davidson J, D'Amore R, Elkins C, Herald K, Incorvia M, Johnson A, Karol R, Seibert R, Slavejkov A, Wagner K (2001) Providing clarity and a common language to the fuzzy front end. Res Technol Manage 44(2):46–55

Koen P, Ajamian GM, Boyce S, Clamen A, Fisher E, Fountoulakis S, Johnson A, Puri P, Seibert R (2002) Fuzzy-front end/effective methods, tools and techniques. In: Belliveau P, Griffin A, Somermeyer S (eds) The PDMA ToolBook for new product development. Wiley, New York, pp 2–35

Kogut B, Metiu A (2001) Open-source software development and distributed innovation. Oxford Econ Policy 17:248–264

Kohler T, Füller J, Matzler K, Stieger D (2011) Co-creation in virtual worlds: the design of the user experience. MIS Quart 35(3):773–788

Kohn S (2012) Von der Idee zum Konzept/Erfolgsfaktoren in den frühen Phasen des Innovationsprozesses. Harland media, Lichtenberg

Korba P, Baumgartner F, Völlmin B, Manunatha AP (2012) Integration and management of PV-battery systems in the grid. In: 27th European photovoltaic solar energy conference and exhibition, Frankfurt, 24–28 Sept 2012

Koschatzky K (2001) Networks in innovation research and innovation policy – an introduction. In: Koschatzky K, Kulicke M, Zenker A (eds) Innovation networks: concepts and challenges in the European perspective. Physica Verlag, Heidelberg

Kosow H, Gaßner R (2008) Methoden der Zukunfts- und Szenarioanalyse. Überblick, Bewertung und Auswahlkriterien, vol 103, WerkstattBericht Nr. IZT – Institut für Zukunftsstudien und Technologiebewertung, Berlin

Kotler P, Keller KL (2006) Marketing management. Pearson Education, Prentice Hall, New Jersey

Kotler P, Armstrong G, Saunders J, Wong V (2003) Grundlagen des marketing. Pearson Education, München

Kozinets RV (1998) On netnography: initial reflections on consumer research investigations of cyberculture. Adv Consum Res 25(1):366–371

Krippendorff K (2004) Content analysis: an introduction to its methodology. Sage, Thousand Oaks

Krippendorff K (2011) Principles of design and a trajectory of artificiality. J Prod Innov Manage 28(3):413

Kurylko DT (2005) Moonraker project seeks marketing savvy for VW. Automot News Eur 10 (17):22

Lakhani KR, Jeppesen LB (2007) Getting unusual suspects to solve R&D puzzles. Harv Bus Rev 85(5):30–32

Lakhani KR, Garvin DA, Lonstein E (2011) TopCoder (A): developing software through crowdsourcing (TN). Harvard Business School Case, 610–032

Lamnek S (2005) Qualitative sozialforschung. Psychologie Verlag, München

Lane PJ, Lubatkin M (1998) Relative absorptive capacity and interorganizational learning. Strateg Manage J 19(5):461–477

Lane PJ, Koka BR, Pathak S (2006) The reification of absorptive capacity: a critical review and rejuvenation of the construct. Acad Manage Rev 31(4):833–863

Laursen K, Salter A (2006) Open for innovation: the role of openness in explaining innovation performance among UK manufacturing firms. Strateg Manage J 27(2):131–150

Leifer LJ, Steinert M (2011) Dancing with ambiguity: causality behavior, design thinking, and triple-loop-learning. Inform Knowl Syst Manage 10(1):151–173

Leonard D, Rayport JF (1997) Spark innovation through emphatic design. Harv Bus Rev 75 (6):102–113

Leonard-Barton D (1995) Wellsprings of knowledge. Harvard Business School Press, Boston

Lerner J, Tirole J (2005) The scope of open source licenses. J Law Econ Organ 21(1):20–56

Levinthal DA, March JG (1993) The myopia of learning. Strateg Manage J 14:95–112

Linder J, Cantrell S (2000) Changing business models: surveying the landscape. Working paper. Technical report, Accenture Institute for Strategic Change

Littler D, Leverick F, Bruce M (1995) Factors affecting the process of collaborative product development: a study of UK manufacturers of information and communications technology products. J Prod Innov Manage 12(1):16–32. doi:10.1111/1540-5885.1210016

Loch C, Terwiesch C, Thomke S (2001) Parallel and sequential testing of design alternatives. Manage Sci 47(5)

Lockwood T (2009) Design thinking: integrating innovation, customer experience, and brand value. Allworth Press, New York

Loewy R, Weseloh HA (1953) Hässlichkeit verkauft sich schlecht, Die Erlebnisse des erfolgreichsten Formgestalters unserer Zeit. Econ Verlag, Düsseldorf

Mabogunje A (1997) Measuring conceptual design performance in mechanical engineering: a question based approach. Doctoral Dissertation, Stanford University, Stanford

MacCormack A (2001) How internet companies build software. MIT Sloan Manage Rev 42(2):75–84

Magee GB (2005) Rethinking invention: cognition and the economics of technological creativity. J Econ Behav Organ 57(1):29–48

Magretta J (2002) Why business models matter. Harv Bus Rev 80(5):86–92

Malhotra NK (2009) Basic marketing research. A decision-making approach. Pearson Education, Upper Saddle River

Manyika J, Chui M, Bughin J, Brown B, Dobbs R, Roxburgh C, Byers Hung A (2011) Big data: the next frontier for innovation, competition, and productivity. McKinsey Global Institute

Marple D (1961) The decisions of engineering design. IEEE Transactions on Engineering Management, June, 8:55–71

ME310 (2011) EXPE 2011 (class report)

ME310 (2011, January 30) ME310. Retrieved from http://www.stanford.edu/group/me310/me310 2010/about.html

Meffert H, Bruhn M (2003) Dienstleistungsmarketing. Grundlagen – Konzepte – Methoden. Gabler Verlag, Wiesbaden

Michalko M (2006) Thinkertoys: a handbook of creative-thinking techniques, 2nd edn. Ten Speed Press, Berkeley

Millard A (1990) Edison and the business of innovation. John Hopkins University Press, Baltimore

Miller BP (2012) Quick brainstorming activities for busy managers: 50 exercises to spark your team's creativity and get results fast. AMACOM, New York

Minx E, Kollosche I (2009) Kontingenz und zyklische Zukunftsbetrachtung. In: Popp R, Schüll E (eds) Zukunftsforschung und Zukunftsbetrachtung. Springer, Berlin, pp 167–174

Mockus A, Fielding RT, Herbsleb JA (2000) Case study of open source software development: the apache server. In: Proceedings of the 22nd international conference on software engineering, Limerick, Ireland, pp 263–272

Mohr JJ, Sengupta S, Slater SF (2010) Marketing of high-technology products and innovations. Pearson Education, Upper Saddle River

Montgomery D (1991) Design and analysis of experiments. Wiley, New York

Montoya-Weiss MM, O'Driscoll TM (2000) From experience: applying performance support technology in the fuzzy front end. J Prod Innov Manage 17(2):143–161. doi:10.1111/1540-5885.1720143

Morris M, Schindehutte M, Allen J (2005) The entrepreneur's business model: toward a unified perspective. J Bus Res 58(6):726–735

Mowery DC (2009) Plus ca change: industrial R&D in the "third industrial revolution". Ind Corp Change 18(1):1–50

Mowery DC, Oxley JE, Silverman BS (1996) Strategic alliances and interfirm knowledge transfer. Strateg Manage J 17(Special Issue):77–91

Mowery DC, Oxley JE, Silverman BS (1998) Technological overlap and interfirm cooperation: implications for the resource-based view of the firm. Res Policy 27(5):507–523

Mühlbacher H, Füller J, Huber L (2011) Online forum discussion-based forecasting of new product market performance. J Res Manage 33(3):221–234

Müller-Seitz G, Reger G (2009) Is open source software living up to its promises? Insights for open innovation management from two open source software-inspired projects. R&D Manage 39(4):372–381. doi:10.1111/j.1467-9310.2009.00565.x

Müller-Stewens G, Lechner C (2005) Strategisches management. Wie strategische Initiativen zum Wandel führen. Schäffer-Poeschel Verlag, Stuttgart

Murphy SA, Kumar V (1997) The front end of new product development: a Canadian survey. R&D Manage 27(1):5–16

Myers DG (2004) Exploring psychology. Worth Publishers, New York
Nachtwey J, Mair J (2008) Design ecology! Neo-Grüne Markenstrategien. Verlag Hermann Schmidt, Mainz
Nijssen EJ, Hillebrand B, Vermeulen PAM (2005) Unraveling willingness to cannibalize: a closer look at the barrier to radical innovation. Technovation 25(12):1400–1409
Nonaka I, Takeuchi H (1995) The knowledge-creating company. Oxford University Press, New York
Nooteboom B (1992) Towards a dynamic theory of transactions. J Evol Econ 2(4):281–299
Nooteboom B (1999) Inter-firm alliances: analysis and design. Routledge, London
Nooteboom B (2004) Inter-firm collaboration, learning and networks: an integrated approach. Routledge, London
Nooteboom B, Van Haverbeke W, Duysters G, Gilsing V, Van Den Oord A (2007) Optimal cognitive distance and absorptive capacity. Res Policy 36(7):1016–1034
Nowotny H (2008) Insatiable curiosity. Innovation in a fragile future (trans: Mitch Cohen) The MIT Press, Cambridge
Ogawa S, Piller FT (2006) Reducing the risks of new product development. MIT Sloan Manage Rev 47(Winter):65–72
Osterwalder A (2004). The business model ontology – a proposition in a design science approach, Universite de Lausanne
Osterwalder A, Pigneur Y (2010) Business model generation: a handbook for visionaries, game changers, and challengers. Wiley, Hoboken
Osterwalder A, Pigneur Y, Tucci C (2005) Clarifying business models: origins, present, and future of the concept. Commun Ass Inform Syst (CAIS) 16(1):1–25
Page SE (2007) The difference – how the power of diversity creates better groups, firms, schools, and societies. Princeton University Press, Princeton
Patton MQ (2002) Qualitative research & evaluation methods. Sage, Thousand Oaks
Paustian C (2001) Better products through virtual customers. MIT Sloan Manage Rev 42(3):14–16
Perez-Freije J (2008) Design of organizational controls for managing innovation. Dr. Kovač, Hamburg
Peruzzi M (2011) Open innovation – Kundeneinbindung im fuzzy front end der voestalpine Stahl GmbH, 8. Forum Innovation, Wien
Piller FT, Walcher D (2006) Toolkits for idea competitions: a novel method to integrate users in new product development. R&D Manage 36(3):307–318. doi:10.1111/j.1467-9310.2006.00432.x
Poetz MK, Schreier M (2012) The value of crowdsourcing: can users really compete with professionals in generating new product ideas? J Prod Innov Manage 29(2):245–256. doi:10.1111/j.1540-5885.2011.00893.x
Porter M (1980) Competitive strategy. Free Press, New York
Porter AL, Ashton WB, Clar G, Coates JF, Cuhls K, Cunningham SW, Ducatel K, van der Duin P, Cunningham SW, Ducatel K, van der Duin P, Georgehiou L, Gordon T, Linstone H, Marchau V, Massari G, Miles I, Mogee M, Salo A, Scapolo F, Smits R, Thissen W, Group TFAMW (2004) Technology futures analysis: toward integration of the field and new methods. Technol Forecast Soc Change 71(3):287–303
Poskela J, Martinsuo M (2009) Management control and strategic renewal in the front end of innovation. J Prod Innov Manage 26(6):671–684
PRL (2011, January 30) Product realization laboratory. Retrieved from http://www.stanford.edu/group/prl/prl site/
Quinn JB (1985) Managing innovation: controlled chaos. Harv Bus Rev, May–June, 63(3):73–84
Rams D (1994a) Design und Verantwortung. Spektrum der Wissenschaft 1994(4), 107. Heidelberg: Spektrum der Wissenschaft Verlagsgesellschaft
Rams D (1994b) Weniger, aber besser. Less but better. Jo Klatt Design + Design Verlag, Hamburg

Raymond ES (1999) The cathedral and the bazaar. Musings on Linux and open source by an accidental revolutionary. O'Reilly & Associates, Sebastopol

Reger G (2001) Technology foresight in companies: from an indicator to a network and process perspective. Technol Anal Strateg Manage 13(4):533–553

Reid SE, De Brentani U (2004) The fuzzy front end of new product development for discontinuous innovations: a theoretical model. J Prod Innov Manage 21(3):170–184. doi:10.1111/j.0737-6782.2004.00068.x

Reitzig MG, Sorenson O (2010) Intra-organizational provincialism. Available at SSRN: http://ssrn.com/abstract=1552059 or http://dx.doi.org/10.2139/ssrn.1552059

Rindfleisch A, Moorman C (2001) The acquisition and utilization of information in new product alliances: a strength-of-ties perspective. J Market 65(2):1–18

Ritter T, Gemünden HG (2003) Network competence: its impact on innovation success and its antecedents. J Bus Res 56(9):745–755

Rohrbeck R (2010a) Corporate foresight: towards a maturity model for the future orientation of a firm. Physica-Verlag/Springer, Heidelberg

Rohrbeck R (2010b) Harnessing a network of experts for competitive advantage – technology scouting in the ICT industry. R & D Manage 40(2):169–180

Rohrbeck R, Gemünden HG (2011) Corporate foresight: its three roles in enhancing the innovation capacity of a firm. Technol Forecast Soc Change 78(2):231–243

Rohrbeck R, Schwarz JO (2013) The value contribution of strategic foresight: insights from an empirical study of large European Companies. Technological forecasting and social change, forthcoming, available at SSRN, http://ssrn.com/abstract=2194787

Rohrbeck R, Hölzle K, Gemünden HG (2009) Opening up for competitive advantage – how deutsche telekom creates an open innovation ecosystem. R & D Manage 39(4):420–430

Rohrbeck R, Konnertz L, Knab S (2013) Collaborative business modelling for systemic and sustainable innovations. Int J Technol Manage, forthcoming

Ross BH (1989) Distinguishing types of superficial similarities: different effects on the access and use of earlier problems. J Exp Psychol Learn Mem Cogn 15(3):456–468

Rothwell R (1994) Towards the fifth-generation innovation process. Int Market Rev 11(1):7

Rowley T, Behrens D, Krackhardt D (2000) Redundant governance structures: an analysis of structural and relational embeddedness in the steel and semiconductor industries. Strateg Manage J 21(3):369–386

Rudek S (2008) Organisation der Verkaufsförderung bei Konsumgüterherstellern. Gabler Verlag, Wiesbaden

Rudzinski C (2009) Informationsmärkte: der Unterschied, der einen Unterschied macht. Revue für Postheroisches Manage 4:90–95

Sandmeier P, Jamali N, Kobe C, Enkel E, Gassmann O, Meier M (2004) Towards a structured and integrative front-end of product innovation. 2004. In: R&D management conference (RADMA), Lisabon

Sandmeier P, Jamali N (2007) Eine praktische Strukturierungs-Guideline für das Management der frühen Innovationsphase. In: Herstatt C, Verworn B (eds), Management der frühen Innovationsphasen. Grundlagen – Methoden – Neue Ansätze. GWV Fachverlage GmbH, Wiesbaden, pp 339–356

Schief M, Buxmann P (2012) Business models in the software industry. In: Hawaii international conference on system sciences, Maui, HI, USA, pp 3328–3337

Schweitzer F, Gaubinger K, Gassmann O (2012) A three-track process model of the front end of innovation in the machine building industry. In: Proceedings of the XXIII ISPIM conference, Barcelona, pp 17–20 June 2012

Schweitzer FM, Buchinger W, Gassmann O, Obrist M (2012b) Crowdsourcing – leveraging innovation through online idea competitions. Res Technol Manage 55(3):32–38. doi:10.5437/08956308X5503055

Schwery A, Raurich VF (2004) Supporting the technology-push of a discontinuous innovation in practice. R&D Manage 34(5):539–552. doi:10.1111/j.1467-9310.2004.00361.x

Send H, Friesike S (2013) Participation in online co-creation: assessment and review of motivations. HIIG Working Paper Series
Shafer SM, Smith HJ, Linder JC (2005) The power of business models. Bus Horiz 48:199–207
Shippey KC (2009) A short course in international intellectual property rights. World Trade Press, Novato
Silber A (2007) Schnittstellenmanagement im CRM-Prozess des Industriegütervertriebs. Modellbasierte Analyse und Gestaltung der Verbesserungspotentiale. Deutscher Universitäts-Verlag, Wiesbaden
Simon HA (1969) The sciences of the artificial, 2nd edn. MIT Press, Cambridge
Simons R (1995) Levers of control: how managers use innovative control systems to drive strategic renewal. Harvard Business School Press, Boston
Sirkin DM, Sonalkar N, Jung M, Leifer LJ (2009) Lowering barriers to distributed design research collaboration. In: Proceedings of the 17th international conference on engineering design (ICED'09), vol. 9, pp 279–286
Slama A, Korell M, Warschat J, Ohlhausen P (2006) Auf dem Weg zu schnelleren innovationsprojekten. In: Bullinger H-J (ed) Fokus innovation. Carl Hanser Verlag, München
Slaughter RA (1997) Developing and applying strategic foresight. ABN Rep 5(10):13–27
Sloane P (2012) A guide to open innovation and crowdsourcing: advice from leading experts. Kogan Page, London
Smith M, Hansen F (2002) Managing intellectual property: a strategic point of view. J Intell Cap 3 (4):366–374
Smith PG, Reinertsen DG (1991) Developing products in half the time. Van Nostrand Reinhold, New York
Smith M, Busi M, Ball P, Van der Meer R (2008) Factors influencing an organisations ability to manage innovation/a structured literature review and conceptual model. Int J Innov Manage 12 (4):655–676
Sobek D, Ward A, Liker J (1999) Toyota's principles of set-based concurrent engineering. Sloan Manage Rev 40(2):67–83
Song XM, Parry ME (1996) What separates Japanese new product winners from losers. J Prod Innov Manage 13(5):422–439
Soukhoroukova A, Spann M, Skiera B (2012) Sourcing, filtering, and evaluating new product ideas: an empirical exploration of the performance of idea markets. J Prod Innov Manage 29 (1):100–112
Stats IW (2012) Internet world stats – usage and population statistics. http://www.internet-worldstats.com/stats.htm. Accessed 22 Nov 2012
Steiner G (2003) Kreativitätsmanagement: Durch Kreativität zur innovation. In: Strebel H (ed) Innovations- und Technologiemanagement. WUV Universitätsverlag, Wien
Steinmüller K (1997) Grundlagen und Methoden der Zukunftsforschung: Szenarien, Delphi, Technikvorausschau, vol 21, WerkstattBericht. SFZ Sekretariat für Zukunftsforschung, Gelsenkirchen
Stuart TE (1998) Network positions and propensities to collaborate: an investigation of strategic alliance formation in a high-technology industry. Adm Sci Q 43(3):668–698
Surowiecki J (2004) The wisdom of crowds: why the many are smarter than the few and how collective wisdom shapes business, economies, societies and nation. Abacus, London
Tang JC (1989) Toward an understanding of the use of shared workspaces by design teams. Stanford University, Stanford
Tang JC (1991) Findings from observational studies of collaborative work. Int J Man machine Stud 34(2):143–160
Tang JC, Leifer LJ (1991) An observational methodology for studying group design activity. Res Eng Des 2(4):209–219
Tapscott D, Williams AD (2006) Wikinomics – how mass collaboration changes everything. Penguin Group, New York

Teece DJ (2007) Explicating dynamic capabilities: the nature and microfoundations of (sustainable) enterprise performance. Strateg Manage J 28(13):1319–1350
Teece DJ (2010) Business models, business strategy and innovation. Long Range Plann 43 (2–3):172–194
ter Wal A, Criscuolo P, Salter A (2011) Absorptive capacity at the individual level: an ambidexterity approach to external engagement. Paper presented at the DRUID 2011, Copenhagen
Terwiesch C, Xu Y (2008) Innovation contests, open innovation, and multiagent problem solving. Manage Sci 54(9):1529–1543
Thomke S (1998) Managing experimentation in the design of new products. Manage Sci 44 (6):743–762
Thomke S (2003) Experimentation matters: unlocking the potential of new technologies for innovation. Harvard Business School Press, Boston
Thomke S, Fujimoto T (2000) The effect of front-loading problem-solving on product development performance. J Prod Innov Manage 17(2):128–142
Thomke S, Reinertsen D (2012) Six myths of product development. Harv Bus Rev, May issue
Thomke S, von Hippel E, Franke R (1998) Modes of experimentation: an innovation process and competitive variable. Res Policy 27:315–332
Thomke S, Holzner M, Gholami T (1999) The crash in the machine. Sci Am 280:92–97
Thumm N (2001) Management of intellectual property rights in European biotechnology firms. Technol Forecast Soc Change 67(2–3):259–272
Tidd J, Bessant J (2009) Managing innovation. Wiley, West Sussex
Tischner U, Schmincke E, Rubik F, Prösler M (2000) Was ist EcoDesign? vol 13. Verlag Form, Frankfurt/Main
Trott P (2012) Innovation management and new product development. Prentice Hall, London
Ulrich KT, Eppinger SD (2008) Product design and development. McGraw-Hill, New Jersey
Ulwick AW (2005) What customers want. McGraw-Hill, New York
UrbanDictionary (2011) UrbanDictionary–DARPAhard. Retrieved from http://www.urbandictionary.com/define.php?term=DARPA%20hard
Van Aken JE, Weggeman MC (2000) Managing learning in informal innovation networks: overcoming the Daphne Dilemma. R&D Manage 30(2):139–149
Van Atta RH, Cook A, Gutmanis I, Lippitz MJ, Lupo J, VA IFDAA (2003) Transformation and transition: DARPA's role in fostering an emerging revolution in military affairs. Volume 2- Detailed Assessments
Verworn B (2005) Die frühen Phasen der Produktentwicklung. Deutscher Universitätsverlag, Wiesbaden
Verworn B (2009) A structural equation model of the impact of the "fuzzy front end" on the success of new product development. Res Policy 38(10):1571–1581
Verworn B, Herstatt C (1999) Approaches to the "Fuzzy Front End" of innovation. Arbeitspapier Nr. 2
Verworn B, Herstatt C (2007) Bedeutung und Charakteristika der frühen Phasen des Innovationsprozesses. In: Herstatt C, Verworn B (eds) Management der frühen Innovationsphasen, Grundlagen – Methoden – Neue Ansätze. GWV Fachverlage GmbH, Wiesbaden, pp 4–19
Verworn B, Herstatt C, Nagahira A (2008) The fuzzy front end of Japanese new product developmentprojects: impact on success and differences between incremental and radical projects. R&D Manage 38(1):1–19
Veugelers M, Bury J, Viaene S (2010) Linking technology intelligence to open innovation. Technol Forecast Soc Change 77(2):335–343
Vianna M, Vianna Y, Adler I, Lucena B, Russo B (2012) Design thinking: business innovation. MJV Press, Rio de Janeiro
von Hippel E (1986) Lead users: a source of novel product concepts. Manage Sci 32(7):791–805. doi:10.1287/mnsc.32.7.791 DOI:10.1287/mnsc.32.7.791#_self
von Hippel E (2005) Democratizing innovation. MIT Press, Cambridge

von Hippel E, Tyre MJ (1995) How 'learning by doing' is done: problem identification in novel process equipment. Research Policy 24(1), Cambridge

von Hippel E, Franke N, Prügl R (2008) "Pyramiding": efficient identification of rare subjects. MIT Sloan School of Management Working Paper, pp 4719–08

von Hippel E, Franke N, Prügl RW (2009) Pyramiding: efficient search for rare subjects. Res Policy 38(9):1397–1406. doi:10.1016/j.respol.2009.07.005

Walker OC, Mullins JW (2011) Marketing strategy: a decision-focuses approach. McGraw Hill, New York

Ward TB (2004) Cognition, creativity, and entrepreneurship. J Bus Venturing 19(2):173

Wasserman S, Faust K (1994) Social network analysis: methods and applications. Cambridge University Press, Cambridge

Weiblen T, Giessmann A, Bonakdar A, Eisert U (2012) Leveraging the software ecosystem: towards a business model framework for marketplaces. In: Proceedings of the 9th international joint conference on e-business and telecommunications (ICETE), Rome, Italy, pp 187–193

Werani T, Prem C (2009) Produktkonzeption. In: Gaubinger K, Werani T, Rabl M (eds) Praxisorientiertes innovations- und Produktmanagement. Gabler, Wiesbaden, pp 101–113

Wheelwright SC, Clark KB (1992) Revolutionizing product development – quantum leaps in speed, efficiency, and quality. The Free Press, New York

Wikipedia (2012) About wikipedia. http://en.wikipedia.org/wiki/Main_Page. Accessed 31 Dec 2012

WIPO (2011) Statistics on patents, trademarks, and industrial designs. http://www.wipo.int/ipstats/en/statistics/

Wirtschaftswoche (11/2010) Vom Träumer zum Aufräumer, 6. Düsseldorf: Handelsblatt GmbH

Wuyts S, Colombo MG, Dutta S, Nooteboom B (2005) Empirical tests of optimal cognitive distance. J Econ Behav Organ 58(2):277–302

Zahra SA, George G (2002) Absorptive capacity: a review, reconceptualization, and extension. Acad Manage Rev 27(2):185–203

Zhang Q, Doll WJ (2001) The fuzzy front end and success of new product development: a causal model. Eur J Innov Manage 4(2):95–112

Zott C, Amit R (2008) The fit between product market strategy and business model: implications for firm performance. Strateg Manage J 29(1):1–26

Zott C, Amit R, Massa L (2011) The business model: recent developments and future research. J Manage 37(4):1019–1042

# Index

**A**
Absorptive capacity, 124, 303
　individual-level, 118, 119
　potential, 114, 115, 118, 124
Analogical thinking, 50, 51, 54–58

**B**
Bilateral relationships, 110, 114
Bisociation method, 165–166
BMI. *See* Business model innovation (BMI)
Brainstorming, 20, 49, 82, 128, 153, 161, 166, 168–169, 172, 173, 222, 230, 245, 253
Business ideas, 122, 233, 235, 238, 240, 245
Business model innovation (BMI), 227, 229, 269–273
　map, 92
　navigator, 97

**C**
CATWOE, 173–174
Causality, 13, 141, 143–145, 155
Change management, 257, 270
Client centers, 252, 253
Co-creation, 33
Cognitive distance, 13, 111, 112, 114, 115, 117, 118, 120, 124
Collaboration platform, 276
Collaborative innovation, 227, 252
Collaborative insourcing, 210
Collective notebook method, 172–173
Commercialization, 11, 30, 48, 104, 225, 255, 289, 290, 292, 294
Concept writing skills, 240
Concurrent engineering, 10, 261
Consumer business, 241
Consumer insights, 32, 41, 223, 238, 239, 241
Consumer markets, 243
Controlling, 8, 10, 17, 50, 263, 292
Corporate culture, 185, 187, 196, 197, 199, 251, 293, 308
Corporate foresight, 59, 60, 67, 70–73, 229
Corporate social responsibility, 184
Corporate venturing, 229
Creative potential, 9, 17, 34, 40, 166, 188
Creativity methods, 219, 222
Creativity techniques, 27, 49, 50, 160, 177, 217, 308
Cross-industry innovation(s), 11, 49, 52–58, 110, 114
Crowdfunding, 78, 186
Crowdsourcing, 11, 12, 31, 42, 63, 75–87, 210, 211, 230, 238, 243, 244, 246–249, 277, 297, 303
Customer active paradigm, 34
Customer insights, 6, 32
Customer integration, 31–38, 40–42, 44–48, 201, 277, 290, 293, 294
Customer pain points, 195

**D**
Deep reasoning questions, 149, 154
Design
　strategic, 13, 183–185, 187, 189, 191,
　sustainable, 179–185, 191
　thinking, 93, 141–177, 189, 269–270, 272, 273, 275, 306

**E**
Early innovation phase, 9, 10, 13, 17, 114, 124, 208–212, 264–267, 301
Economic viability, 269–273
Effectiveness, 267, 275, 289
Effectiveness and efficiency, 264, 267, 275, 289
Energy storage systems, 201, 205
Engineering design, 138, 141, 143, 144

Enterprise 2.0, 229–231
Evaluation process, 5, 241
Expert innovation journey, 213, 214, 217–219
External assessment, 24, 26

**F**
Fact finding missions, 266
First-of-a-kind projects (FOAK), 255
Five 'Whys,' 175
FOAK. *See* First-of-a-kind projects (FOAK)
Focus group, 37, 42, 45, 221–225, 279
Front-end activities
  early, 5, 19, 21
  later, 5
Future_bizz, 122–123
Future workshops, 35, 276, 278–279

**G**
Gallery method, 171–172
Generative design questions, 149, 154
Global technology outlook (GTO), 253, 254
GTO. *See* Global technology outlook (GTO)

**I**
ICT. *See* Information and communication technology (ICT)
Idea generation, 6, 7, 11, 21, 26, 27, 82, 202, 213, 214, 217, 219, 222, 224, 225, 240, 245, 264, 290
Idea maturation, 27
Ideation jam, 230
Ideation platforms, 76, 230
Incubator, 211, 233–235
Industry Solutions Labs (ISL), 11, 251–255
Information and communication technology (ICT), 16, 109, 253
Information market, 295–298
InnoGate process, 237
Innovation culture, 46, 57, 195, 309
Innovation essentials, 266
Innovation flexibility approach, 116
Innovation opportunities, 10, 19, 24, 62, 65, 67, 73, 123, 266
Innovation partnerships, 231, 253–255
Innovation strategy, 12, 21, 30, 50, 207, 296
Innovator, 6, 69, 79–84, 92, 105
Intellectual property, 12, 13, 76, 81, 99, 100, 102, 103, 105–107, 123, 205, 228
Intuition, 166, 169
ISL. *See* Industry Solutions Labs (ISL)

**J**
Joint development, 119, 211, 218, 252

**K**
Knowledge integration capability, 119, 124
Knowledge transfer, 33, 47, 48, 114, 124

**L**
Lead user, 26, 33, 36, 41, 42, 44, 46, 69, 203, 231, 279, 290
Life cycle, 100
  patent, 100, 101, 106, 107
  of technologies, 13, 100
Life cycle of technologies, 100
Life sciences, 209, 210, 234, 275

**M**
Management Cockpit, 14, 282, 284, 286
Manufacturer active paradigm, 34, 35
Market research tools, 33, 37, 47
Method 6-3-5, 171
Mind map, 166
Morphological box, 170
Multilateral network, 119

**N**
Needs information, 32–35, 42, 43, 46, 225
Netnography, 6, 41, 47
Network analysis, 111, 112
Network culture, 120, 121, 124
New product development, 4, 5, 11, 21, 35, 102, 198, 237, 309
NexGen Systems, 18–19
Not invented here (NIH) syndrome, 46, 92, 93, 118, 119

**O**
Online community, 279
Online competition, 221–225
Open foresight, 295–299
Open ideation, 230
Open innovation, 11, 12, 23, 31, 73, 83, 111, 227–231, 243, 295, 297, 298, 303, 305, 307
Open source, 41, 45, 47, 77, 146, 244, 246
Opponent, 43, 69, 71
Out come driven innovation, 39

# Index

**P**
Patent portfolio, 99, 100, 104–107
Patent risk, 82
Pearlfinder network, 113
Peer production, 248
Performance measurement, 30, 123
Product portfolio, 207, 240, 258, 262
Prognosis, 176
Prosumer, 33
Prototyping, 10, 16, 141, 142, 145, 147–150, 153, 174–175, 272, 302, 304, 307
Provocation technique, 174

**R**
Resource efficiency approach, 116
Revealing strategic intent, 81
Risk of followers, 81

**S**
Scenario thinking, 68, 298
Scope of action, 7
Scouting, 11, 63, 64, 71, 73, 106, 117, 228, 229
Scout network, 63
Shadow labor force, 243
Similarity, 51, 55, 161, 167
Six thinking hats, 164–165
Social media, 12, 186, 222, 235, 279
Solution information, 32–34, 36, 44, 225
Solver, 76, 79
Spider meeting, 163–164
Stage-gate process, 9, 17, 21, 28, 209–211, 266, 268
Success factor, 14, 16, 75, 184, 218–219, 255, 289, 296, 305, 308
Sustainable design, 179–185, 189–191
Sustainable innovation, 13, 181, 182, 185–189, 265, 305
Sweet spot, 6
Synectics, 160–161

**T**
Technology development (TD), 10, 20–23, 26, 66, 89, 290
Think tank, 255
TILMAG, 161–162
Toolkits, 31, 33, 40, 41, 47, 222, 224
Trend collection, 277–278
Trend evaluation, 278
Trial-and-error, 127, 138
TRIZ, 65, 167–168

**U**
Uncertainty, 5, 8, 15, 17, 21, 32, 37, 44, 50, 60, 64, 67, 68, 71, 130, 141, 142, 189, 190, 213, 259, 264, 301, 305
User interviews, 175–176

**V**
Value network, 86, 281–283, 285, 286
Venture capital, 34, 229, 233, 235
Virtual solutions, 302

**W**
Weak link chain, 246
Web 2.0, 41, 61, 73, 87, 229
Wind tunneling, 295–297
Wisdom of the crowd, 12, 81, 295
World café, 176, 177

The manufacturer's authorised representative in the EU is Springer Nature Customer Service Centre GmbH, Europaplatz 3, 69115 Heidelberg, Germany. If you have any concerns regarding our products, please contact ProductSafety@springernature.com

Printed and bound by CPI Group (UK) Ltd, Croydon, CR0 4YY

23/03/2026

02076668-0015